Magnetics, Dielectrics, and Wave Propagation with MATLAB® Codes

Future microwave, wireless communication systems, computer chip designs, and sensor systems will require miniature fabrication processes in the order of nanometers or less as well as the fusion of various material technologies to produce composites consisting of many different materials. This requires distinctly multidisciplinary collaborations, implying that specialized approaches will not be able to address future world markets in communication, computer, and electronic miniaturized products.

Anticipating that many students lack specialized simultaneous training in magnetism and magnetics, as well as in other material technologies, *Magnetics, Dielectrics, and Wave Propagation with MATLAB® Codes* avoids application-specific descriptions, opting for a general point of view of materials per se. Specifically, this book develops a general theory to show how a magnetic system of spins is coupled to acoustic motions, magnetoelectric systems, and superconductors. Phenomenological approaches are connected to atomic-scale formulations that reduce complex calculations to essential forms and address basic interactions at any scale of dimensionalities. With simple and clear coverage of everything from first principles to calculation tools, the book revisits fundamentals that govern magnetic, acoustic, superconducting, and magnetoelectric motions at the atomic and macroscopic scales, including superlattices.

Constitutive equations in Maxwell's equations are introduced via general free energy expressions which include magnetic parameters as well as acoustic, magnetoelectric, semiconductor, and superconducting parameters derived from first principles. More importantly, this book facilitates the derivation of these parameters, as the dimensionality of materials is reduced toward the microscopic scale, thus introducing new concepts. The deposition of ferrite films at the atomic scale complements the approach toward the understanding of the physics of miniaturized composites. Thus, a systematic formalism of deriving the permeability or the magnetoelectric coupling tensors from first principles, rather than from an ad hoc approach, bridges the gap between microscopic and macroscopic principles as applied to wave propagation and other applications.

Magnetics, Dielectrics, and Wave Propagation with MATLAB® Codes

Second Edition

Carmine Vittoria

CRC Press
Taylor & Francis Group
Boca Raton London New York

CRC Press is an imprint of the
Taylor & Francis Group, an **informa** business

Second edition published 2024
by CRC Press
2385 Executive Center Drive, Suite 320, Boca Raton, FL 33431

and by CRC Press
4 Park Square, Milton Park, Abingdon, Oxon, OX14 4RN

CRC Press is an imprint of Taylor & Francis Group, LLC

ISBN: 9781032555683 (hbk)
ISBN: 9781032555690 (pbk)
ISBN: 9781003431244 (ebk)

DOI: 10.1201/9781003431244

Typeset in Times
by codeMantra

Sempre Reginella,
my wife Anmarie

Contents

Preface

By now, there are many books that describe the phenomenon of magnetic resonance and/or wave propagation in magneto-dielectric materials. Most descriptions are phenomenological and specialized to specific applications. The problems with such descriptions are twofold: (1) Nowadays, most students with an engineering or scientific background are not trained in magnetism or magnetics to be able to appreciate specialized presentations, phenomenological, or otherwise. (2) Modern trends in microwave applications are toward miniature devices, nanosized structures, multifunctional applications, composites of various types of materials, etc. As such, there is a need for multidisciplinary collaborations to meet modern needs in microwave technology. Specialized approaches will not be able to address system requirements of the future. We believe it is time to revisit the fundamentals that govern the phenomenon of magnetic resonance and wave propagation in magneto-dielectric materials. The objective of this book is to be able to connect phenomenological approaches to comprehensive microscopic formulations in order to gain a new physical perspective as related to modern trends in microwave technology. The presentation must be inclusive rather than exclusive. It must be done in a way to invite participation of nonexperts. Future systems require the participation of a variety of experts with different backgrounds.

This book provides a comprehensive description of wave propagation in magneto-dielectric materials. The magnetic state of a magnetic ion, for example, is considered at the atomic scale, and a mathematical link to wave propagation is provided at the macroscopic scale. Qualitative and quantitative arguments are presented to calculate magnetic parameters from first principles. A clear understanding of the origin of magnetic parameters from first principles is very important, as these parameters are to be included in a large-scale thermodynamic quantity referred to as the free energy. However, more importantly, it is to be able to calculate these parameters as the dimensionality of materials is reduced toward the microscopic scale. The conduit between the free energy and Maxwell's equations is, of course, the permeability and permittivity tensors, which are shown to be derived from the free energy. As an example, we present a systematic way of deriving the permeability tensor of the most practical magnetic materials, cubic and hexagonal crystal structures. This approach is rather simple, but very general. Effectively, this book bridges the gap between microscopic and macroscopic principles as applied to wave propagation.

In Chapter 1, both the MKS and CGS systems of units are introduced in order to allow cross-fertilization of ideas between engineers and physicists. The purpose of Chapters 2–5 is to build the background for engineering students and to help them to apply fundamental ideas. We introduce basic definitions and terminologies by considering a macroscopic wire loop. We then shrink the loop to atomic scale and still maintain the same definitions! We reverse the shrinking process by considering the magnetic potential energy or the free magnetic energy of a macroscopic body. In Chapter 6, we start with the free magnetic energy and begin to develop the conduit for wave propagation. While most or all electromagnetic books in the past introduce constitutive equations as a complementary set of equations to Maxwell's equations, we introduce them via the free energy. A unified approach is developed by which Maxwell's equations are coupled to the free energy of a solid without ever introducing constitutive relationships by an ad hoc procedure. For example, in Chapter 9, we consider wave propagation in ordered magnetic materials by coupling the free magnetic energy to Maxwell's equations. Clearly, once a free energy is specified or formulated from first principles, as for example in Chapter 7, one may readily consider wave propagation in any media. This may, of course, include ferroelectric materials as well. The step-by-step procedure of unifying Maxwell's equations with the free energy via the equation of motion represents the essence of this book. This procedure allows for both scalar and tensorial permeability and permittivity quantities. As such, these constants must all obey the Kramer–Kronig relations, as described in Chapter 8.

Chapters 9–12 deal with specialized applications toward planar composite configurations in which electromagnetic waves propagate in a specific direction. This was done on purpose in order to contrast this work from previous books in which waveguides were chosen as the medium of propagation. This book focuses on the understanding of electromagnetic wave propagation in films, multilayers, and planar devices, since this represents the modern technological trend. Usually, in other books, field variables from Maxwell's equations are substituted into the equation of motion resulting in a dispersion relation. In Chapter 9, we reverse the order of substitution by introducing a k-dependent permeability tensor derived from the equations of motion into Maxwell's equations. The advantage of this approach is that it allows for proper identification of purely magnetic waves (spin waves) and purely electromagnetic waves, as well as the mixture of the two types of waves. This calculational method is sufficiently general to include k-dependent permittivity tensors derived from corresponding equations of motion. Since we are dealing with thin-film configurations, Chapter 11 introduces electromagnetic boundary conditions, which are purely magnetic, as well as the usual Maxwell's boundary conditions. Finally, in Chapter 12, we formulate electromagnetic propagation in multilayer superlattice composites. The essence of this approach is that, however complex the mode of propagation in a given medium, the algebra always reduces to the simplest form of a 2×2 matrix, since there is only one electromagnetic electric field and one magnetic field at the surface.

This book may serve three purposes: (1) It may introduce a novice to the field of magnetism, magnetics (technical magnetism), and wave propagation in very anisotropic media. Clearly, some mathematical and scientific background would help the novice. (2) It may serve as a textbook for a regular college-level course at the senior (somewhat motivated) level or first-year graduate school level. One semester may cover the first five chapters and another semester the remaining five. It is to be noted that problems assigned at the end of each chapter have solutions. This is not to encourage cheating, but rather to encourage students to ask the "what if" question. For example, all the solutions are based on a premise. The question is, how would the solution look like if the original premise is modified or amended? It is like taking a "mental" derivative of the given solutions. In summary, students are encouraged to take the solutions to a higher level of understanding. The content of the appendices only serves to complement the text or provide a specialized result derived from the text. There was no attempt to provide a complete compendium of references. Rather, whenever possible, a book reference was cited to help the novice. (3) It may serve as a complementary source of information to an expert or practitioner carrying out research in the field. We believe our approach, or point of view, is rather unique and worthy of attention of our peers.

Carmine Vittoria
Professor Emeritus
Northeastern University
Boston, Massachusetts

For MATLAB® and Simulink® product information, please contact:
The MathWorks, Inc.
3 Apple Hill Drive
Natick, MA, 01760-2098 USA
Tel: 508-647-7000
Fax: 508-647-7001
E-mail: info@mathworks.com
Web: www.mathworks.com

Preface to the Second Edition

In the Second edition, the following topics have been added: (1) Deposition of single crystal ferrite films at the atomic scale; (2) the phenomena of magnetoelectricity in hexaferrite films; and (3) quantitative microscopic model for magnetic relaxation in ordered magnetic materials. The combination of past (previous edition) and the above topics offers a comprehensive theoretical and practical approach to the field of magnetism and applications thereof, as in new artificial ferrite materials, nano-composites and nano-structures, superlattices, amorphous magnetic films, ferrite film preparation, new theoretical formulations at the microscopic and macroscopic levels, inter-facing with non-magnetic materials, new technologies, etc.

The ATLAD (alternate target laser ablation deposition) technique (Chapter 4) was deployed in the early 1990s in order to deposit single crystal ferrite films at the atomic scale. It allowed for the first time to tinker with "mother nature" to place magnetic ions in a unit cell not condoned by the natural growth process. However, it took about 20 years of development to hone the technique to the point it became a viable one in depositing cations at any site in a crystal unit cell. The original patent disclosure was submitted in 1993, but issued 25 years later. The ATLAD technique was utilized to enhance the magnetoelectric linear coupling in hexaferrites (Chapter 10) by placing magnetic cations in a special site within the unit cell. The impact of being able to "surgically" alter the make-up of a crystal unit cell is truly immense to modern technologies in which the application of voltage or magnetic field can efficiently affect the properties of a film.

Since Landau's proposed phenomenological mechanism for magnetic relaxation in ordered magnetic materials in 1935, his mechanism is now addressed in terms of a spin–lattice microscopic model in quantitative agreement with measured relaxation times of insulating and metallic magnetic materials. This is an important consideration in setting physical limitations on the use of ordered magnetic materials in modern-day technologies.

Acknowledgments

When I entered graduate school at Yale University, I had pre-conceived notions about magnetism. Professor W. P. Wolf erased every one of them and planted seeds for new and beautiful ideas about magnetism. I want to thank Professor Wolf for taking the time and patience to do that. Also, I want to acknowledge the help of my graduate students: Jianwei Wang, Soack Dae Yoon, Anton Geiler, Xu Zuo, Hessam Izadkhah, Saba Zare, Rezaoul Karim, and Tomokazu Sakai. Their selfless hard work made it possible for the new edition.

Author

Dr. Carmine Vittoria's career spans 50–55 years in academia and government research establishments. His approach to scientific endeavors has been to search for the common denominator or thread that links the various sciences to make some logical sense. His fields of study include physics, electrical engineering, ceramics, metallurgy, surface or interfaces, and nanocomposite films. His interests in science range from the physics of particle–particle interaction at the atomic scale to nondestructive evaluation of bridge structures, from EPR of a blood cell to electronic damage in the presence of gamma rays, from design of computer chips to radar systems, and from microscopic interfacial structures to thin-film composites. The diversity and seriousness of his work and his commitment to science are evident in the ~500 publications in peer-reviewed journals, ~25 patents, and three other scientific books. Dr. Vittoria is also the author of a nonscientific books on soccer for children; memoirs: "Bitter Chicory to Sweet espresso," "Once Upon a Hill," and "Hidden in Plain Sight". He is a life fellow of the IEEE (1990) and an APS fellow (1985). He has received research awards and special patent awards from government research laboratories.

Dr. Vittoria was appointed to a professorship position in 1985 in the Electrical Engineering Department at Northeastern University, and he was awarded the distinguished professorship position in 2001 and a research award in 2007 by the College of Engineering.

In addition, he was cited for an outstanding teacher award by the special need students at Northeastern University. His teaching assignments included electromagnetics, antenna theory, microwave networks, wave propagation in magneto-dielectrics, magnetism and superconductivity, electronic materials, microelectronic circuit designs, circuit theory, electrical motors, and semiconductor devices.

1 Review of Maxwell Equations and Units

MAXWELL EQUATIONS IN MKS SYSTEM OF UNITS

Maxwell's equations in MKS are written as follows:

$$\vec{\nabla} \times \vec{E} = -\frac{\partial \vec{B}}{\partial t}, \tag{1.1}$$

$$\vec{\nabla} \times \vec{H} = \vec{J} + \frac{\partial \vec{D}}{\partial t}, \tag{1.2}$$

$$\vec{\nabla} \cdot \vec{D} = \rho, \tag{1.3}$$

and

$$\vec{\nabla} \cdot \vec{B} = 0, \tag{1.4}$$

where

$$\vec{J} = \sigma \vec{E},$$

with J being the current density in the medium. The above equation is sometimes referred to as one of the constitutive equations. We don't do that here because it can easily be combined with another constitutive equation, as we shall see later.

The two remaining constitutive equations are written as

$$\vec{B} = \mu_0 \left(\vec{H} + \vec{M} \right) = \mu \vec{H} \tag{1.5}$$

and

$$\vec{D} + \varepsilon_0 \vec{E} + \vec{P} = \varepsilon \vec{E}, \tag{1.6}$$

where
  ε is the permittivity constant
  μ is the permeability constant

DOI: 10.1201/9781003431244-1

The units of each field quantity are defined as follows:

J = Current density (A/m²)
E = Electric field intensity (V/m)
B = Magnetic flux density (Wb/m²)
H = Magnetic field intensity (A/m)
D = Electric displacement (C/m²)
ρ = Charge density (C/m³)
σ = Conductivity (mhos/m)
M = Magnetization (A/m)
P = Electric polarization (C/m²)
μ_0 = Permeability of free space = $4\pi \times 10^{-7}$ H/m
ε_0 = Dielectric constant of free space = $(1/36\pi) \times 10^{-9}$ F/m

ε and μ may be defined in terms of their respective susceptibilities χ or

$$\mu = \mu_0 \left(1 + \chi_m\right) \text{ (H/m)}$$

and

$$\varepsilon = \varepsilon_0 \left(1 + \chi_e\right)(\text{F/m}),$$

where χ_m and χ_e are the magnetic and electric susceptibilities, respectively.

MAJOR AND MINOR MAGNETIC HYSTERESIS LOOPS

We have assumed in the above relations that both μ and ε are scalars, but complex quantities. This implies losses or dissipation in a magneto-dielectric medium. Let's examine the ramifications of complex μ. Let's consider the magnetic response to a magnetic field excitation consisting of DC and time-varying components. For simplicity, the time-varying field may be either linearly or circularly polarized. Let's express the field as follows:

$$H = H_0 + h(t),$$

where
H_0 is the static or DC magnetic field component
$h(t)$ is the time-varying field

The magnetic response to H may be represented in terms of hysteresis loops where the magnetization, M, is plotted as a function of H (see Figure 1.1). We will model a DC hysteresis loop in Chapter 5. We have superimposed the time response in the DC hysteresis plot in order to illustrate the type of hysteretic loop responses that may be excited in a magnetic medium.

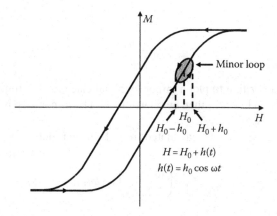

FIGURE 1.1   Major and minor magnetic hysteresis loops.

A linearly polarized time-varying field may be expressed as follows:

$$h(t) = h_0 \cos \omega t$$

and the magnetization time response as

$$m(t) = m_0 \cos(\omega t - \varphi).$$

For the sake of generality, a phase angle φ is arbitrarily introduced between m and h. As we shall see soon, note that the polarity of this phase angle is not so arbitrary. In complex notation, the above equations become

$$h = h_0$$

and

$$m = m_0 e^{-j\varphi}.$$

The complex magnetic susceptibility for time-varying fields, χ, is defined as

$$\chi = \chi' - j\chi'' = \frac{m}{h} = \frac{m_0}{h_0} e^{-j\varphi}.$$

Equating the two above equations yields

$$\chi' = \chi_0 \cos \varphi,$$

$$\chi'' = \chi_0 \sin \varphi,$$

and

$$\chi_0 = \frac{m_0}{h_0}.$$

Now, we are in a position to plot a minor loop and calculate the time rate of energy loss. The dissipation loss will then be related to χ''. The minor loop is basically a plot of $m(t)$ versus $h(t)$, see Figure 1.2.

The power loss, P_D, may be calculated from the minor loop response as

$$P_D = \left(\frac{\omega}{2}\right)V\mu_0 \int h \; dm,$$

where
  $V \equiv$ volume of magnetic sample
  $\omega = 2\pi f$
  $f =$ frequency (Hz)

The factor of 2 is a result of averaging sinusoidal fields over time. Clearly, the integration over the minor loop gives rise to a net power loss each time the loop traces one complete cycle, losing some net energy per second. The above integration simplifies to

$$P_D = \left(\frac{\omega}{2}\right)V\mu_0 (ab) m_0 h_0,$$

where
  $a =$ major axis of minor loop $= \sqrt{2}\cos(\varphi/2)$
  $b =$ minor axis of the minor loop $= \sqrt{2}\sin(\varphi/2)$

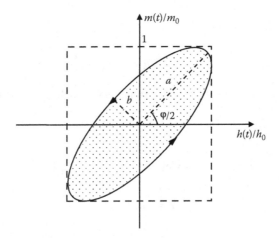

FIGURE 1.2   Minor magnetic hysteresis loop.

After some simple algebraic manipulations, we obtain

$$P_D = \left(\frac{\omega}{2}\right) V \mu_0 \chi'' h_0^2 = \left(\frac{\omega}{2}\right) V \mu'' h_0^2,$$

since

$$\mu = \mu' - j\mu'' = \mu_0 \left(1 + \chi' - j\chi''\right).$$

It is noted that dissipative power can only be a positive quantity. Hence, both χ'' and μ'' must necessarily be greater than zero which implies that the polarity of φ cannot simply be assumed to be arbitrary.

Let's now consider the same analysis for a circularly polarized excitation, since it is a common excitation field in many experimental situations. We may write the circular polarized field in complex notation as

$$h^{\pm} = h_0 e^{\pm j\omega t}.$$

The (+) sign corresponds to right-hand circular (RHC) and (−) to left-hand circular (LHC) polarized excitation fields. The magnetic response will also be circularly polarized

$$m^{\pm} = m_0^{\pm} e^{-j\varphi_{\pm}} \left(e^{\pm j\omega t}\right),$$

where

$$m_0^+ \neq m_0^-$$

and

$$\varphi_+ \neq \varphi_-.$$

As in the case of the linearly polarized excitation, the response lags in phase by an amount $\varphi_{\pm}$ depending on the sense of rotation of the circularly polarized excitation, see Figure 1.3. Furthermore, the amplitudes $\left(m_0^{\pm}\right)$ are usually different. It is simple to show that the magnetic response is of the same form as in the linear case or

$$m^{\pm} = \chi^{\pm} h_0 e^{\pm j\omega t},$$

where

$$\chi^{\pm} = \left(\chi'\right)^{\pm} - j\left(\chi''\right)^{\pm} = \chi_0^{\pm} \cos \varphi_{\pm} - j\chi_0^{\pm} \sin \varphi_{\pm}$$

and

$$\chi_0^{\pm} = \frac{m_0^{\pm}}{h_0}.$$

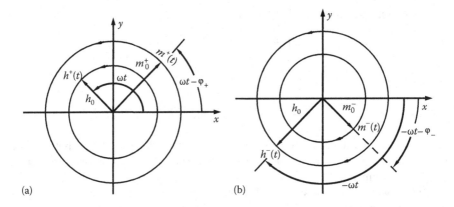

(a)   (b)

FIGURE 1.3   (a) RHC polarization excitation and response. (b) LHC polarization excitation and response.

Thus, the loss factors are uniquely defined as

$$\left(\chi''\right)^{\pm} = \chi_0^{\pm}\sin\varphi_{\pm}.$$

As such, the average power loss calculated from minor loop excitations must result in

$$P_D^{\pm} = \left(\frac{\omega}{2}\right)\mu_0 V\left(\chi''\right)^{\pm}h_0^2.$$

In most cases of interest, $P_D^{+} \neq P_D^{-}$, since the magnetic response depends on the sense of polarization.

TENSOR AND DYADIC QUANTITIES

A scalar permeability may be represented by a (3×3) diagonal tensor or matrix of the form

$$[\mu] = \begin{bmatrix} \mu_{xx} & 0 & 0 \\ 0 & \mu_{yy} & 0 \\ 0 & 0 & \mu_{zz} \end{bmatrix},$$

where

$$\mu_{ii} = \frac{B_i}{H_i}, \ i = x, y, \text{ and } z \text{ coordinate axes.}$$

with B_i representing either the static or dynamic component of the magnetic flux density along the i-direction. There are two cases of interest to consider: (1) static H_i and (2) dynamic $H_i = h_i$. The designation h_i is to denote small field dynamic field excitations. It is understood that h_i is a complex quantity. It may be related to a sinusoidal time dependence using the connection

$$h_i(t) = \text{Re}\left[h_i e^{j\omega t}\right].$$

1. Static H_i:
   For a semi-infinite isotropic magnetic medium, whereby H_i is collinear with B_i,

   $$\mu_{xx} = \mu_{yy} = \mu_{zz} = \mu.$$

   For finite-size magnetic samples of various shapes, magnetically anisotropic collinearity may no longer be assumed. B_i is usually interpreted as being the projection of the B field or the magnetization M along H_i. Thus,

   $$\mu_{ii} = \mu_0 \left(\frac{H_i + M_i}{H_i}\right), \; M_i = M \cos\alpha_i,$$

   where α_i is the angle between the direction of M and H_i. In most experimental cases,

   $$\mu_{xx} \neq \mu_{yy} \neq \mu_{zz},$$

   since α's are different for each direction of H. For non-collinear situations, the tensor for the permeability can no longer be represented by a diagonal tensor. The off-diagonal elements may be calculated from the following:

   $$\mu_{ij} = \mu_0 \left(\frac{M_j}{H_i}\right), \; j \neq i$$

   since $M_j \neq 0$. In summary, the permeability tensor is of this form for non-collinear situations:

   $$[\mu] = \begin{bmatrix} \mu_{xx} & \mu_{xy} & \mu_{xz} \\ \mu_{yx} & \mu_{yy} & \mu_{yz} \\ \mu_{zx} & \mu_{zy} & \mu_{zz} \end{bmatrix}.$$

   The diagonal terms can be measured using vibrating sample magnetometer (VSM) techniques. However, the off-diagonal elements are rather difficult to measure, but possible.

2. Dynamic Field Excitations, h_i:
   Again, for a semi-infinite magnetic medium and no bias DC magnetic field or random magnetization direction, the dynamic permeability is isotropic or

   $$\mu_{xx} = \mu_{yy} = \mu_{zz} = \mu.$$

The introduction of a static external magnetic field polarizes the magnetic medium such that the precessional motion of the magnetic moment and the direction of the dynamic drive field relative to the static magnetization can only make the dynamic permeability tensor non-diagonal. However, there is one special case when it is diagonal and that is for the drive dynamic field to be collinear with the static saturation magnetization. The argument is that once saturation is achieved for any direction of the static magnetic field or magnetization, more magnetic fields of any kind along the static magnetization direction are not going to increase the magnetization or polarization beyond saturation values. As such, the diagonal elements are simply

$$\mu_{xx} = \mu_{yy} = \mu_{zz} = \mu_0.$$

In general, for arbitrary direction of the exciting field, the permeability tensor for the dynamic components of the magnetization can be rather complex. One of the objectives of this book is to bring a systematic order on how to derive dynamic permeability tensors for some very anisotropic magnetic materials (see Chapter 5, for example). For now, let's treat the derivation as a mathematical exercise. Mathematically, an incremental dB along a given direction (say the x-axis) may be expressed as follows:

$$dB_x = \mu_0 dH_x + \mu_0 \left(\frac{\partial M_x}{\partial H_x} dH_x + \frac{\partial M_x}{\partial H_y} dH_y + \frac{\partial M_x}{\partial H_z} dH_z \right). \qquad (1.7)$$

Let's be specific and define the dynamic components as follows:

$$dB_i = b_i,$$

$$dM_i = m_i,$$

$$dH_i = h_i, \quad i = x, y, \text{ and } z,$$

and

$$\frac{\partial M_i}{\partial H_j} \Rightarrow \frac{\partial m_i}{\partial h_j} \quad \text{for dynamic fields.}$$

The above equation holds for $i \neq j$ as well as $i = j$, and it implies a nonlinear theory for the derivation of the permeability tensor. Usually, the relationships between the m's and h's are assumed to be linear. Clearly, b_i, h_i, and m_i are dynamic fields that infer time dependence. In summary, Equation 1.7 may be written in tensor form as

$$\begin{bmatrix} b_x \\ b_y \\ b_z \end{bmatrix} = \mu_0 \left\{ \begin{bmatrix} 1 & 0 & 0 \\ 0 & 1 & 0 \\ 0 & 0 & 1 \end{bmatrix} + \begin{bmatrix} \dfrac{\partial m_x}{\partial h_x} & \dfrac{\partial m_x}{\partial h_y} & \dfrac{\partial m_x}{\partial h_z} \\ \dfrac{\partial m_y}{\partial h_x} & \dfrac{\partial m_y}{\partial h_y} & \dfrac{\partial m_y}{\partial h_z} \\ \dfrac{\partial m_z}{\partial h_x} & \dfrac{\partial m_z}{\partial h_y} & \dfrac{\partial m_z}{\partial h_z} \end{bmatrix} \right\} \begin{bmatrix} h_x \\ h_y \\ h_z \end{bmatrix}. \tag{1.8}$$

In a compact form, Equation 1.8 may be written as

$$[b] = \mu_0 \left\{ [I] + [\chi_m] \right\} [h], \tag{1.9}$$

where
$[I]$ = unit matrix
$[\chi_m]$ = magnetic susceptibility tensor whose elements are

$$\chi_{ij} = \frac{\partial m_i}{\partial h_j}. \tag{1.10}$$

The permeability tensor may also be defined as

$$[\mu] = \mu_0 \left\{ [I] + [\chi_m] \right\} = \begin{bmatrix} \mu_{xx} & \mu_{xy} & \mu_{xz} \\ \mu_{yx} & \mu_{yy} & \mu_{yz} \\ \mu_{zx} & \mu_{zy} & \mu_{zz} \end{bmatrix}. \tag{1.11a}$$

Equation 1.9 may be written in vector form or

$$\vec{b} = \ddot{\mu} \cdot \vec{h}, \tag{1.11b}$$

where $\ddot{\mu}$ is a dyadic vector and is expressed as follows:

$$\ddot{\mu} = \mu_{xx}\vec{a}_x\vec{a}_x + \mu_{xy}\vec{a}_x\vec{a}_y + \mu_{yx}\vec{a}_y\vec{a}_x + \mu_{yy}\vec{a}_y\vec{a}_y + \mu_{yz}\vec{a}_y\vec{a}_y + \mu_{zx}\vec{a}_z\vec{a}_x + \mu_{zy}\vec{a}_z\vec{a}_y + \mu_{zz}\vec{a}_z\vec{a}_z.$$

The other constitutive relationship in tensor and dyadic forms is given below:

$$[D] = [\varepsilon][e] \tag{1.12}$$

and

$$\vec{D} = \ddot{\varepsilon} \cdot \vec{e}, \tag{1.13}$$

where

$$[\varepsilon] = \varepsilon_0 \left\{ [I] + [\chi_e] \right\} = \begin{bmatrix} \varepsilon_{xx} & \varepsilon_{xy} & \varepsilon_{xz} \\ \varepsilon_{yx} & \varepsilon_{yy} & \varepsilon_{yz} \\ \varepsilon_{zx} & \varepsilon_{zy} & \varepsilon_{zz} \end{bmatrix}, \tag{1.14}$$

$$[I] = \begin{bmatrix} 1 & 0 & 0 \\ 0 & 1 & 0 \\ 0 & 0 & 1 \end{bmatrix},$$

$$(\chi_e)_{ij} = \left(\frac{1}{\varepsilon_0} \right) \frac{\partial p_i}{\partial e_j}, \tag{1.15}$$

$p_i \equiv$ dynamic electric polarization along the i-direction
$i, j \equiv x, y,$ and z coordinate axes

In summary, the value of each matrix element of [ε] and [μ] is proportional to the induced electric and magnetic polarization, respectively. The existence of the off-diagonal matrix elements is due to anisotropic inductions of polarizations in a magneto-dielectric medium.

MAXWELL EQUATIONS IN GAUSSIAN SYSTEM OF UNITS

Maxwell equations are written in Gaussian units, which is part of several electromagnetic unit systems within CGS. The Gaussian system of units is based on electrostatic (esu) and electromagnetic (emu) systems of units. In appendix, conversions between systems of units are provided.

$$\vec{\nabla} \times \vec{E} = -\frac{1}{c} \frac{\partial \vec{B}}{\partial t}, \tag{1.16}$$

$$\vec{\nabla} \times \vec{H} = \frac{4\pi\sigma}{c} \vec{E} + \frac{1}{c} \frac{\partial \vec{D}}{\partial t}, \tag{1.17}$$

$$\vec{\nabla} \cdot \vec{D} = 4\pi\rho, \tag{1.18}$$

and

$$\vec{\nabla} \cdot \vec{B} = 0. \tag{1.19}$$

The constitutive equations in CGS units become

$$\vec{B} = \vec{H} + 4\pi\vec{M} = \ddot{\mu} \cdot \vec{H} \tag{1.20}$$

and

$$\vec{D} = \vec{E} + 4\pi\vec{P} = \ddot{\varepsilon} \cdot \vec{E}. \tag{1.21}$$

In tensor form, we may write

$$[\mu] = [I] + 4\pi[\chi_m]$$

and

$$[\varepsilon] = [I] + 4\pi[\chi_e].$$

In CGS system of units, $\mu_0 = \varepsilon_0 = 1$. Thus, the matrix elements of $[\mu]$, for example, are dimensionless as implied in Equations 1.20 and 1.21. One convenient result of using the Gaussian system of units is that the characteristic impedance of free space, Z_0, is simply 1, since

$$Z_0 = \sqrt{\frac{\mu_0}{\varepsilon_0}} = 1 \quad \text{(Gaussian units)}$$

and
$Z_0 = 120\pi$ ohms (MKS units).
Let's now assign units to each quantity in Maxwell's equations:

E = statvolt/cm
B = G (Gauss)
H = Oe (Oersted)
M = emu/cm^3
P = statcoulomb/cm^2
D = statvolt/cm
ρ = statcoulomb/cm
σ = 1/s
$c = 3 \times 10^{10}$ cm/s

EXTERNAL, SURFACE, AND INTERNAL ELECTROMAGNETIC FIELDS

Maxwell's equations represent a set of equations that relate the internal electromagnetic fields in a magneto-dielectric medium to the polarization fields of that medium. It establishes fundamental relationships between these sets of fields. Polarization fields are the result of local interactions between the electromagnetic fields and the medium. A convenient way to relate these internal field interactions in terms of surface electromagnetic fields is via the so-called Poynting integral relation. The reader is referred to standard textbooks for its derivation. The Poynting integral equation in MKS units is then

$$\oint_s (\vec{E}_s \times \vec{H}_s^*) \cdot d\vec{S} = -j\omega \int_V dV \left(\vec{B} \cdot \vec{H}^* - \vec{D}^* \cdot \vec{E} - j\frac{\sigma}{\omega}|\vec{E}|^2 \right). \tag{1.22}$$

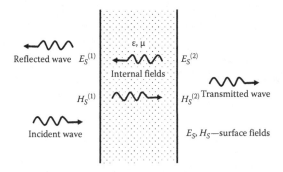

FIGURE 1.4 Electromagnetic external, surface, and internal fields.

The left-hand side contains only surface or external fields ($\vec{E}_S$ and $\vec{H}_S$), and the right-hand side contains only internal fields ($\vec{E}$, $\vec{H}$, $\vec{B}$, and $\vec{D}$). The surface fields may be expressed in terms of external reflected and transmission fields (see Figure 1.4). The object of an experimentalist is to monitor changes in surface fields in order to decipher the physical state of the magneto-dielectric medium. Equation 1.22 provides that connection or relationship between the fields.

For a scalar dielectric constant, Equation 1.22 may be rewritten as follows:

$$\oint_s \left(\vec{E}_S \times \vec{H}_S^* \right) \cdot d\vec{S} = -j\omega \int_V dV \left(\vec{B} \cdot \vec{H}^* - \varepsilon_V^* \left| \vec{E} \right|^2 \right), \tag{1.23}$$

where $\varepsilon_V^* = \varepsilon^* + j(\sigma/\omega)$.

The left-hand side is a surface integral over an enclosed volume and it is in units of watts. Without loss of generality, we may write

$$-Z_S \left| \vec{H}_S \right|^2 A = -j\omega A \int dy \left(\vec{B} \cdot \vec{H}^* - \varepsilon^* \left| \vec{E} \right|^2 \right),$$

where

$Z_S = E_S/H_S$
$A \equiv$ area
$y \equiv$ coordinate normal to surface (arbitrary)

We have assumed that the surface fields are uniform over the surface and propagation is along the y-direction. Also, for simplicity and without loss of generality, let's assume an enclosed volume whose shape is that of a very thick slab such that all surface fields on the far side of the slab vanish. Equivalently, we may assume asymmetrical excitation. Thus, solving for Z_S, we obtain

$$Z_S = j\omega \int dy \left[\mu_0 \left(\frac{\left| \vec{H} \right|^2 + \vec{M} \cdot \vec{H}^*}{\left| \vec{H}_S \right|^2} \right) - \varepsilon^* \frac{\left| \vec{E} \right|^2}{\left| \vec{H}_S \right|^2} \right], \tag{1.24}$$

where $\vec{B} = \mu_0 (\vec{H} + \vec{M})$, MKS.

We define an effective internal permeability, $\mu_{V|}$, and impedance, Z_V, as

$$\mu_{V|} = \frac{\mu_0}{t} \int dy \frac{\left(|\vec{H}|^2 + \vec{M} \cdot \vec{H}^* \right)}{|\vec{H}_S|^2} \tag{1.25}$$

and

$$|Z_V|^2 = \frac{1}{t} \int dy \frac{|\vec{E}|^2}{|\vec{H}_S|^2}. \tag{1.26}$$

This requires that $\vec{E}$ and $\vec{H}$ be solved in terms of $\vec{H}_S$ via the Maxwell boundary and spin (if necessary) boundary conditions at the surface. We will postpone spin boundary conditions for Chapter 9. Equation 1.24 simplifies to the following:

$$Z_S = j\omega \left(\mu_V - \frac{\varepsilon^*}{|Y_V|^2} \right) t, \tag{1.27}$$

where
$Y_V = 1/Z_V$
$t =$ slab thickness

Let

$$Z_S = R_S + jX_S,$$

$$\mu_V = \mu' - j\mu'',$$

and

$$\varepsilon = \varepsilon' - j\varepsilon''.$$

Then,

$$R_S = \omega t \left(\mu'' + \frac{\varepsilon''}{|Y_V|^2} \right)$$

and

$$X_S = \omega t \left(\mu' - \frac{\varepsilon'}{|Y_V|^2} \right).$$

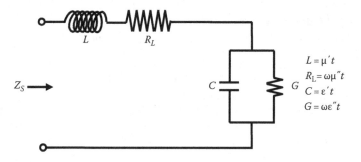

FIGURE 1.5 Equivalent circuit for Z_S.

Clearly, losses are additive, as reflected in R. An equivalent circuit for Z_S is shown in Figure 1.5. It is interesting to note that too many books on electromagnetic theory contain a factor of ε in Equation 1.22 as a result of assuming

$$\vec{D}^* = \varepsilon \vec{E}^*,$$

implying that $\varepsilon^* = \varepsilon$.

All materials are lossy including air. Hence, the omission of taking the complex conjugate of ε leads to nonconservation of energy or power in the application of the Poynting integral. Equivalently, it implies active circuit networks rather than passive networks in magneto-dielectric media, and this conclusion is incorrect.

Finally, if we are to represent a medium by a set of circuit parameters, there cannot be more than four parameters, as shown in Figure 1.5. The Poynting integral equation states that the net time rate of change of electromagnetic energy within a given volume is equal to the negative of the total work done by the fields on the medium. The beauty of the Poynting relationship is that any changes in the stored energy or the potential energy within a medium manifest itself as a change in the surface fields. In fact, it is exactly this principle that allows experimentalists to be able to characterize the properties of microwave materials. Interaction between internal fields and the medium is included in the stored energies. Finally, surface fields may be related to external fields located far from the medium by simple application of electromagnetic boundary conditions at the surface.

PROBLEMS

1.1 Show that using the tensor or dyadic forms for permeability leads to the same B field.

1.2 Derive expression for skin depth in MKS system of units. You may assume a TEM wave propagating in a medium with the following parameters: $\sigma = 5.8 \times 10^7$ mhos/m, $\mu = \mu_0$, and $f = 9 \times 10^9$ Hz in order to calculate the skin depth.

1.3 In engineering practice, such as with motors and generators, B is generally given as a function of H. Given the following measured parameters, determine the differential permeability tensor:

$$B_x = a_1 + a_2 H_x + a_3 H_y + a_5 H_x^2 + a_6 H_y^2 + \cdots$$

$$B_y = b_1 + b_2 H_x + b_3 H_y + b_5 H_x^2 + b_6 H_y^2 + \cdots$$

$$B_z = c_1 + c_2 H_z + c_3 H_z^2 + \cdots$$

1.4 A unit cell has the normalized dimensions $a = 2$, $b = 3$, and $c = 4$. Identify the unit cell. Assume no coupling between orthogonal axes, where a, b, and c are along x, y, and z, respectively, and assume linear effect of dimensions upon D when E is applied. Find the permittivity tensor.

1.5 Is $\int \vec{B} \cdot \vec{H} dV = \int \vec{H} \cdot \vec{B} dV$, if $[\mu]$ is not diagonal?

1.6 Redo Problem 1.2, but use Maxwell's equations in CGS system of units.

1.7 Write the Lorenz equation for a charge in a magnetic field and electric field in MKS and CGS units.

1.8 Express Poynting vector integral in real time and calculate power dissipation using the time representation.

APPENDIX 1.A: CONVERSION OF UNITS

There are several CGS systems of units. The Gaussian system of units is quite common in the scientific community. This system is based on the electrostatic (esu) and electromagnetic (emu) systems of units. In Tables 1.A.1 and 1.A.2, conversions between the MKS and Gaussian, esu and emu systems of units are given. We have utilized for the most of this book the Gaussian system. The conversions are straightforward except for the unit of magnetic dipole moment, since it has not been agreed upon. The reader is referred to Bleaney and Bleaney (1965) and Morrish (1965) listed in the reference section for much of the information in Tables 1.A.1–1.A.4. In this book, the magnetic dipole moment is defined as follows:

$$m = IdS, \tag{1.A.1}$$

where
  $I \equiv$ current (A)
  $dS \equiv$ incremental area (m^2)

In MKS, the magnetic moment is in units of A m^2. However, other authors (see Bleaney and Bleaney, 1965) prefer to write m as

$$m = \mu_0 IdS. \tag{1.A.2}$$

In Equation A.1.2, m is in units of Wb m. Equation A.1.1 implies the following definition of magnetic flux density:

$$B = \mu_0 (H + M), \tag{1.A.3}$$

TABLE 1.A.1

Conversion from MKS to Gaussian System of Units

Quantity	MKS	Conversion	Gaussian
Length	M	100	Cm
Mass	Kg	1000	G
Time	S	1	S
Density	kg/m^3	10^{-3}	g/cm^3
Force	N	10^5	Dyne
Work (energy)	J	10^7	Erg
Charge	C	1/10	statcoulomb
Potential (voltage)	V	1/300	statvolt
Current	A	1/10	statampere
Electric displacement, D	C/m^2	$12\pi \times 10^5$	statvolt/cm
Electric field, E	V/m	$10^{-4}/3$	statvolt/cm
Polarization, P	C/m^2	3×10^5	statcoulomb/cm^2
Resistance, R	Ohm	$10^{-11}/9$	s/cm
Conductivity, σ	mho/m	9×10^9	1/s
Inductance, L	H	$10^{-11}/9$	stathenry
Capacitance, C	F	9×10^{11}	Cm
Magnetic density, B	Wb/m^2	10^4	G
Magnetic field, H	A/m	$4\pi \times 10^{-3}$	Oe
Magnetic flux, Φ	Wb	10^8	maxwell
Magnetization, M^a	Wb/m^2	$10^4/4\pi$	emu/cm^3
Magnetization, M^b	A/m	10^{-3}	emu/cm^3
Magnetic dipole moment, m^b	A m^2	10^3	Emu
Magnetoelectric linear coupling (α)	s/m	3×10^8	Dimensionless

[a] See definition of M in Equation 1.A.4.
[b] See definition of M in Equation 1.A.3.

TABLE 1.A.2

Conversion to emu and esu Systems of Units

Quantity	MKS	Emu	esu
Charge	C	1/10	3×10^9
Current	A	1/10	3×10^9
Voltage	V	10^8	1/300
Power	W	10^7	10^7
Resistance	ohm	10^9	$10^{-11}/9$
Inductance	H	10^9	$10^{-11}/9$
Capacitance	F	10^{-9}	9×10^{11}

TABLE 1.A.3
Conversion for Various Units of Energy

eV	cm^{-1}	K	J	Cal
1	0.8066×10^4	1.1605×10^4	1.6021×10^{-19}	3.8291×10^{-20}
1.2398×10^{-4}	1	1.4387	1.9862×10^{-23}	4.7400
0.8617×10^{-4}	0.6951	1	1.3853×10^{-23}	3.2995×10^{-24}
0.6242×10^{19}	0.5034×10^{23}	0.7219×10^{23}	1	0.23995
2.6116×10^{19}	2.1064×10^{23}	3.0307×10^{23}	4.1840	1

TABLE 1.A.4
Physical Constants

Electron charge	1.6021×10^{-19} C
Velocity of light	2.9979×10^{10} cm/s
Acceleration of gravity	9.8066 m/s^2
Planck's constant	6.6252×10^{-34} J s
Boltzmann's constant	1.3804×10^{-23} J/deg
Avogadro's number	6.0249×10^{23} (1/g mol)
Electron mass	9.1083×10^{-31} kg
Bohr magneton	9.2732×10^{-21} emu
	1.1653×10^{-29} Wb m
Bohr's radius	5.2917×10^{-9} cm
Rydberg's constant	3.2898×10^{15} (1/s)
Fine structure constant	1/137

where
$M = m/V$ (A/m)
$V \equiv$ volume

Equation A.1.2 implies the following definition for B:

$$B = \mu_0 H + M. \tag{1.A.4}$$

In Equation A.1.4, M is defined in units of Wb/m^2, since B and M are of the same units. The definition of B in Equation A.1.4 is consistent with the definition in the form of its counterpart, the electric field density or electric displacement D. So, there is some arbitrariness about the definition of B, but we are going to use the definition of B, as defined in Equation A.1.3. This choice makes M and H to be of the same units (A/m), which is the convention throughout the world.

REFERENCES

B.I. Bleaney and B. Bleaney, *Electricity and Magnetism*, Clarendon Press, Oxford, UK, 1965.
J.D. Jackson, *Classical Electrodynamics*, John Wiley & Sons, New York, 1975.
A.H. Morrish, *The Physical Principles of Magnetism*, Wiley, New York, 1965.

SOLUTIONS

1.1   From Equation 1.11a,

$$[B] = [\mu][H],$$

$$
\begin{bmatrix} B_x \\ B_y \\ B_z \end{bmatrix} = \begin{bmatrix} \mu_{xx} & \mu_{xy} & \mu_{xz} \\ \mu_{yx} & \mu_{yy} & \mu_{yz} \\ \mu_{zx} & \mu_{zy} & \mu_{zz} \end{bmatrix} \begin{bmatrix} H_x \\ H_y \\ H_z \end{bmatrix} = \begin{bmatrix} \mu_{xx}H_x & \mu_{xy}H_y & \mu_{xz}H_z \\ \mu_{yx}H_x & \mu_{yy}H_y & \mu_{yz}H_z \\ \mu_{zx}H_x & \mu_{zy}H_y & \mu_{zz}H_z \end{bmatrix}.
$$

Therefore,

$$
\begin{cases} B_x = \mu_{xx}H_x + \mu_{xy}H_y + \mu_{xz}H_z \\ B_y = \mu_{yx}H_x + \mu_{yy}H_y + \mu_{yz}H_z \\ B_z = \mu_{zx}H_x + \mu_{zy}H_y + \mu_{zz}H_z \end{cases}
\tag{S.1.1}
$$

From Equation 1.11b,

$$\vec{B} = \ddot{\mu} \cdot \vec{H},$$

$$
B_x\vec{a}_x + B_y\vec{a}_y + B_z\vec{a}_z = \begin{pmatrix} \mu_{xx}\vec{a}_x\vec{a}_x + \mu_{xy}\vec{a}_x\vec{a}_y + \mu_{xz}\vec{a}_x\vec{a}_z \\ +\mu_{yx}\vec{a}_y\vec{a}_x + \mu_{yy}\vec{a}_y\vec{a}_y + \mu_{yz}\vec{a}_y\vec{a}_z \\ +\mu_{zx}\vec{a}_z\vec{a}_x + \mu_{zy}\vec{a}_z\vec{a}_y + \mu_{zz}\vec{a}_z\vec{a}_z \end{pmatrix} \cdot \left(H_x\vec{a}_x + H_y\vec{a}_y + H_z\vec{a}_z \right)
$$

$$
= \left(\mu_{xx}H_x + \mu_{xy}H_y + \mu_{xz}H_z \right)\vec{a}_x + \left(\mu_{yx}H_x + \mu_{yy}H_y + \mu_{yz}H_z \right)\vec{a}_y + \left(\mu_{zx}H_x + \mu_{zy}H_y + \mu_{zz}H_z \right)\vec{a}_z.
$$

Therefore,

$$
\begin{cases} B_x = \mu_{xx}H_x + \mu_{xy}H_y + \mu_{xz}H_z \\ B_y = \mu_{yx}H_x + \mu_{yy}H_y + \mu_{yz}H_z \\ B_z = \mu_{zx}H_x + \mu_{zy}H_y + \mu_{zz}H_z \end{cases}
\tag{S.1.2}
$$

Equations 1.11a and 1.11b are the same.

1.2 From Maxwell's equations in MKS system of units,

$$\left\{ \begin{array}{l} \nabla \times \vec{E} = -\dfrac{\partial \vec{B}}{\partial t} \\[2mm] \nabla \times \vec{H} = \vec{J} + \dfrac{\partial \vec{D}}{\partial t} \\[2mm] \nabla \cdot \vec{D} = \rho \\[2mm] \nabla \cdot \vec{B} = 0 \end{array} \right. ,$$

and two constitutive equations $\vec{D} = \varepsilon \vec{E}, \ \vec{B} = \mu \vec{H}$.

We rewrite Maxwell's equations in the time-harmonic form, and the medium obeys Ohm's law $\vec{J} = \sigma \vec{E}$ and $\rho = 0$:

$$\left\{ \begin{array}{l} \nabla \times \vec{E} = -j\omega\mu\vec{H} \\[1mm] \nabla \times \vec{H} = \sigma\vec{E} + j\omega\varepsilon\vec{E} \\[1mm] \nabla \cdot \vec{E} = 0 \\[1mm] \nabla \cdot \vec{H} = 0 \end{array} \right. ,$$

$$\nabla \times \left(\nabla \times \vec{E} \right) = -j\omega\mu\nabla \times \vec{H},$$

$$\nabla \left(\nabla \cdot \vec{E} \right) - \nabla^2 \vec{E} = -j\omega\mu \left(\sigma\vec{E} + j\omega\varepsilon\vec{E} \right),$$

$$\nabla^2 \vec{E} + \omega^2 \varepsilon\mu \left(1 - j\frac{\sigma}{\omega\varepsilon} \right) \vec{E} = 0.$$

For TEM wave, we assume

$$\vec{E} = E_0 e^{-jkz} \vec{a}_x,$$

$$k^2 = \omega^2 \varepsilon\mu \left(1 - j\frac{\sigma}{\omega\varepsilon} \right),$$

$$k = \sqrt{\omega^2 \varepsilon\mu \left(1 - j\frac{\sigma}{\omega\varepsilon} \right)}.$$

For highly conducting medium, $\sigma \gg \omega\varepsilon$

$$k \approx \sqrt{\omega\mu\sigma}\sqrt{-j} = \sqrt{\frac{\omega\mu\sigma}{2}}(1-j).$$

Let $k = \beta - j\alpha$

$$\alpha = \sqrt{\frac{\omega\mu\sigma}{2}}.$$

Skin depth in MKS units:

$$\delta = \frac{1}{\alpha} = \sqrt{\frac{2}{\omega\mu\sigma}},$$

$$\delta = \sqrt{\frac{2}{2\pi \times 9 \times 10^9 \times 4\pi \times 10^{-7} \times 5.8 \times 10^7}}$$

$$= 0.697 \times 10^{-6} \, \text{m} = 0.697 \mu\text{m}$$

1.3   From definition, the element of differential permeability

$$\mu_{ij}^{AC} = \frac{\partial B_i}{\partial H_j},$$

where

$$i, j = x, y, z$$

$$\mu_{xx}^{AC} = a_2 + 2a_5 H_x$$

$$\mu_{xy}^{AC} = a_3 + 2a_6 H_y$$

$$\mu_{xz}^{AC} = 0$$

$$\mu_{yx}^{AC} = b_2 + 2b_5 H_x$$

$$\mu_{yy}^{AC} = b_3 + 2b_6 H_y$$

$$\mu_{yz}^{AC} = 0$$

$$\mu_{zx}^{AC} = 0$$

$$\mu_{zy}^{AC} = 0$$

$$\mu_{zz}^{AC} = c_2 + 2c_3 H_z$$

Therefore, the differential permeability tensor is

$$[\mu]^{AC} = \begin{bmatrix} a_2 + 2a_5 H_x & a_3 + 2a_6 H_y & 0 \\ b_2 + 2b_5 H_x & b_3 + 2b_6 H_y & 0 \\ 0 & 0 & c_2 + 2c_3 H_z \end{bmatrix}.$$

1.4   Reference book:
(See Jackson, 1975)
Instead of a simple cubic lattice, we discuss the case with an ortho-rhombic lattice. The total field at any given molecule in the medium is $\vec{E} + \vec{E}_i$, where $\vec{E}$ is the macroscopic electric field with the relation $\vec{P} = \chi_e \vec{E}$, and $\vec{E}_i$ is the internal field from the neighboring molecules. The internal field $\vec{E}_i$ can be rewritten as the difference of two terms:

$$\vec{E}_i = \vec{E}_{near} - \vec{E}_P,$$

where $\vec{E}_{near}$ is the molecule field due to other dipoles $\vec{E}_P$ is the average field from the polarization $\vec{P}$

Equation 4.18 in Jackson's book shows that the average electric field that comes from $\vec{P}$ is

$$\vec{E}_P = -\frac{4\pi}{3}\vec{P}.$$

The field due to the molecules nearby is determined by the specific atomic configuration and locations of the nearby molecules. The positions of the dipoles $\vec{P}$ are given by the coordinates $\vec{r} = 2l\vec{a}_x + 3m\vec{a}_y + 4n\vec{a}_z$ along the axes x, y, and z. According to Equation 4.13 in Jackson's book, the field at the origin due to all the dipoles is

$$\vec{E}_{near} = \sum_{l,m,n} \frac{3(\vec{p} \cdot \vec{r})\vec{r} - r^2 \vec{p}}{r^5}.$$

The x-component of the field can be written in the form

$$E_x^{near} = \sum_{l,m,n} \frac{3(2lp_x + 3mp_y + 4np_z)2l - (4l^2 + 9m^2 + 16n^2)p_x}{(4l^2 + 9m^2 + 16n^2)^{5/2}}.$$

Since the indices run equally over positive and negative values, the cross terms involving mlp_y and nlp_z vanish.

$$E_x^{near} = \sum_{l,m,n} \frac{3(4l^2)-(4l^2+9m^2+16n^2)}{(4l^2+9m^2+16n^2)^{5/2}} P_x$$

$$= \sum_{l,m,n} \frac{3(4l^2)-(4l^2+9m^2+16n^2)}{(4l^2+9m^2+16n^2)^{5/2}} 24P_x = \beta P_x.$$

The y-component of the field:

$$E_y^{near} = \sum_{l,m,n} \frac{3(9m^2)-(4l^2+9m^2+16n^2)}{(4l^2+9m^2+16n^2)^{5/2}} P_y$$

$$= \sum_{l,m,n} \frac{3(9m^2)-(4l^2+9m^2+16n^2)}{(4l^2+9m^2+16n^2)^{5/2}} 24P_y = \beta P_y.$$

The z-component of the field:

$$E_z^{near} = \sum_{l,m,n} \frac{3(16n^2)-(4l^2+9m^2+16n^2)}{(4l^2+9m^2+16n^2)^{5/2}} P_z$$

$$= \sum_{l,m,n} \frac{3(16n^2)-(4l^2+9m^2+16n^2)}{(4l^2+9m^2+16n^2)^{5/2}} 24P_z = -(\alpha+\beta) P_z.$$

Therefore, $\vec{E}_{near}$ is related to the $\vec{P}$ through a traceless tensor $\ddot{S}_{\alpha\beta}$ that has the symmetry properties of the lattice:

$$\vec{E}_{near} = \ddot{S}_{\alpha\beta} \cdot \vec{P},$$

$$[S_{\alpha\beta}] = \begin{bmatrix} \alpha & 0 & 0 \\ 0 & \beta & 0 \\ 0 & 0 & -\alpha-\beta \end{bmatrix}.$$

In the case of a simple cubic lattice, it's easy to prove $\alpha = \beta = 0$.
The internal field can be written as

$$\vec{E}_i = \ddot{S}_{\alpha\beta} \cdot \vec{P} + \frac{4\pi}{3} \vec{P}.$$

The polarization vector $\vec{P}$ was defined as

$$\vec{P} = N\vec{p}_{mol},$$

where N is the volume density $\vec{p}_{mol}$ is the average dipole moment of the molecules

In our case,

$$N = \frac{1}{2 \times 10^{-8} \times 3 \times 10^{-8} \times 4 \times 10^{-8}}, \ \vec{p}_{mol} = \vec{p}.$$

This dipole moment is approximately proportional to the electric field acting on the molecule. This gives

$$\vec{p} = \gamma_{mol}\left(\vec{E} + \vec{E}_i\right),$$

γ_{mol} is called *molecular polarizability*, which characterizes the response of the molecules to an applied field.

Finally, combining the above three equations, we get

$$\vec{P} = N\gamma_{mol}\left(\vec{E} + \vec{\vec{S}}_{\alpha\beta}\vec{P} + \frac{4\pi}{3}\vec{P}\right).$$

Solving for $\vec{P}$ in terms of $\vec{E}$ and using the definition that $[\varepsilon] = [I] + 4\pi[\chi_e]$, we find

$$[\varepsilon] = \begin{bmatrix} 1 + \dfrac{4\pi}{\dfrac{1}{N\gamma_{mol}} - \dfrac{4\pi}{3} - \alpha} & & \\ & 1 + \dfrac{4\pi}{\dfrac{1}{N\gamma_{mol}} - \dfrac{4\pi}{3} - \beta} & \\ & & 1 + \dfrac{4\pi}{\dfrac{1}{N\gamma_{mol}} - \dfrac{4\pi}{3} + \alpha + \beta} \end{bmatrix}.$$

1.5

$$[B] = [\mu][H],$$

where

$$[\mu] = \begin{bmatrix} \mu_{xx} & \mu_{xy} & \mu_{xz} \\ \mu_{yx} & \mu_{yy} & \mu_{yz} \\ \mu_{zx} & \mu_{zy} & \mu_{zz} \end{bmatrix} \text{ is non-diagonal.}$$

$$[B] = \begin{bmatrix} \mu_{xx} & \mu_{xy} & \mu_{xz} \\ \mu_{yx} & \mu_{yy} & \mu_{yz} \\ \mu_{zx} & \mu_{zy} & \mu_{zz} \end{bmatrix} \begin{bmatrix} H_x \\ H_y \\ H_z \end{bmatrix} = \begin{bmatrix} \mu_{xx}H_x & \mu_{xy}H_y & \mu_{xz}H_z \\ \mu_{yx}H_x & \mu_{yy}H_y & \mu_{yz}H_z \\ \mu_{zx}H_x & \mu_{zy}H_y & \mu_{zz}H_z \end{bmatrix},$$

$$\vec{B} \cdot \vec{H} = [B]^T [H] = \left(\mu_{xx}H_xH_x + \mu_{xy}H_yH_x + \mu_{xz}H_zH_x \right) + \left(\mu_{yx}H_xH_y + \mu_{yy}H_yH_y + \mu_{yz}H_zH_y \right)$$

$$+ \left(\mu_{zx}H_xH_z + \mu_{zy}H_yH_z + \mu_{zz}H_zH_z \right),$$

$$\vec{H} \cdot \vec{B} = [H]^T [B] = \left(\mu_{xx}H_xH_x + \mu_{xy}H_xH_y + \mu_{xz}H_xH_z \right) + \left(\mu_{yx}H_yH_x + \mu_{yy}H_yH_y + \mu_{yz}H_yH_z \right)$$

$$+ \left(\mu_{zx}H_zH_x + \mu_{zy}H_zH_y + \mu_{zz}H_zH_z \right),$$

$$\int \vec{B} \cdot \vec{H} dv = \int \vec{H} \cdot \vec{B} dv,$$

because $H_iH_j = H_jH_i$, where $i, j = x, y, z$.

1.6 From Maxwell's equations in CGS system of units,

$$\begin{cases} \nabla \times \vec{E} = -\dfrac{1}{c}\dfrac{\partial \vec{B}}{\partial t} \\[2mm] \nabla \times \vec{H} = \dfrac{4\pi\sigma}{c}\vec{E} + \dfrac{1}{c}\dfrac{\partial \vec{D}}{\partial t}, \\[2mm] \nabla \cdot \vec{D} = 4\pi\rho \\[2mm] \nabla \cdot \vec{B} = 0 \end{cases}$$

and two constitutive equations $\vec{D} = \varepsilon\vec{E}, \vec{B} = \mu\vec{H}$.

We rewrite Maxwell's equations in the time-harmonic forms, and the medium obeys Ohm's law $\vec{J} = \sigma\vec{E}$ and $\rho = 0$:

$$\begin{cases} \nabla \times \vec{E} = -j\dfrac{\omega\mu}{c}\vec{H} \\[2mm] \nabla \times \vec{H} = \dfrac{4\pi\sigma}{c}\vec{E} + j\dfrac{\omega\sigma}{c}\vec{E}, \\[2mm] \nabla \cdot \vec{E} = 0 \\[2mm] \nabla \cdot \vec{H} = 0 \end{cases}$$

$$\nabla\left(\nabla \times \vec{E} \right) = -j\dfrac{\omega\mu}{c}\nabla \times \vec{H}$$

$$\nabla\left(\nabla\cdot\vec{E}\right)-\nabla^2\vec{E}=-j\frac{\omega\mu}{c}\left(\frac{4\pi\sigma}{c}\vec{E}+j\frac{\omega\sigma}{c}\vec{E}\right),$$

$$\nabla^2\vec{E}+\frac{\omega^2\varepsilon\mu}{c^2}\left(1-j\frac{4\pi\sigma}{\omega\varepsilon}\right)\vec{E}=0$$

For TEM wave, we assume

$$\vec{E}=E_0e^{-jkz}\vec{a}_x,$$

$$k^2=\frac{\omega^2\varepsilon\mu}{c^2}\left(1-j\frac{4\pi\sigma}{\omega\varepsilon}\right),$$

$$k=\sqrt{\frac{\omega^2\varepsilon\mu}{c^2}\left(1-j\frac{4\pi\sigma}{\omega\varepsilon}\right)}.$$

For highly conducting medium, $\sigma\gg\omega\varepsilon$,

$$k\approx\sqrt{\frac{4\pi\omega\mu\sigma}{c^2}}\sqrt{-j}=\sqrt{\frac{2\pi\omega\mu\sigma}{c^2}}\left(1-j\right).$$

Let $k=\beta-j\alpha$

$$\alpha=\frac{\sqrt{2\pi\omega\mu\sigma}}{c}.$$

Skin depth in CGS units:

$$\delta=\frac{1}{\alpha}=\frac{c}{\sqrt{2\pi\omega\mu\sigma}}=\frac{c}{2\pi\sqrt{\sigma\mu f}}$$

$$\delta=0.697\times10^{-4}\,\text{cm}=0.697\mu\text{m}.$$

1.7   In MKS units,

$$\vec{F}=q\left(\vec{E}+\vec{v}\times\vec{B}\right).$$

In CGS units,

$$\vec{F}=q\left(\vec{E}+\frac{\vec{v}}{c}\times\vec{B}\right).$$

1.8   Poynting vector integral equation in real time (in MKS) is

$$\oint_s\left(\vec{E}_s\times\vec{H}_s\right)\cdot d\vec{S}=-\int_V dV\left(\vec{H}\cdot\frac{\partial\vec{B}}{\partial t}+\vec{E}\cdot\frac{\partial\vec{D}}{\partial t}+\vec{J}\cdot\vec{E}\right).$$

The power dissipated in a magnetic medium is

$$P_D = \frac{1}{2} \int_V dV \vec{H} \cdot \frac{\partial \vec{B}}{\partial t}.$$

The averaging is over time, and H and B are collinear. Thus,

$$H = h_0 \cos \omega t$$

and

$$B = \mu_0 \left(h_0 \cos \omega t + m_0 \cos(\omega t - \varphi) \right).$$

Substituting the above relations for H and B and averaging over time, we obtain

$$H \frac{\partial B}{\partial t}\bigg|_t = \frac{\mu_0}{T} \int dt \left\{ -\frac{\omega}{2} \Big[h_0^2 \sin(2\omega t) + m_0 h_0 \left(\sin(2\omega t - \varphi) - \sin(\varphi) \right) \Big] \right\}$$

$$= \frac{\omega}{2} \mu_0 m_0 h_0 \sin \varphi,$$

where T = period (s).

Averaging sin $(2\omega t)$ and sin $(2\omega t - \varphi)$ with time ranging from 0 to T yields zero, but

$$\frac{1}{T} \int_0^T \sin(\varphi) dt = \sin(\varphi).$$

Then,

$$P_D = \frac{\omega}{2} V \mu_0 m_0 h_0 \sin \varphi,$$

where

$$\mu'' = \mu_0 \frac{m_0}{h_0} \sin \varphi.$$

The factor of 2 is a result of time averaging. Hence, we arrive at the final result that

$$P_D = \frac{\omega}{2} V \mu'' h_0^2,$$

where V is the volume.

This is to be compared with the previous results in the text.

2 Classical Principles of Magnetism

HISTORICAL BACKGROUND

Year[a]	Discovery
5000 BC	Magnetite was discovered in China
2400 BC	Navigation by compass
500 BC	Lodestone (magnetite)
1700	Chemical battery
1800	Generation of magnetic induction field
1850	Magnetic field generated from magnetized bodies
1900	Development of modern quantum mechanics
1913	Discovery of metal superconductors
1930	Development of magnetic materials
1946	Nuclear magnetic resonance
1947	Semiconductors
1950	Oxide ferrites
1965	Development of oxide films
1986	Oxide superconductors

[a] Approximate.

FIRST OBSERVATION OF MAGNETIC RESONANCE

From a historical perspective, the discovery of nuclear magnetic resonance probably had the greatest impact on the use of ferrites at microwave frequencies. The phenomenon of magnetic resonance was predicted around the early 1930s. As microwave sources became available during the 1940s, there were many efforts then to observe electron magnetic resonance as well as nuclear magnetic resonance. Perhaps the development of the radar had something to do with these efforts. Microwave sources were available during and after World War II. Gorter mentioned to the author in 1970 that he planned to measure electron paramagnetic resonance on a magnetic sulfide compound very early in the development of paramagnetic salts, but unfortunately the linewidth was as broad as the "Mississippi River" to have made a meaningful measurement. However, the first verified experiment on magnetic resonance in which nuclear spins were in precessional motion about the magnetic field was conceived in 1947 using standard resonant R–L–C tank circuits, as shown in Figure 2.1.

DOI: 10.1201/9781003431244-2

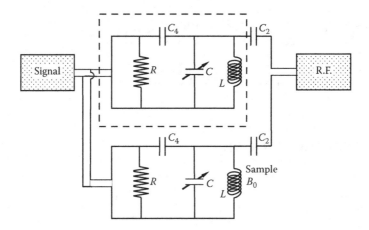

FIGURE 2.1   Nuclear magnetic resonance experiment.

Tap water from the Charles River in Boston, MA was used for testing, and absorption was monitored by adjusting the two tank circuits as in an electrical bridge arrangement in which the voltages at the output of the two tank circuits were nulled out at frequencies well above and below the magnetic resonance frequency. The bridge technique was extended later at microwave frequencies in which electron spins precessed at resonance.

DEFINITION OF MAGNETIC DIPOLE MOMENT

In this chapter, we wish to develop the basic principles of magnetism based on Maxwell's equations and classical mechanics. The only concept in magnetism that has no simple analog to classical mechanics is the concept of exchange interaction between electrons. We will treat this electrostatic interaction between electrons as a magnetic interaction, where the interaction is represented as the dot product between magnetic spins and the magnetic spins are represented as classical vectors. We will discuss this topic in detail later; for the moment, let's define the magnetic dipole moment classically. Assume an enclosed electrical wire loop carrying current placed in a uniform magnetic induction field, $\vec{B}$. We divide the cross area of the loop into small rectangular incremental strips, see Figure 2.2.

The force acting on an incremental current within $d\vec{l_1}$ is by Lorentz law:

$$d\vec{F_1} = Id\vec{l_1} \times \vec{B}.$$

The direction of $d\vec{F_1}$ is into the paper (see Figure 2.2) and $d\vec{F_2}$ is out of the paper. The magnitude of $d\vec{F_1}$ is $\left| d\vec{F_1} \right| = Idl_1 B \sin\phi_1$, where ϕ_1 is the angle between $d\vec{l_1}$ and $\vec{B}$ and is equal to $\phi_1 = \sin^{-1}(dy/dl_1)$. Thus, $dy = dl_1 \sin\phi_1$.

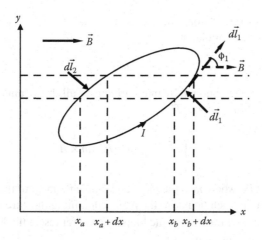

FIGURE 2.2 A current-carrying wire loop in the presence of a B field.

Summarizing,

$$\left|d\vec{F}_1\right| \cong IBdy,$$

and

$$\left|d\vec{F}_2\right| \cong IBdy.$$

The torque acting on the incremental current element at position (1) is

$$d\vec{T}_1 = \vec{x}_{ab} \times d\vec{F}_1 = \vec{x}_{ab} \times \left(Id\vec{l}_1 \times \vec{B}\right),$$

where

$$\vec{x}_{ab} = \left(x_b - x_a\right)\vec{a}_x. \tag{2.1}$$

with $(x_b - x_a)$ being the length of the incremental rectangular strip along the x-axis, as shown in Figure 2.2. We can write

$$d\vec{T}_1 = Id\vec{S} \times \vec{B},$$

since

$$d\vec{S} = \vec{x}_{ab} \times d\vec{l}_1, \tag{2.2}$$

and $d\vec{S}$ is a vector normal to the plane containing the incremental area. The magnitude of $d\vec{S}$ is the area of the rectangular incremental strip. The resultant vector

$d\vec{S} \times \vec{B}$ or the torque axis is along the y-axis. The magnetic dipole moment is defined for the incremental rectangular strip as

$$d\vec{m} = Id\vec{S}. \tag{2.3}$$

The total torque may be obtained by integrating over all the rectangular incremental strips contained within the loop. Thus,

$$\vec{T} = \int_s d\vec{T}_1 = \int_s d\vec{m} \times \vec{B} = \vec{m} \times \vec{B}. \tag{2.4}$$

We may write $\vec{m} = I\vec{S}$, where $\vec{m}$ is the magnetic dipole moment of the coil. The directions of $\vec{m}$ and $\vec{S}$ are perpendicular to the plane containing the wire loop. The unit of $\left|\vec{S}\right|$ is m², and it is simply the area of the loop. We can represent the loop rotation by a vector $\vec{m}$, as illustrated in Figure 2.3.

The magnitude of $\vec{T}$ is

$$\left|\vec{T}\right| = mB\sin\theta = T \tag{2.5}$$

Clearly, the application of a $\vec{B}$ field to a closed coil carrying current I induces mechanical motion or torque with the axis of rotation perpendicular to both $\vec{B}$ and the vector $\vec{S}$ or $\vec{m}$. Since there is motion, there exists potential to do work. The incremental potential energy, dU, required to rotate the coil by an amount $d\theta$ is

$$dU = Td\theta,$$

$$dU = mB\sin\theta d\theta,$$

where $d\theta$ is the amount of angular rotation of the loop.

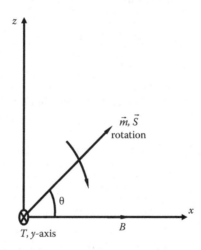

FIGURE 2.3   Magnetic moment in the presence of a B field.

The total potential energy to rotate the coil is then (by integration over θ)

$$U = -mB\cos\theta = -\vec{m} \cdot \vec{B} = \text{Am}^2\text{Wb} / \text{m}^2 = \text{J.} \tag{2.6}$$

This is recognized as the Zeeman magnetizing energy or magnetic potential energy in units of joules (J). In general, the magnetic potential energy of a coil carrying current I is

$$U = -\int d\vec{m} \cdot \vec{B} = -I\int d\vec{S} \cdot \vec{B}.$$

Remembering that $\vec{\nabla} \cdot \vec{B} = 0$, there exists a vector that satisfies the vector equation

$$\vec{\nabla} \cdot \left(\vec{\nabla} \times \vec{A}\right) = 0.$$

The vector $\vec{A}$ is recognized as the vector magnetic potential. Thus,

$$\vec{B} = \vec{\nabla} \times \vec{A}. \tag{2.7}$$

Utilizing Equation 2.7, we obtain

$$U = -I\int d\vec{S} \cdot \left(\vec{\nabla} \times \vec{A}\right). \tag{2.8}$$

The surface integral may be expressed into a line integral by invoking the Stokes theorem or

$$\int d\vec{S} \cdot \left(\vec{\nabla} \times \vec{A}\right) = \oint_c \vec{A} \cdot d\vec{l}.$$

Finally, we have

$$U = -I\oint_c \vec{A} \cdot d\vec{l}.$$

The current density J (a/m^2) may be related to $Id\vec{l}$ by the relation

$$Id\vec{l} = \vec{J}dV.$$

Equivalently, U may be related to a current density J of the wire: $U = -\int_V \left(\vec{A} \cdot \vec{J}\right)dV$, where the integration is over the volume of the current-carrying loop of wire.

Making use of $\vec{A} = \left(\vec{B} \times \vec{r}\right)/2$, U may be written as follows:

$$U = -\frac{1}{2}\int_V \left[\left(\vec{B} \times \vec{r}\right) \cdot \vec{J}\right]dV,$$

which is the same as

$$U = -\vec{B} \cdot \int_V \frac{1}{2}\left[\vec{r} \times \vec{J}\right]dV.$$

By putting the above relation into the form $U = -\vec{B}\left(\vec{B} \times \vec{r}\right)\int dm$, we identify $d\vec{m}$ as follows:

$$d\vec{m} = \frac{1}{2}\left[\vec{r} \times \vec{J}\right]dV. \tag{2.9}$$

Integration is carried along the path enclosed by the current-carrying wire. Clearly, $\vec{m}$ is a vector normal to the plane containing $\vec{r}$ and $\vec{J}$.

Assume a magnetic moment equal to the Bohr magneton, $\mu_B = 9.27 \times 10^{-24}$ A m², and an orbiting electron with radius equal to Bohr's radius, $r_B = 0.52 \times 10^{-10}$ m. Let's ask the question: what is the current and velocity of the electron based on classical arguments? Clearly, it is a hypothetical question because the analysis is purely classical, but the physical parameters assumed are derived from quantized arguments of quantum mechanics. Nevertheless, let's see where it leads. By definition, the current I may be calculated from

$$I = \frac{m}{\pi r^2} = \frac{\mu_B}{\pi r_B^2},$$

which yields $I = 0.56 \times 10^{-3}$ A.

The current may be related to the velocity, v, as follows:

$$I = \frac{dq}{dt} = \frac{\Delta q}{\Delta t} = \frac{|e|}{2\pi r_B / v}$$

or

$$v = \frac{2\pi r_B}{|e|}I = 1.14 \times 10^6 \text{ m/s},$$

which is well below the velocity of light. One may conclude that the orbital motion of an electron can be described classically without resorting to relativistic corrections. However, spin motion may be explained only in terms of Dirac's relativistic equations as well as spin–orbit interaction in an atom.

MAGNETOSTATICS OF MAGNETIZED BODIES

In the previous section, we have introduced the definition of magnetic dipole moment. A basic question to consider is whether the existence of $\vec{m}$ implies a $\vec{B}$ field. Let's address this question and assume a circular coil as before, as shown in Figure 2.4. We have from Maxwell's equations that (in the absence of displacement currents)

$$\vec{\nabla} \times \vec{H} = \vec{J}.$$

Since

$$\vec{B} = \vec{\nabla} \times \vec{A},\ A\ \text{in wb/m}, \tag{2.10}$$

we may write

$$\vec{\nabla} \times \vec{B} = \mu_0 \vec{J}$$

or

$$\vec{\nabla} \times \left(\vec{\nabla} \times \vec{A} \right) = \mu_0 \vec{J}$$

which gives

$$\nabla \left(\vec{\nabla} \cdot \vec{A} \right) - \nabla^2 \vec{A} = \mu_0 \vec{J}. \tag{2.11}$$

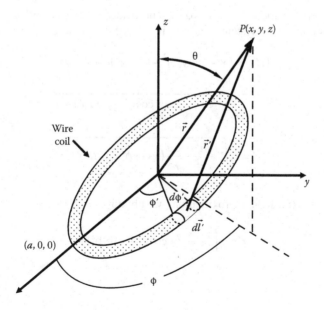

FIGURE 2.4   Observation point P relative to the coil.

For static fields, the Lorentz condition states that

$$\vec{\nabla}\left(\vec{\nabla}\cdot\vec{A}\right)=0$$

so that

$$\nabla^2\vec{A}=-\mu_0\vec{J}. \tag{2.12}$$

The solution to (2.12) is

$$\vec{A}=\frac{\mu_0}{4\pi}\int\frac{\vec{J}\vec{r}'}{r'}dV', \tag{2.13}$$

where
the prime denotes a source point
$\vec{A}$ assumes the direction of $\vec{J}$
Let's rewrite Equation 2.13 in terms of I. Use the relation

$$\vec{J}dV'=I d\vec{l}'.$$

Then,

$$\vec{A}=\frac{\mu_0}{4\pi}\int\frac{I d\vec{l}'}{r'}, \tag{2.14}$$

where $d\vec{l}'=-a d\phi'\sin\phi'\vec{a}_x-a d\phi'\cos\phi'\vec{a}_y$. The reader is referred to Figure 2.4 for a definition of coordinate system. Since dl' is defined, let's define r'. Write

$$\left(r'\right)^2=\left(x-a\cos\phi'\right)^2+\left(y-a\sin\phi'\right)^2+z^2$$

$$r'=\sqrt{x^2+y^2+z^2-2a\left(x\cos\phi'+y\sin\phi'\right)+a^2},$$

$$r'=r\sqrt{1+\frac{a^2}{r^2}-\frac{2a}{r^2}\left(x\cos\phi'+y\sin\phi'\right)}.$$

Assuming $r\gg a$ (far-field approximation), we obtain

$$r'\cong r\left[1-\frac{a}{r^2}\left(x\cos\phi'+y\sin\phi'\right)\right]$$

or

$$r'\cong r-\frac{a}{r}\left(x\cos\phi'+y\sin\phi'\right).$$

Finally, we have a simple relation for $1/r'$

$$\frac{1}{r'} \cong \frac{1}{r} + \frac{a}{r^3}\left[x\cos\phi' + y\sin\phi'\right].$$

Substituting $\vec{dl'}$ and r' into Equation 2.14, we obtain

$$\vec{A} = \frac{\mu_0}{4\pi}\int I\left(-ad\phi'\sin\phi'\vec{a}_x + ad\phi'\cos\phi'\vec{a}_y\right)\left[\frac{1}{r} + \frac{a}{r^3}\left(x\cos\phi' + y\sin\phi'\right)\right].$$

Writing $\vec{A}$ into components, we obtain

$$A_x = -\frac{\mu_0}{4\pi}a^2 I\int_0^{2\pi}\left[\frac{1}{r} + \frac{x\cos\phi' + y\sin\phi'}{r^3}\right]\sin\phi'd\phi',$$

$$A_x = -\frac{\mu_0}{4\pi}\left[\pi a^2 I\left(\frac{y}{r^3}\right)\right],$$

$$A_y = -\frac{\mu_0}{4\pi}a^2 I\int_0^{2\pi}\left[\frac{1}{r} + \frac{x\cos\phi' + y\sin\phi'}{r^3}\right]\cos\phi'd\phi',$$

and

$$A_y = \frac{\mu_0}{4\pi}\left[\pi a^2 I\left(\frac{x}{r^3}\right)\right].$$

A convenient and compact form of $\vec{A}$ is

$$\vec{A} = \frac{\mu_0}{4\pi r^3}\left(\vec{m}\times\vec{r}\right),$$

which yields

$$\vec{A} = \frac{\mu_0 m\sin\theta}{4\pi r^2}\vec{a}_\phi.$$

Since

$$\vec{m} = m\vec{a}_z,$$

$$\vec{r} = x\vec{a}_x + y\vec{a}_y + z\vec{a}_z,$$

$\vec{a}_\phi \cong$ unit vector in the circumferential direction or azimuth direction,

and

$$m = \pi a^2 I.$$

The angle between $\vec{r}$ and the z-axis is θ.

It is interesting to note that $\vec{A}$ for an electromagnetic radiating wire coil at far field is given as

$$\vec{A} = \frac{\mu_0}{4} a^2 I e^{-jkr} \left[\frac{jk}{r} + \frac{1}{r^2} \right] \sin\theta \vec{a}_\varphi,$$

where $k = \omega\sqrt{\mu_0 \mathcal{J}_0}$.

Clearly, in the limit that $\omega \to 0$ (DC case), we have exactly the same result as above. The field at point $P(x, y, z)$ may simply be calculated from

$$\vec{B} = \vec{\nabla} \times \vec{A}.$$

Thus, the field at P due to a magnetic moment along the z-direction becomes

$$\vec{B} = \frac{\mu_0 m}{4\pi r^3} \left[\frac{3xz}{r^2} \vec{a}_x + \frac{3yz}{r^2} \vec{a}_y - \left(\frac{3y^2}{r^2} + \frac{3x^2}{r^2} - 2 \right) \vec{a}_z \right]. \tag{2.15}$$

In Figure 2.5, the field $\vec{B}$ is plotted about the magnetic moment, $\vec{m}$, pointing in the z-direction.

Here, we need to exercise some caution in plotting $\vec{B}$ near $\vec{m}$. In fact, at the $\vec{m}$ site $(x=y=z=0)$, Equation 2.15 yields

$$\vec{B} = \infty.$$

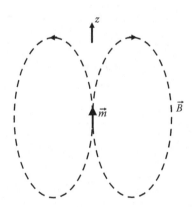

FIGURE 2.5 $\vec{B}$-field about a dipole magnetic moment, $\vec{m}$.

This is an incorrect result since we expect $\vec{B}$ to be parallel to $\vec{m}$ at $x=y=z=0$. The problem is that one cannot use a far-field calculation to infer a near-field result. Let's now derive a near-field result at the origin. Equation 2.14 is sufficiently general to calculate $\vec{A}$ for this case. Thus, $\vec{A}$ is given as follows:

$$\vec{A} = -\frac{\mu_0 I}{2} \vec{a}_\varphi.$$

yielding the $\vec{B}$ field at $x=y=z=0$ as

$$\vec{B} = \mu_0 \frac{I}{2a} \vec{a}_z = \mu_0 \left(\frac{2}{3} \frac{m}{V} \right) \vec{a}_z,$$

where $V = (4/3)\pi a^3$.

If we adopt the convention for the definition of the $\vec{B}$ field as

$$\vec{B} = \mu_0 \left(\vec{H} + \vec{M} \right),$$

we arrive at the conclusion that $\vec{H}$ must be for the example of a single wire loop

$$\vec{H} = -\mu_0 \frac{1}{3} \frac{m}{V} \vec{a}_z = -\mu_0 \frac{M}{3} \vec{a}_z$$

and

$$M = \frac{m}{V}.$$

This result is a bit misleading or disingenuous, because it is not physical. For example, if we apply a direct current to a circular wire coil, we can readily calculate the $\vec{B}$ field at the center of coil, without including any "demagnetizing" field or "Lorentzian" field. Ampere's law will account for it. However, if we have an isolated magnetic impurity in the shape of a sphere within a non-magnetic medium, indeed, there will be a demagnetizing field within the spherical impurity in the order of

$$\vec{H} = -\frac{\vec{M}}{3}, \text{MKS}.$$

The pattern of the $\vec{B}$ field around a magnetic dipole moment is very similar to the magnetic field around the earth. We are led to the conclusion that the earth is one "large" magnetic dipole moment. The reader is referred to the "References" section. In the appendix, we provide quantitative arguments for the magnetic field generation on earth. The model in the appendix was first presented as a series of class lectures in 1980. The physical picture of that model is summarized in Figure 2.6.

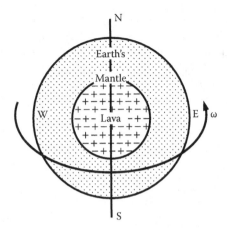

FIGURE 2.6   Model for the generation of the earth's magnetic field.

Net currents generated within the melting lava flow from west to east in order to generate fields with polarity shown in Figure 2.6. There are many models to explain the field about the earth.

Our interest here is on the magnetism of magneto-dielectric media rather than geophysics. The reader is referred to the appendix for more details about the model.

Now let's get back to magnetics in general. Let's consider N magnetic dipoles in a medium and consider its potential energy. The Zeeman interaction energy between two magnetic dipole moments is

$$U_i = -\vec{m}_i \cdot \vec{B},$$

where

$\vec{m}_i$ is the magnetic dipole moment at site i
U_i is the magnetic potential energy at site i
$\vec{B}$ is the field generated by the dipole moment at site j and is equal to

$$\vec{B} = \frac{\mu_0 m_j}{4\pi r_{ij}^3} \vec{D}_{ij},$$

where

$$\vec{D}_{ij} = \left[\vec{a}_x \frac{3x_{ij}z_{ij}}{r_{ij}^2} + \vec{a}_y \frac{3y_{ij}z_{ij}}{r_{ij}^2} - \left(\frac{3x_{ij} + 3y_{ij}^2}{r_{ij}^2} - 2 \right) \vec{a}_z \right],$$

$$r_{ij} = \left| \vec{r}_i - \vec{r}_j \right|,$$

and

$$q_{ij} = |q_i - q_i|, \quad q \equiv x, y, \text{ and } z.$$

For simplicity, we have assumed $\vec{m}_j$ to be along the z-axis. For $N-1$ dipoles, the pair interactions are summed over all of the jth sites. Thus,

$$U_i = -\frac{\mu_0}{4\pi} \sum_{j}^{N-1} \left(\frac{m_j}{r_{ij}^3} \right) \vec{m}_i \cdot \vec{D}_{ij}.$$

The (−) sign is a reminder that the basic dipole–dipole interaction is of the Zeeman type. The dipole–dipole interaction as outlined above leads to magnetic potential energies, such as the demagnetizing energy or magnetostatic energy.

ELECTROSTATICS OF ELECTRIC DIPOLE MOMENT

It is no surprise that we can describe the electric displacement field, $\vec{D}$, about an electric dipole moment, $\vec{p}$, in the same way as $\vec{B}$ near $\vec{m}$, since in free space,

$$\vec{\nabla} \cdot \vec{D} = 0$$

which implies

$$\vec{D} = -\vec{\nabla} \times \vec{F}. \tag{2.16}$$

where $\vec{F}$ is the analog of $\vec{A}$ and is referred to the vector electric potential with units of C/m. The (−) sign is arbitrary. Since

$$\vec{\nabla} \cdot \left(-\vec{\nabla} \times \vec{F} \right) = \vec{\nabla} \cdot \vec{D} = 0.$$

Thus, we can utilize all of the arguments presented so far to calculate $\vec{B}$. Specifically, we may write for $\vec{F}$ as follows:

$$\vec{F} = -\frac{1}{4\pi r^3} \left(\vec{p} \times \vec{r} \right). \tag{2.17}$$

As in the case of $\vec{A}$, the above expression applies for far-field evaluation of $\vec{D}$. Let's assume similarly

$$\vec{p} = p\vec{a}_z,$$

where $p = qd$ (see Figure 2.7).

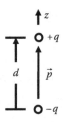

FIGURE 2.7   Electric dipole moment, $\vec{p}$.

As before (see section on Magnetostatics of Magnetized Bodies)

$$\vec{F} = -p\frac{\sin\theta}{4\pi r^2}\vec{a}_\varphi.$$

The far field for $\vec{D}$ may be calculated from

$$\vec{D} = -\vec{\nabla} \times \vec{F}$$

and

$$\vec{E} = -\frac{1}{\int_0}\vec{\nabla} \times \vec{F}\ \left(\text{free space}\right).$$

Figure 2.8 displays the far-field plot of $\vec{E}$ about $\vec{p}$, where $\vec{E}$ is of similar form as $\vec{B}$ or

$$\vec{E} = \frac{p}{4\pi\int_0 r^3}\left[\vec{a}_x\frac{3yz}{r^2} + \vec{a}_y\frac{3yz}{r^2} - \vec{a}_z\left(\frac{3\left(x^2+y^2\right)}{r^2} - 2\right)\right].$$

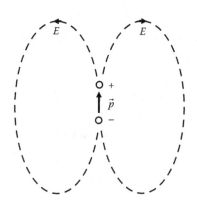

FIGURE 2.8   Electric field, $\vec{E}$, about electric dipole moment, $\vec{P}$.

Again, the above expression applies for far field. Finally, the potential energy of an electric dipole moment, $\vec{p}$, in an external electric field, $\vec{E}_{ex}$, is given below (analogous to the magnetic case):

$$U_e = -\vec{p} \cdot \vec{E}_{ex} = \text{CmV/m} = J.$$

RELATIONSHIP BETWEEN B AND H FIELDS

A convenient and compact expression for $\vec{A}$ is

$$\vec{A} = \frac{\mu_0}{4\pi r^3}(\vec{m} \times \vec{r}) = \frac{\mu_0}{4\pi}\vec{m} \times \frac{\vec{r}}{r^3} = -\frac{\mu_0}{4\pi}\vec{m} \times \vec{\nabla}\left(\frac{1}{r}\right).$$

It is noted, for example, that there is no component of $\vec{A}$ along the $\vec{m}$-direction, as demonstrated in the above representation. Rewrite $\vec{A}$ as

$$\vec{A} = \frac{\mu_0}{4\pi}\vec{\nabla} \times \frac{\vec{m}}{r}.$$

Using $\vec{B} = \vec{\nabla} \times \vec{A}$, we obtain

$$\vec{B} = \frac{\mu_0}{4\pi}\vec{\nabla} \times \left(\nabla \times \frac{\vec{m}}{r}\right)$$

and

$$\vec{B} = \frac{\mu_0}{4\pi}\vec{\nabla}\left(m \cdot \vec{\nabla}\left(\frac{1}{r}\right)\right),$$

since $\nabla^2(1/r) = 0$. This says that if a medium can be represented by $\vec{m}(r)$, one can then calculate the field about that medium. A magnetized body may be represented by a magnetization vector, $\vec{M}$, at a point P in the body, as shown in Figure 2.9.

The magnetization may be related to the magnetic moment by dividing the moment by the incremental volume. Both M_x and I may be spatially dependent. Consider the following at point P:

$$m_x = M_x dxdydz. \tag{2.18}$$

Notice that the unit of m_x is m^2 and M_x a/m. By definition, $m_x = I\,dz\,dy$, and therefore, $I\,dz\,dy = M_x\,dx\,dy\,dz$, which results in

$$I = M_x dx. \tag{2.19}$$

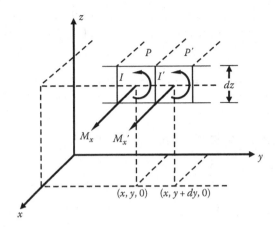

FIGURE 2.9   Magnetic current.

The current I' located at $y + dy$ may be expressed in terms of I using Taylor series expansion or

$$I' = I + \frac{\partial I}{\partial y} dy + \frac{1}{2} \frac{\partial^2 I}{\partial y^2} (dy)^2 + \text{higher order terms.} \qquad (2.20)$$

Substituting Equation 2.19 into Equation 2.20, we obtain

$$I' = I + \frac{(\partial M_x dx) dy}{\partial y} + \dots \qquad (2.21)$$

For small increments of y,

$$\frac{I - I'}{dx\, dy} \cong -\frac{\partial M_x}{\partial y}. \qquad (2.22)$$

The units of the left-hand side of Equation 2.22 are a/m² which by definition are the current density along the z-direction or

$$J_z = -\frac{\partial}{\partial y} M_x.$$

By considering similar current loops in the xz plane, we would find the other component of J_z which is equal to $(\partial/\partial x) M_y$. Thus, we write

$$J_z = \frac{\partial}{\partial x} M_y - \frac{\partial}{\partial y} M_x. \qquad (2.23)$$

By considering the three orthogonal directions of $\vec{J}$, we may summarize simply as

$$\vec{J}_m = \vec{\nabla} \times \vec{M}. \tag{2.24}$$

The subscript m on $\vec{J}$ indicates magnetic current. This is not to be confused with conduction or displacement current. It is the current that gives rise to a magnetic moment as in circular current, for example.

Let us consider Maxwell's equation containing the conduction current explicitly:

$$\vec{\nabla} \times \vec{H} = \vec{J}_\sigma, \tag{2.25}$$

where J_σ is the conduction current density. Adding $\vec{\nabla} \times \vec{M}$ on both sides of Equation 2.25 yields

$$\vec{\nabla} \times \left(\vec{H} + \vec{M} \right) = \vec{J}_\sigma + \vec{J}_m. \tag{2.26}$$

Multiplying the above equation by μ_0, we obtain

$$\vec{\nabla} \times \mu_0 \left(\vec{H} + \vec{M} \right) = \mu_0 \left(\vec{J}_\sigma + \vec{J}_m \right),$$

where

$$\vec{B} = \mu_0 \left(\vec{H} + \vec{M} \right). \tag{2.27}$$

Therefore,

$$\vec{\nabla} \times \vec{B} = \mu_0 \left(\vec{J}_\sigma + \vec{J}_m \right).$$

$\vec{B}$ may be defined in terms of a dyadic permeability, $\ddot{\mu}$:

$$\vec{B} = \ddot{\mu} \cdot \vec{H}.$$

Then, Equation 2.27 becomes

$$\vec{\nabla} \times \left(\ddot{\mu} \cdot \vec{H} \right) = \mu_0 \left(\vec{\nabla} \times \vec{H} + \vec{\nabla} \times \vec{M} \right).$$

Rearranging, we have

$$\vec{\nabla} \times \left(\ddot{\mu} \cdot \vec{H} - \mu_0 \left(\vec{H} + \vec{M} \right) \right) = 0.$$

We introduce a dyadic susceptibility such that

$$\vec{M} = \ddot{\chi} \cdot \vec{H}.$$

Finally, we have

$$\vec{\nabla} \times \left(\left[\ddot{\mu} - \mu_0 \left(\ddot{I} + \ddot{\chi} \right) \right] \cdot \vec{H} \right) = 0,$$

concluding that

$$\ddot{\mu} = \mu_0 \left(\ddot{I} + \ddot{\chi} \right). \tag{2.28}$$

It is noted that

$$\ddot{I} \cdot \vec{H} = \left(\vec{a}_x \vec{a}_x + \vec{a}_y \vec{a}_y + \vec{a}_z \vec{a}_z \right) \cdot \left(H_x \vec{a}_x + H_y \vec{a}_y + H_z \vec{a}_z \right),$$

$$\ddot{I} \cdot \vec{H} = H_x \vec{a}_x + H_y \vec{a}_y + H_z \vec{a}_z = \vec{H},$$

$$\ddot{I} \cdot \vec{H} = \vec{H}.$$

The analog to $\ddot{I}$ is the unit tensor. Thus, a relationship between permeability and susceptibility dyadics or tensors has been established.

GENERAL DEFINITION OF MAGNETIC MOMENT

It is well known that the orbital motion implies angular momentum. In this section, we wish to establish a connection between magnetic moment and angular momentum. We had from previous arguments that

$$\vec{m} = \frac{1}{2} \oint \left(\vec{r} \times \vec{J} \right) dV. \tag{2.29}$$

A simple rearrangement shows that

$$\vec{m} = \frac{1}{2} \oint \left(\vec{r} \times I d\vec{l} \right), \tag{2.30}$$

where $d\vec{l}$ is the incremental length over which current flows (Figure 2.10).

Let's define the incremental charge, dq, over dl as $I d\vec{l} = \vec{v} dq$, where v is the linear velocity of the incremental charge. Substituting the above into (2.30), one obtains

$$\vec{m} = \frac{1}{2} \int \left(\vec{r} \times \vec{v} \right) dq = \frac{1}{2m_q} \int \left(\vec{r} \times \vec{p} \right) dq, \tag{2.31}$$

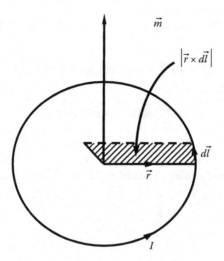

FIGURE 2.10 Graphic representation of Equation 2.30. The magnitude of $d\vec{m}$ is 1/2 of shaded area.

where

$\vec{m}_q$ = mass of charge
$p = mv$ = linear momentum
Define angular momentum as $\vec{G} = \vec{r} \times \vec{p}$. Thus, Equation 2.31 may be rewritten as

$$\vec{m} = \frac{1}{2m_q}\int \vec{G}dq = \frac{\vec{G}}{2m_q}\int dq$$

or simply

$$\vec{m} = \frac{q}{2m_q}\vec{G}.$$

Let $\gamma = q/2m_q \equiv$ gyromagnetic ratio. Let's evaluate γ in MKS and CGS systems of units.

$$\gamma = e/2m = \frac{1.602 \times 10^{-19}\,\mathrm{C}}{2 \times 9.1090 \times 10^{-31}\,\mathrm{kg}} = 8.7934 \times 10^{10}\,\mathrm{C/kg},$$

$$\gamma = 8.7934 \times 10^{10}\,\frac{\mathrm{rad/\,s}}{\mathrm{Wb/\,m}^2} = 2\pi \times 1.4 \times 10^{10}\,\frac{\mathrm{Hz}}{\mathrm{Wb/\,m}^2},\ \mathrm{MKS}$$

$$\gamma = 2\pi \times 1.4 \times 10^{10}\,\frac{\mathrm{Hz}}{10^4\mathrm{G}} = 2\pi \times 1.4 \times 10^6\,\frac{\mathrm{Hz}}{\mathrm{Oe}},\ \mathrm{CGS}.$$

$$\mu_B \equiv \text{Bohr magneton} = \gamma\hbar = 9.273 \times 10^{-24}\,\text{A m}^2, \text{ MKS}$$

$$\mu_B = 9.273 \times 10^{-21}\,\text{statmpere cm}^2 = 9.273 \times 10^{-21}\,\text{emu}$$

$$= 9.273 \times 10^{-21}\,\text{Gcm}^3/4\pi, \text{ CGS}$$

The reader should be aware that sometimes in this book we use the symbol β for the Bohr magneton for convenience. Finally, $\vec{m}$ can be related to angular momentum as follows:

$$\vec{m} = \gamma\vec{G}. \tag{2.32}$$

For a single particle, we have the following:

$$q = +e, \text{ proton}\left(\gamma = \frac{e}{2m_p}\right),$$

$$q = -e, \text{ electron}\left(\gamma = \frac{-|e|}{2m_e}\right),$$

where

$m_q = m_p = $ mass of proton
$m_q = m_e = $ mass of electron

CLASSICAL MOTION OF THE MAGNETIC MOMENT

Let's go back to the original problem of m in the presence of a $\vec{B}$ field. Torque is related to both $\vec{m}$ and $\vec{B}$ as

$$\vec{T} = \vec{m} \times \vec{B} \tag{2.33}$$

The time-rate change of angular momentum is by definition the torque or

$$\vec{T} = \frac{d\vec{G}}{dt}.$$

Since $\vec{m} = \gamma\vec{G}$, Equation 2.33 becomes finally

$$\frac{1}{\gamma}\frac{d\vec{m}}{dt} = \vec{m} \times \vec{B}. \tag{2.34}$$

Equation 2.34 is valid for macroscopic as well as microscopic magnetic media. Thus, Equation 2.34 forms the basis for the microwave properties of magnetic materials provided $\vec{B}$ is known within the magnetized media. We will develop in the next few chapters a systematic theoretical formulation in determining $\vec{B}$ internal to a magnetized media (see Chapter 5).

This equation of motion is applicable to a coil in motion as well as a charged particle in the presence of a $\vec{B}$ field. It is a very general equation. For spin motion or for relativistic motions, γ is modified:

$$\gamma = g \frac{e}{2m_e},$$

where $g = 2.0013$ for free electron (J. Schwinger). For most practical materials, $g \sim 2$.

Let's now consider the equation of motion of a magnetic moment $\vec{m}$ in the presence of a static $\vec{B}$ field along the z-axis, for example. Using the equation of motion for $\vec{m}$, one obtains

$$\frac{1}{\gamma} \frac{dm_x}{dt} = m_y B$$

and

$$\frac{1}{\gamma} \frac{dm_y}{dt} = -m_x B.$$

Also,

$$\frac{1}{\gamma} \frac{dm_z}{dt} = 0, \; m_z = \text{constant}.$$

If the moment is induced by the electron charge, it is proper to change γ to $-|\gamma|$, since the charge of the electron is $-|e|$. Thus, the equations of motion yield

$$\frac{1}{|\gamma|} \frac{dm_x}{dt} = -m_y B$$

and

$$\frac{1}{|\gamma|} \frac{dm_y}{dt} = m_x B.$$

Assume solution for m_x as $m_x = A \cos \omega t$, then, we have

$$\frac{\omega}{|\gamma|} A \sin \omega t = m_y B$$

and

$$\left[\left(\frac{\omega}{\gamma}\right)^2 - B^2\right] A\cos\omega t = 0.$$

In order to obtain a nontrivial solution, we require that

$$\frac{\omega}{|\gamma|} = B,$$

$$m_x = A\cos\omega t,$$

and

$$m_y = A\sin\omega t.$$

This means that the DC field, B, has induced an angular motion with an angular velocity of ω. Thus, the normal mode solution or the natural way the magnetic moment is set in motion by the B field is a circular motion. The solutions imply that $m_x^2 + m_y^2 = A^2$, which is the equation of a circle in the (m_x, m_y) plane. Thus, the magnetic moment rotates in a counterclockwise motion or precessional motion around B in a circular motion. This is the normal mode solution or natural way for the notion of the magnetic moment. The angular velocity of the motion is ω and is directly proportional to $|\gamma| B$. There is motion only in the presence of B.

$$\omega = |\gamma| B.$$

This is sometimes referred to as the Larmor frequency.

The amplitude A is determined by the excitation applied to the system or to the magnetic moment. At $T=0$, where for example m is at equilibrium and there is no excursion from equilibrium, no torque is generated, and therefore, no precessional motion of m. Since there is no damping, the magnetic moment will precess forever once its motion has been initiated by an external perturbation. However, every magnetic system is lossy. In general, the amplitude, A, is maximum only at resonance ($\omega/|\gamma|=B$). For frequencies away from the resonance frequency, the oscillation amplitudes are smaller than at resonance. The excitations may come in the form of thermal fluctuations, strain fields, electromagnetic transients, or even mechanical excitations. In short, a small disturbance is needed to move the magnetic moment away from its equilibrium position. The B field specifies the angular frequency at which the motion is potentially perpetual.

In a lossy medium, an electromagnetic field is needed to keep m away from its equilibrium position. In a sense, the drive field balances the losses of the medium so that A is the result of this balance. Henceforth, we will ignore the sign of γ and be mindful in the future that the actual natural motion is opposite of what we will calculate in the remaining chapters, especially in systems where the magnetic moments are derived from electronic motion.

Let's calculate the magnetic resonance frequency for an external DC field of $B=0.32$ wb/m². We will do this in two systems of units: MKS and CGS. The magnetic resonance frequency obeys the following relation: $f=(\gamma/2\pi)B$.

MKS	CGS
$\dfrac{\gamma}{2\pi}=g\times1.4\times10^{10}\,\dfrac{\text{Hz}}{\text{Wb/m}^2}$	$=g\times1.4\times10^{6}\,\dfrac{\text{Hz}}{\text{Oe}}$
$g=2$	$=2$
$B=0.32\text{Wb}/\text{m}^2$	$=3,200$ G
$f=2.8\times10^{10}\,\dfrac{\text{Hz}}{\text{Wb}/\text{m}^2}\times0.32\text{Wb}/\text{m}^2$	$=2.8\times10^{6}\,\dfrac{\text{Hz}}{\text{Oe}}\times3,200$ G
$f=8.96$ GHz	$=8.96$ GHz +

PROBLEMS

2.1 Assume that a current loop is canted 30° to a B field and that the current is constant through the loop and the radius is also fixed. Calculate the torque and plot the magnetic potential energy qualitatively.

2.2 Calculate the force acting on a compass needle 2 in. long assuming the following parameters: radius of needle$=0.1$ mm; $B=1$ Wb/m²; and earth's field$=1/4$ Oe. Assume that the initial position of needle axis is at 45° with respect to the earth's field direction.

2.3 Evaluate the vector potential $\vec{A}$ for a current loop of radius, $r=2$ cm, and with a current $I=2$ A. Calculate the B field.

2.4 Show that for a uniform B field that

$$\vec{A}=\frac{1}{2}\left(\vec{B}\times\vec{r}\right).$$

2.5 Show that in the limit of $a>r'\vec{B}$, it becomes

$$\vec{B}=\frac{\mu_0 m}{2\pi r^3}\,\vec{a}_z.$$

2.6 The earth's field is 0.6 Oe at the poles. Assume the earth's radius to be 4,000 km. Calculate the equivalent current within the earth to generate such a field, and hence, m.

2.7   Calculate the magnetic moment of a current loop using Equation 2.30. Assume $I=2$ A and $r=2$ cm.

2.8   Calculate the precessional resonance frequency of a superconducting ring in space. You may assume that the earth's field is ~0.01 Oe, the radius of the ring is $1/\sqrt{\pi}$ cm, and the thickness of the ring is 100 Å. Also, assume a density of 10 g/cm³. The key to this problem is to determine g from the weight of the ring.

2.9   Evaluate γ in MKS and CGS systems of units for an electron. Assume $g=2$.

2.10   For a simple model of an atom where the electron moves in a circular orbit around the nucleus, determine the classical expression for the magnetic moment.

2.11   Develop an expression for the Bohr magneton by quantizing the angular momentum.

2.12   Using the value of the Bohr magneton, calculate the magnitude of the angular velocity and tangential velocity of an orbiting electron (assume a radius of 1 Å). Is this a realistic value?

APPENDIX 2.A

Based on the enormous internal current generated in the earth's lava, let's estimate the charge density, n, to support this amount of current generated (see Problem 2.6). Starting with the definition of magnetic moment,

$$m = \int \pi x^2 J(x)\,dxdz,$$

where

$J(x) = env = enx\omega = 2\pi enx/T$
$\omega = 2\pi/T = $ earth's rotation frequency
$T = $ period $= 86{,}400$ s
$n = $ carrier density
$e = $ electron charge
$x = r\sin\theta$
$dx\,dz = r\,d\theta\,dr$

The limits of integration on r are from 0 to R and on θ are from π to 0. The result of integration is

$$m = \frac{8\pi^2 enR^5}{15T} = I\pi R^2.$$

Solving for n, we obtain

$$n = \frac{15IT}{8\pi eR^3}.$$

Assuming $I = 2 \times 10^9$ A (see Problem 2.6) and $R \sim 4,000$ km, we obtain

$$n \approx 10^7 / \text{cm}^3 \text{carriers.}$$

BIBLIOGRAPHY

B.I. Bleaney and B. Bleaney, *Electricity and Magnetism*, Clarendon Press, Oxford, UK, 1965.
S. Chikazumi, *J. Phys. Soc. Jpn.* **5**, 327, 1950.
G.A. Glatzmaier, The geodynamo, www.es.ucsc.edu/~glatz/geodynamo.html
L.D. Landau and E.M. Lifshitz, *Mechanics*, Pergamon Press, Oxford, U.K., 1960.
L. Neel, *J. Phys. Radium* **12**, 339, 1951; **13**, 249, 1952.
P. Weiss, *J. Phys.* **6**, 667, 1907.
W.P. Wolf, *Phys. Rev.* **108**, 1152, 1957.

SOLUTIONS

2.1   We assume that the radius and the current of the ring are a and I. The magnetic moment is given by $|\vec{m}| = \pi a^2 I$.
The torque is calculated using

$$\vec{\tau} = \vec{m} \times \vec{B} = |\vec{m}| |\vec{B}| \sin\left(\frac{\pi}{3}\right) \vec{a}_\perp,$$

$$= \pi a^2 I |\vec{B}| \sin\left(\frac{\pi}{3}\right) \vec{a}_\perp,$$

where $\vec{a}_\perp \equiv$ unit vector perpendicular to both $\vec{m}$ and $\vec{B}$.
Therefore,

$$|\vec{\tau}| = \frac{\pi a^2 I |\vec{B}| \sqrt{3}}{2}.$$

The potential energy is given by

$$U = -\vec{m} \cdot \vec{B} = -|\vec{m}| |\vec{B}| \cos\left(\frac{\pi}{3}\right),$$

$$= \pi a^2 I |\vec{B}| \cos\left(\frac{\pi}{3}\right).$$

Therefore,

$$U = -\frac{\pi a^2 I |\vec{B}|}{2}.$$

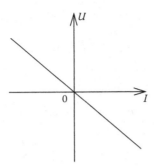

FIGURE S2.1   Plot of potential energy versus I.

2.2   The magnetic energy is $U = -\vec{m} \cdot \vec{B} = |\vec{m}||\vec{B}|\cos\theta$.
The torque is given by

$$|\vec{\tau}| = \frac{\partial U}{\partial \theta} = |\vec{m}||\vec{B}|\sin\theta = lF \ .$$

Therefore, F is

$$|\vec{F}| = \frac{|\vec{m}||\vec{B}|\sin\theta}{l} = \frac{|\vec{m}||\mu_0\vec{H}|\sin\theta}{l},$$

$$= \frac{\mu_0|\vec{M}||\vec{H}|V}{l}\sin\theta,$$

$$= \mu_0|\vec{M}||\vec{H}|\pi r^2 \sin\theta,$$

$$= \frac{1}{16\sqrt{2}} \times 10^{-3}\,N,$$

$$|\vec{F}| = 3.13 \times 10^{-5}\,N.$$

2.3   Maxwell's equation in this case is $\nabla \times \vec{H} = \vec{J}$. The vector potential is given by

$$\vec{A} = \frac{\mu_0}{4\pi}\int \frac{\vec{J}(\vec{r})dV'}{r'},$$

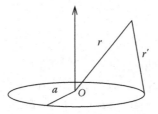

FIGURE S2.3 Geometry of the problem (2.3).

where $\vec{J} dV' = I d\vec{l}\,'$.

$$\vec{A} = \frac{\mu_0}{4\pi} \int \frac{I d\vec{l}\,'}{r'},$$

where $dl' = a d\phi' \left(\cos\phi' \vec{a}_{y'} - \sin\phi' \vec{a}_{x'} \right)$.

$$\left(r'\right)^2 = \left(x - a\cos\phi'\right)^2 + \left(y - a\sin\phi'\right)^2 + \left(z'\right)^2.$$

Use far zone approximation, $r \gg a$:

$$\frac{1}{r'} = \frac{1}{r} + \frac{a}{r^3} \left(x\cos\phi' + y\sin\phi' \right),$$

$$\vec{A} = \frac{\mu_0 I}{4\pi} \int a d\phi' \left(\cos\phi' \vec{a}_{y'} - \sin\phi' \vec{a}_{x'} \right) \left\{ \frac{1}{r} + \frac{a}{r^3} \left(x\cos\phi' + y\sin\phi' \right) \right\},$$

$$\therefore \vec{A} = -\frac{\mu_0}{4\pi} \pi a^2 I \left(\frac{y}{r^3} \vec{a}_x - \frac{x}{r^3} \vec{a}_y \right) = -\frac{8\pi \times 10^{-11}}{r^3} \left(y\vec{a}_x - x\vec{a}_y \right),$$

$$\therefore \vec{B} = \nabla \times \vec{A} = \frac{8\pi \times 10^{-11}}{r^3} \left\{ \frac{3xz}{r^2} \vec{a}_x + \frac{3yz}{r^2} \vec{a}_y - \left(\frac{3y^2}{r^2} + \frac{3x^2}{r^2} - 2 \right) \vec{a}_z \right\}.$$

2.4 Let $\vec{B} = B_x \hat{a}_x + B_y \hat{a}_y + B_z \hat{a}_z$. Assume that $\vec{A} = (1/2)\left(\vec{B} \times \vec{r}\right)$.
By the use of determinant, we can calculate A as follows:

$$\vec{A} = \frac{1}{2} \begin{vmatrix} \vec{a}_x & \vec{a}_y & \vec{a}_z \\ B_x & B_y & B_z \\ x & y & z \end{vmatrix} = \frac{1}{2}\left[\left(B_y z - B_z y\right)\vec{a}_x + \left(B_z x - B_x z\right)\vec{a}_y + \left(B_x y - B_y x\right)\vec{a}_z \right].$$

Take the curl of the above equation:

$$\vec{B} = \nabla \times \vec{A} = \frac{1}{2} \begin{vmatrix} \vec{a}_x & \vec{a}_y & \vec{a}_z \\ \dfrac{\partial}{\partial x} & \dfrac{\partial}{\partial y} & \dfrac{\partial}{\partial z} \\ B_y z - B_z y & B_z x - B_x z & B_x y - B_y x \end{vmatrix}$$

$$= \frac{1}{2}\left[(B_x + B_x)\vec{a}_x + (B_y + B_y)\vec{a}_y + (B_y + B_z)\vec{a}_z\right]$$

$$= B_x \vec{a}_x + B_y \vec{a}_y + B_z \vec{a}_z$$

$$= \vec{B}.$$

This shows us that $\vec{A} = (1/2)(\vec{B} \times \vec{r})$.

2.5   From Problem 2.3, we have that

$$\vec{B} = \frac{\mu_0 |\vec{m}|}{4\pi r^3}\left[\frac{3xz}{r^2}\vec{a}_x + \frac{3yz}{r^2}\vec{a}_y + \left\{2 - \frac{3}{r^2}(x^2 + y^2)\right\}\vec{a}_z\right].$$

For $x = y = 0$,

$$\therefore \vec{B} = \frac{\mu_0 |\vec{m}|}{4\pi r^3}\vec{a}_z.$$

2.6   We define the radius of earth as a. We again use the answer of Problem 2.3 in this question.

$$|\vec{B}| = \frac{\mu_0 I a^2}{2\pi a^3} = \frac{\mu_0 I}{2\pi a}, \ a = 4{,}000 \text{ km},$$

$$I = \frac{2\pi a B}{\mu_0} = 2 \times 10^9 \text{ A}.$$

2.7   The magnetic moment is given by Equation 2.30:

$$\vec{m} = \frac{1}{2}\int (\vec{r} \times \vec{J})d^3 r.$$

Use the fact that $\vec{J}d^3 r = I d\vec{l}$ (where ds is a line segment), then we have

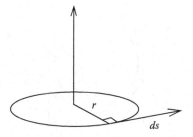

FIGURE S2.7 Geometry of the problem (2.7)

$$\vec{m} = \frac{1}{2}\int\left(\vec{r}\times d\vec{l}\right),$$

$$\therefore \vec{r}\times d\vec{l} = rdl\vec{a}_{\perp},$$

$$\therefore |\vec{m}| = \frac{1}{2}\int_{0}^{r} lrdl = \pi r^{2}l,$$

$$\therefore |\vec{m}| = \pi r^{2}I = 0.0025 \ A \ m^{2},$$

where $I = 2$ A and $r = 2$ cm.

2.8 From Figure S2.8,

$$|\vec{m}| = \left|\vec{G}\right|\gamma = aN_{s}m_{s}v_{s}\gamma = \pi a^{2}J_{s}(tw)$$

$$= a^{2}\pi\frac{N_{s}}{V}Q_{s}v_{s}(tw)$$

$$= a^{2}\pi\frac{N_{s}}{(tw)2\pi a}Q_{s}v_{s}(tw).$$

If $\gamma = Q_{s}/2m_{s}$, then $g_{s} = 1$.
We have

$$\left(\frac{\omega}{\gamma}\right) = H,$$

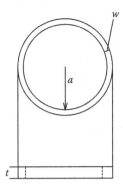

FIGURE S2.8   Geometry of the problem (2.8)

where H is the earth's field.

$$\frac{2\pi f}{g_s\left(2\pi \times 1.4 \times 10^6\right)} = H.$$

Therefore, $f = 1.4 \times 10^4$ Hz.

2.9 γ is given by $\gamma = g(e^-/2m)$, where $g = 2$ for electron

a. γ_{MKS} in MKS system:

$$\gamma_{MKS} = 2 \times \frac{e^-}{2m_e} = \frac{1.6021 \times 10^{-19}\ \text{C}}{9.1091 \times 10^{-31}\ \text{kg}} = 1.76 \times 10^{11}\ \text{C}/\text{kg}.$$

b. γ_{CGS} in CGS system:

$$\gamma_{CGS} = 2 \times \frac{e^-}{2m_e} = \frac{1.6021 \times 10^{-20}}{9.1091 \times 10^{-28}} = 1.7588 \times 10^7\ \text{Hz/Oe}.$$

$$\frac{\gamma_{CGS}}{2\pi} = 2.8 \times 10^6\ \text{Hz/Oe} = g \times 1.4 \times 10^6\ \text{Hz/Oe}.$$

2.10   The velocity is given by $v = 2\pi r/T$, where v is the velocity of electron, r is the radius of the orbit, and T is the period of electron. The current I is given by

$$I = \frac{ev}{2r\pi}.$$

Therefore, the magnetic moment $\vec{m}$ is calculated as follows:

$$|\vec{m}| = \pi r^2 I = \left(\frac{ev}{2r\pi}\right)r^2\pi = \frac{1}{2}erv.$$

We calculate the angular momentum L:

$$L = m_e rv \quad (m_e : \text{mass of electron}).$$

Therefore,

$$rv = \frac{L}{m_e},$$

$$|\vec{m}| = \frac{1}{2}e\left(\frac{L}{m_e}\right) = \left(\frac{e}{2m_e}\right)L, \; g = 1.$$

2.11 The magnetic moment $\vec{m}$ is calculated as (Figure S2.11)

$$|\vec{m}| = \frac{qL}{2m_e} = \frac{qv}{2\pi r}A,$$

where
$A = \pi r^2$
v is a velocity of electron
We quantize the angular momentum as follows:

$$L = \hbar n = mvr.$$

The magnetic moment is

$$|\vec{m}| = -\frac{e\hbar}{2m_e}n$$

where $q=-e$ (the charge of electron).
Therefore, the final answer is

$$\mu_B = \frac{e\hbar}{2m_e} = 9.27\times 10^{-24}\, \text{J/T} = 9.27\times 10^{-21}\, \text{erg/Oe}.$$

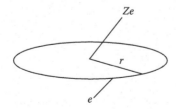

FIGURE S.2.11 Geometry of the problem (2.11)

2.12 Bohr magneton is $\mu_B = \hbar\gamma = e\hbar/m_e$, where m_e is the mass of electron. We can find an equation by Newton's second law:

$$\frac{1}{4\pi\varepsilon_0}\frac{e}{r^2} = m_e r\omega^2,$$

where ω is an angular velocity.

Finally,

$$\omega = \frac{1}{r}\sqrt{\frac{1}{4\pi\varepsilon_0 r}\left(\frac{\mu_B}{\hbar}\right)} \text{ and } v = r\omega = \sqrt{\frac{1}{4\pi\varepsilon_0 r}\left(\frac{\mu_B}{\hbar}\right)}.$$

Therefore, the answer is

$$\omega = 2.81\times10^{25}\,\text{rad}/\text{s and}$$

$$v = 2.81\times10^{15}\,\text{m}/\text{s}.$$

3 Introduction to Magnetism

One of the fundamental questions resolved in the early development of quantum mechanics was whether the electron motion was wavelike or particle-like. The answer was provided by De Broglie who showed that the particle linear momentum, p, was related to an electromagnetic wavelength, λ, as follows: $\lambda = h/p$, where $h =$ Planck's constant $= 6.65 \times 10^{-34}$ J s.

This relationship was important because it says that the motion of a particle can be described in terms of wave propagation. Thus, for example, a particle enclosed in a microwave cavity can be described in terms of an electromagnetic standing mode or as a particle in a box. De Broglie's equation proved to be a great triumph in the early development of quantum physics, especially when applied to atomic physics. Let's apply the De Broglie relationship to a ring resonator. The ring resonator consists of a simple wire loop and is coupled to a microwave source. Further, let's imagine a hypothetical situation in which the loop is so small that very few, perhaps one single electron, is flowing around the wire loop. Thus, the motion is constrained to be circular. Remarkably, it radiates maximum energy at resonance, as determined by $n\lambda = 2\pi r$, where $n = 1, 2, 3, \ldots$, λ is the electrical wavelength of the radiated energy or the wavelength of the particle within the wire loop, and r is the radius of the loop. If we assume that De Broglie's equation is applicable, λ is related to the linear momentum, p, of the few particles "entrapped" in the resonator. Since the motion is circular, the angular momentum, G, is simply $G = pr$. This implies that (applying the De Broglie relationship)

$$2\pi r = \frac{h}{G} rn$$

or

$$G = n\hbar.$$

G is in units of $\hbar$ and discrete. This implies that the magnetic moment of the ring resonator must also be quantized, since G is discrete.

Let's analyze the resonator quantum mechanically. The kinetic energy, E, due to the angular motion is given by

$$E = \frac{p^2}{2m} = \frac{mr^2\omega^2}{2},$$

DOI: 10.1201/9781003431244-3

59

where $p = mr\omega$. The position of the particle can be specified in a probabilistic sense by the wave function ψ, which obeys Schrodinger's equation

$$E\psi = i\hbar \frac{\partial \psi}{\partial t}. \tag{3.1}$$

Since

$$p \rightarrow \hbar \frac{i}{r} \frac{\partial}{\partial \theta},$$

Equation 3.1 may be written as

$$E\psi = -\frac{\hbar^2}{2mr^2} \frac{\partial^2 \psi}{\partial \theta^2} \tag{3.2}$$

Solutions of ψ take the form

$$\psi = A e^{i\Gamma\theta} + B e^{-i\Gamma\theta}$$

where

$$\Gamma^2 = \frac{2mr^2 E}{\hbar^2}.$$

The solution of ψ requires that

$$\psi(\theta + 2\pi = \psi(\theta),$$

which implies that

$$e^{i 2\pi \Gamma} = 1$$

or

$$2\pi\Gamma = 2\pi n$$

or

$$\Gamma = n.$$

Thus,

$$G = n\hbar,$$

Therefore,

$$E = \frac{G^2}{2mr^2} = \frac{(n\hbar)^2}{2mr^2}.$$

The point of this exercise is that De Broglie's relationship implies a nonclassical approach to particles in motion, especially at small scales.

The ramification of a quantized or discrete angular momentum, G, at very small dimensional scales is that the magnetic dipole moment, m, must also be discrete or quantized, since m is proportional to G or

$$m = \gamma G.$$

This means that m at a point in space, or nearly so, must be discrete, and immediately next to it, $m = 0$. This picture is totally in contrast to the classical picture whereby m is assumed to be point by point continuous within a sample. We will construct conceptually, structurally, and mathematically the composition of macroscopic magnetic materials in terms of discrete magnetic moments and show that, in the limit of long wavelength excitations, a classical description of the microwave magnetic properties of materials is sufficient to describe the phenomena in question. To begin with, let's start with the most fundamental building block of a magnetic material: an atom with partially filled 3d, 4f, ..., shells. Our interest here is mostly on the transition metal atoms, since this group of atoms are the building blocks of most of the practical microwave magnetic materials.

ENERGY LEVELS AND WAVE FUNCTIONS OF ATOMS

Let's now consider hydrogenic wave functions of atoms with nucleus charge Ze, where

$$H = -\frac{\hbar^2}{2m_e}\nabla^2 - \frac{Ze^2}{r}.$$

The Hamiltonian H is in terms of the kinetic (first-term) and potential (second-term) energies. Z is the atomic number or number of electrons in the atom.

The Schrodinger equations of this system are

$$H\Psi = E\Psi$$

$$\left[\frac{1}{r^2}\frac{\partial}{\partial r}\left(r^2\frac{\partial}{\partial r}\right) + \frac{1}{r^2\sin\theta}\frac{\partial}{\partial\theta}\left(\sin\theta\frac{\partial}{\partial\theta}\right) + \frac{1}{r^2\sin^2\theta}\frac{\partial^2}{\partial\varphi^2}\right]\Psi$$

$$+ \frac{2m_e}{\hbar^2}\left[E + \frac{Ze^2}{r}\right]\Psi = 0 \tag{3.3}$$

Assuming solution of the form

$$\Psi(r,\theta,\varphi) = R(r)Y(\theta,\varphi)$$

and substituting this solution into Equation 3.1, we obtain

$$
\frac{1}{R(r)}\frac{\partial}{\partial r}\left(r^2\frac{\partial}{\partial r}R(r)\right) + \frac{2m_e}{\hbar^2}\left(E + \frac{Ze^2}{r}\right)r^2
$$

$$
= \frac{1}{Y(\theta,\varphi)\sin\theta}\frac{\partial}{\partial\theta}\left(\sin\theta\frac{\partial Y(\theta,\varphi)}{\partial\theta}\right) - \frac{1}{Y(\theta,\varphi)\sin^2\theta}\frac{\partial^2 Y(\theta,\varphi)}{\partial\varphi^2} \tag{3.4}
$$

Each side can be equated to a constant λ. There are a number of solutions for a given value of λ, but one solution is

$$\lambda = l(l+1), \text{ where } l = 0,1,2,3,\ldots.$$

Thus, we have

$$
\frac{1}{\sin\theta}\frac{\partial}{\partial\theta}\left(\sin\theta\frac{\partial Y(\theta,\varphi)}{\partial\theta}\right) + \frac{1}{\sin^2\theta}\frac{\partial^2 Y(\theta,\varphi)}{\partial\varphi^2} = -l(l+1)Y(\theta,\varphi) \tag{3.5}
$$

and

$$
\frac{1}{r^2}\frac{d}{dr}\left(r^2\frac{d}{dr}R(r)\right) + \frac{2m_e}{\hbar^2}\left(E + \frac{Ze^2}{r}\right)R(r) = \frac{l(l+1)}{r^2}R(r). \tag{3.6}
$$

This is a result of our assumption of separation of variables for the solution of Schrodinger's equation, and the partial differential equation is split into two independent second-order differential equations. One depends only on θ and φ and the other on r. We provide here solutions for both $R(r)$ and $Y(\theta, \varphi)$ and the eigenvalues for the energy as (C. J. Ballhausen)

$$
E_n = -\frac{Z^2 e^2}{2r^2 a_0}, \, n = 1,2,3,\ldots,
$$

where n is referred to as the principal quantum number and it is an integer.

The above eigenvalues could be derived from simple arguments (Bohr). According to Bohr's theory, the Coulomb force of the electron in a central nuclear field equals the centrifugal force in equilibrium so that ($Z=1$, for example)

$$
\frac{e^2}{r^2} = \frac{m_e v^2}{r}.
$$

The total energy in a given orbit of the electron is the sum of the kinetic and potential energies. Making use of the quantization rule for G, we find that

$$E_n = \frac{1}{2} m_e v^2 - \frac{e^2}{r} = -\frac{1}{2} \frac{m_e}{n^2} c^2 \left[\frac{e^2}{\hbar c} \right]^2,$$

$$E_n = -\frac{1}{2} m_e \frac{c^2 \alpha^2}{n^2}, \quad Z = 1,$$

where α = fine structure constant = 1/137. By comparing the two expressions for E_n and for $Z = 1$, the Bohr radius is deduced as

$$a_0 = \frac{\hbar^2}{m_e e^2}.$$

The total energy is negative, corresponding to the fact that the electron is bound to the nucleus and work must be done to remove or strip it from the atom.

Simple case: Hydrogen atom $Z = 1$ ($n = 1$) ground-state energy:

$$\frac{1}{4\pi\varepsilon_0} \frac{e^2}{2a_0}, \quad a_0 = 5.229 \times 10^{-11} \text{m},$$

$$\frac{9}{10^{-9}} \frac{\left(1.6021 \times 10^{-19}\right)^2}{2 \times 5.229 \times 10^{-11}} \text{ J} = 22.0886 \times 10^{-19} \text{ J},$$

$$1 \text{ eV} = 0.80657 \times 10^4 \text{ cm}^{-1},$$

$$1 \text{ eV} = 1.6021 \times 10^{-19} \text{ J},$$

$$\therefore \frac{1}{4\pi\varepsilon_0} \cdot \frac{e^2}{2a_0} = 13.6 \text{ eV}.$$

A word of caution is in order here. For molecular calculations, Slater type wave functions are more useful whereby n and Z are adjusted to take into account electron screening. We will not delve into this. See "References" for this approach. We have omitted the (−) sign in the above value of the ground state.

If the charge of nucleus is $+Ze$, E_n becomes

$$E_n = -\frac{1}{2} m_e \left[\frac{c\alpha Z}{n} \right]^2.$$

The fact that the central field or the attraction force between the electron and the nucleus is three-dimensional means that there are many different ways by which closed orbital motions can be formed. The number of the ways or orbits, N, may be determined in terms of quantum numbers n and l, orbital angular momentum quantum number; ℓ is an integer and is equal to 0, 1, 2, ..., $n-1$. For given n, one may calculate the total number of electrons, N, from the following relationship (spins not included):

$$N = \sum_{l=0}^{n-1} (2l+1) \text{ electronic orbitals.}$$

For example, for $n=2$, $N=4$. This means that there are four different orbital motions for which the total energy of the electrons is the same. We say that the energy is fourfold degenerate. Each orbital is identified by the value of l and also by the value of the angular momentum component quantum number, m_l. The values of m_ℓ are ℓ, $\ell-1$, $\ell-2$, ..., $-\ell$. Each set of n, l, and m_l values represents a specific functional dependence of each orbital on r, θ, and ϕ, respectively. Since each orbital represents a closed orbit, there must be a discrete angular momentum and, therefore, a discrete magnetic moment. The angular momentum of each orbital is given by

$$G_l = \hbar\sqrt{l(l+1)}; \ l = 0,1,2,\ldots,n-1.$$

The magnetic moment amplitude is given as

$$m = \gamma G_l = \gamma\hbar\sqrt{l(l+1)},$$

where
  $\ell \equiv$ orbital quantum number $=0, 1, 2, \ldots, n-1$
  $\gamma = ge/2mc$
  $g=1$

Let's introduce $\beta = \gamma\hbar = $ Bohr magneton, and write m as

$$m = \beta\sqrt{l(l+1)}.$$

SPIN MOTION

There are two degrees of freedom associated with the spin motion. Some suggest using a left- or right-handed motion. The reader is cautioned here in that physicists talk about two degrees of freedom rather than actual spinning or mechanical motion. Dirac's equation is the place to look for the extra degree of freedom in the spin motion. The angular momentum associated with spin motion is

$$G_s = \hbar\sqrt{S(S+1)},$$

where now S is the spin angular momentum quantum number which corresponds to the spin motion and is equal to 1/2 for an electron. The magnetic moment m, is adopting the previous definitions:

$$m = \gamma_s G_s = \gamma_s \hbar \sqrt{S(S+1)}, \; g_s = 2 \qquad \text{or}$$

$$m = g_s \beta \sqrt{S(S+1)}.$$

With $S = 1/2$, it means that there are two degrees of freedom for the spin motion. This is indicated by the value of the component of the spin angular momentum quantum number, m_S. Thus, for $S = 1/2$, $m_s = \pm 1/2$. With the spin motion, the degeneracy is increased by a factor of 2. Thus, the total degeneracy is then (for given n)

$$N = 2 \sum_{l=0}^{n-1} (2l + 1).$$

How do we identify a particular motion of the electron? We either assign the functional dependences on r, θ, and ϕ for a given set of quantum numbers or do it symbolically. Dirac introduced the symbolic notation, as shown below:

$$|\psi = |\, n, l, m_l, S, m_s,$$

where ψ is the wave function of the particle.

The value of n tells us the energy value of the degenerate state and the functional dependence on r. The value of l and m_l tells us a particular angular orbital motion with specific angular momentum, which is a function of θ and ϕ, and m_ℓ is a component quantum number of ℓ. The value of $S = \frac{1}{2}$ tells us that there are two degrees of freedom associated with the spin motion; m_s is a component of S and it quantizes the spin angular momentum into one specific value, $+1/2$, or $-1/2$.

For example, for $n = 2$, there are eight electrons with the same energy value. Any one of the eight electrons can occupy the four orbitals. However, each electron must have only one set of quantum numbers associated with its motion. Thus, we could have the following possibilities of wave functions for $n = 2$:

Electron	Type of Motion
1; 5	$\|\psi = \left\|2,1,1,\dfrac{1}{2},\dfrac{1}{2}; \right\|2,1,1,\dfrac{1}{2},\dfrac{-1}{2}$
2; 6	$\psi = \left\|2,1,0,\dfrac{1}{2},\dfrac{1}{2}; \right\|2,1,0,\dfrac{1}{2},\dfrac{-1}{2}$
3; 7	$\psi = \left\|2,1,-1,\dfrac{1}{2},\dfrac{1}{2}; \right\|2,1,-1,\dfrac{1}{2},\dfrac{-1}{2}$
4; 8	$\|\psi = \left\|2,0,0,\dfrac{1}{2},\dfrac{1}{2}; \right\|2,0,0,\dfrac{1}{2},\dfrac{-1}{2}$

In some books, the shorthand notation is used to indicate the spin state $\left|\frac{1}{2},\frac{1}{2}\right.$ as $|+\rangle$ and $\left|\frac{1}{2},-\frac{1}{2}\right.$ as $|-\rangle$. The electronic states obey the Pauli exclusion principle in which no two electrons can have the same set of quantum numbers.

It is understood that each combination of quantum numbers contained in Dirac bracket actually represents a functional dependence of that electron in orbit. For example, the combinations $\left|3,2,\pm2,\frac{1}{2},\pm\frac{1}{2}\right.$, $\left|3,2,\pm1,\frac{1}{2},\pm\frac{1}{2}\right.$, and $\left|3,2,0,\frac{1}{2},\pm\frac{1}{2}\right.$ representing the 3d transition metal group are shown below.

In this case, $n=3$, $l=2$, $m_l=0, \pm1, \pm2$, $S=1/2$, and $m_s=\pm1/2$. Thus, we may write

$$\left|n,l,m_l,S,m_s = R_{n,l}(r)Y_{l,m_l}(\theta,\phi)\begin{pmatrix} 1 \\ 0 \end{pmatrix}\right.,$$

where

$$\left|m_s=+\frac{1}{2}=\begin{pmatrix} 1 \\ 0 \end{pmatrix}\right.,$$

$$\left|m_s=-\frac{1}{2}=\begin{pmatrix} 0 \\ 1 \end{pmatrix}\right..$$

For the 3d shell electrons, we have that (see References)

$$\left|n,l=\left|3,2=R_{3,2}(r)=\left(\frac{Z}{a_0}\right)^{\frac{3}{2}}\frac{4}{81\sqrt{30}}\left(\frac{Zr}{a_0}\right)^2 e^{-\frac{Zr}{3a_0}}\right.\right., \tag{3.7}$$

$$\left|l,m_l=\left|2,\pm2=Y_{2,\pm2}=\sqrt{\frac{5}{4\pi}}\sqrt{\frac{3}{8}}\sin^2\theta e^{\pm j2\varphi}\right.\right., \tag{3.8}$$

$$\left|l,m_l=\left|2,\pm1=Y_{2,\pm1}=\pm\sqrt{\frac{5}{4\pi}}\sqrt{\frac{3}{2}}\sin\theta\cos\theta e^{\pm j\varphi}\right.\right., \tag{3.9}$$

$$\left|l,m_l=\left|2,0=Y_{2,0}=\sqrt{\frac{5}{4\pi}}\sqrt{\frac{1}{4}}\left(3\cos^2\theta-1\right)\right.\right., \tag{3.10}$$

and $a_0=0.5229\times10^{-8}$ cm = Bohr's radius.

Radial and angular dependence of the wave functions is sketched in Figures 3.1 and 3.2. It is noted that there are 10 possible motions for an electron in the d-shell to occupy. We have ignored electron screening effects from the core electrons and/or electrostatic interactions. The reader is referred to References for more details.

In Figure 3.2, polar plots of $|Y_{l,m_l}|^2$ are shown.

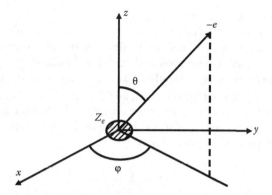

FIGURE 3.1   Coordinate system.

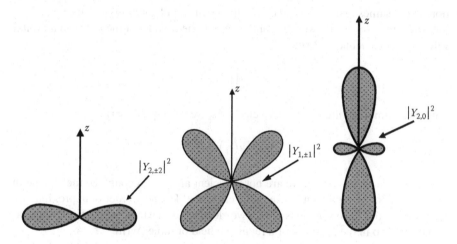

FIGURE 3.2   Possible motions of 3d electrons.

INTRA-EXCHANGE INTERACTIONS

Henceforth, we will only consider electrons in the 3d shell. Clearly, two electrons will interact via the Coulomb interaction:

$$\frac{e^2}{r_{ij}} = \frac{e^2}{\left|\vec{r}_i - \vec{r}_j\right|}.$$

Let's simplify the picture whereby two electrons have only two possible orbital motions and then see where it leads and what conclusions we may draw. After this example, we will examine two electrons interacting in the 3d shell designated simply as $3d^2$ electronic configuration.

Let's assign two different orbitals $u(\vec{r}_1)$ and $v(\vec{r}_2)$ to the two electrons. Thus, $u(\vec{r}_1)$ means that one electron is in orbital u and the other electron in v. However, since the electrons are identical and there are no distinguishable features to identify one from the other, it is equally conceivable to have the possibility of the two electrons switching their orbitals or equivalently make the assignment $u(\vec{r}_2)$ and $v(\vec{r}_1)$. It is understood that $u(\vec{r}_i)$ or $v(\vec{r}_i)$ implies r_i, θ_i, and ϕ_i dependences, where i indicates 1 or 2. We say that the two orbital states are energetically degenerate. Naturally, the two electrons will repel each other when they are in close proximity as expressed in the electrostatic repulsion energy given by

$$H = \frac{e^2}{|\vec{r}_1 - \vec{r}_2|}. \tag{3.11}$$

The interaction will, of course, affect the orbital motion of the electrons as well as the degeneracy. Let's now calculate the effect of the interaction. As a first approximation, one assumes that there is no effect due to this repulsion energy. This means that the wave function of the two electrons is simply the product of the individual orbital wave functions, u and v. Write

$$\psi_1 = u(\vec{r}_1)v(\vec{r}_2).$$

Since we cannot distinguish the two electrons, we can equally write

$$\psi_2 = u(\vec{r}_2)v(\vec{r}_1).$$

We know that ψ_1 and ψ_2 above are not sufficient alone to account for the change in orbitals due to the repulsion between the electrons. The next order of approximation is to say that the wave functions of the two electrons as a whole are a linear combination of the two in the presence of a repulsion field or force. Write

$$\psi = a_1\psi_1 a_2\psi_2. \tag{3.12}$$

We may now proceed to solve the Schrodinger equation:

$$H\psi = E\psi. \tag{3.13}$$

Multiply Equation 3.13 by ψ_1 and integrate over dV_1 and dV_2:

$$\iint \psi_1{}^* H(a_1\psi_1 + a_2\psi_2)dV_1\,dV_2 = \iint \psi_1{}^* E(a_1\psi_1 + a_2\psi_2)dV_1\,dV_2.$$

We can symbolically write the above integrals as

$$a_1 H_{11} + a_2 H_{12} = Ea_1,$$

where

$$H_{11} = \iint u^*(\vec{r}_1)u^*(\vec{r}_2)\left(e^2/|\vec{r}_1 - \vec{r}_2|\right)u(\vec{r}_1)v(\vec{r}_2)dV_1\,dV_2,$$

$$H_{12} = \int\left[\int u^*(\vec{r}_1)\left(e^2/|\vec{r}_1 - \vec{r}_2|\right)v(\vec{r}_1)dV_1\right]v^*(\vec{r}_2)u(\vec{r}_2)dV_2.$$

Notice that $H_{12} \to 0$, if there is no overlap of the two orbits (see term outside square bracket). Let $H_{12}=J=$real quantity. In a similar manner, multiply (3.13) by ψ_2^* and integrate to obtain

$$a_1 H_{21} + a_2 H_{22} = Ea_2,$$

where $H_{21} = \int\left[\int v^*(\vec{r}_1)\left(e^2/|\vec{r}_1 - \vec{r}_2|\right)u(\vec{r}_1)dV_1\right]u^*(\vec{r}_2)v(\vec{r}_2)dV_2$. We require $H_{21} = H_{12}^*$ in order for J to be an observable quantity or real. Finally, let $H_{22}=H_{11}=K$. We summarize the above algebraic set of equations as follows:

$$\begin{bmatrix} H_{11} & H_{12} \\ H_{21} & H_{22} \end{bmatrix}\begin{bmatrix} a_1 \\ a_2 \end{bmatrix} = E\begin{bmatrix} a_1 \\ a_2 \end{bmatrix}$$

or

$$\begin{bmatrix} K-E & J \\ J & K-E \end{bmatrix}\begin{bmatrix} a_1 \\ a_2 \end{bmatrix} = 0.$$

The eigenvalues are $E_{1,2}=K\pm J$. The coefficients a_1 and a_2 may be obtained as follows:

$$(K - E_1)a_1 + Ja_2 = 0.$$

Put $E_1=K+J$ into the above equation and obtain $a_1=a_2$. We also require $\iint \psi^*\psi\,dV_1\,dV_2 = 1$, so that $a_1^2 + a_2^2 = 1$. Thus, $a_1 = a_2 = 1/\sqrt{2}$. For $E_2=K-J$, we have $a_1=-a_2$ and

$$a_1 = 1/\sqrt{2} \text{ and } a_2 = -1/\sqrt{2}.$$

Thus, the corresponding eigen functions are

$$\Psi_{E_1} = \frac{\Psi_1 + \Psi_2}{\sqrt{2}} \text{ and } \Psi_{E_2} = \frac{\Psi_1 - \Psi_2}{\sqrt{2}}.$$

In the overlap region,

$$\Psi_1 \approx \Psi_2$$

so that

$$\Psi_{E_1} \approx \sqrt{2}\Psi_1$$

and

$$\Psi_{E_2} \approx 0.$$

This means that the probability of finding two particles in the overlap region is high for the Ψ_{E_1} eigen function and small for Ψ_{E_2}. Since the nature of the interaction is the repulsion force between two electrons, $E_1 > E_2$. The effect of the repulsion force is to split the degeneracy between the two orbits.

Now let's put the spin functional dependences into Ψ_{E_1} and Ψ_{E_2} since the wave function is made up of both orbital and spin motions. The rule governing the formation of complete electronic wave functions is that they be antisymmetric with respect to interchange of electrons according to Pauli's principle. There are four possible combinations for which the spin angular momentum of each electron may be combined and they are as follows:

Permutation	Electron 1	Electron 2	Spin Orientation
1	α_1	α_2	$\uparrow\uparrow$
2	β_1	β_2	$\downarrow\downarrow$
3	$(\alpha_1\beta_2 + \alpha_2\beta_1)/\sqrt{2}$		$\downarrow\uparrow + \downarrow\uparrow$
4	$(\alpha_1\beta_2 - \alpha_2\beta_1)/\sqrt{2}$		$\downarrow\uparrow - \downarrow\uparrow$

Clearly, the first three combinations of the spin representation are symmetric, but not the last one. This can be checked by interchanging the subscripts 1 and 2. Let's combine the above permutations with Ψ_{E_1} and Ψ_{E_2}. In doing so, we must use it in a way to obey Pauli's rule, which states that no two electrons can have the same set of quantum numbers. Thus, we conclude that

$$\Psi_I = \Psi_{E_2}\alpha_1\alpha_2,$$

$$\Psi_{II} = \Psi_{E_2}\beta_1\beta_2,$$

$$\Psi_{III} = \Psi_{E_2} \frac{\left[\alpha_1\beta_1 + \alpha_2\beta_1\right]}{\sqrt{2}},$$

$$\Psi_{IV} = \Psi_{E_1} \frac{\left[\alpha_1\beta_1 - \alpha_2\beta_1\right]}{\sqrt{2}},$$

where
α denotes the spin state $|+\rangle$
β the state $|-\rangle$

Notice that the Pauli exclusion principle is obeyed automatically, if the two electrons are interchanged by changing the subscripts on ψ's. We can represent the energy-level diagram as follows (see Figure 3.3).

The wave function corresponding to E_2 is threefold degenerate and is often referred to in the literature as the triplet state. The fact that $E_2 < E_1$ implies that the ferromagnetic state is of lower energy, since $S = 1$ (the components of S are $m_s = 1, 0, -1$). The usual designation for the triplet state is ψ_t. The singlet state, ψ_s, which corresponds to E_1 is the antiferromagnetic state, since $S = 0$ ($m_s = 0$) and the spins oppose each other.

The singlet state is higher in energy, because the two particles can coexist in the same region or point. As such, the electrostatic repulsion energy between the two particles is greatest for the singlet state. Clearly, for the ferromagnetic state (ψ_t), the separation between the particles is greater. This can also be understood in terms of the general probability function density for two particles ($P(r_1, r_2)$), which may be expressed as follows (M. Weissbluth):

$$P_2(r_1, r_2) = P_1^\alpha(r_1) P_1^\alpha(r_2) - P_1^{\alpha\alpha}(r_1, r_2) P_1^{\alpha\alpha}(r_2, r_1) + P_1^\alpha(r_1) P_1^\beta(r_2)$$
$$+ P_1^\beta(r_1) P_1^\beta(r_2) - P_1^{\beta\beta}(r_1, r_2) P_1^{\beta\beta}(r_2, r_1) + P_1^\beta(r_1) P_1^\alpha(r_2),$$

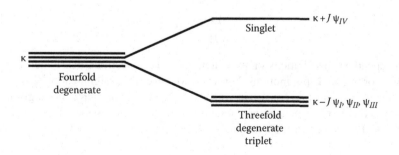

FIGURE 3.3 Energy levels.

where $P_1^\alpha(r_i)$ is the probability density for finding an electron with α-spin at r_i, $i = 1, 2$. $P_1^{\alpha\alpha}(r_1, r_2)$ is the probability density for finding an electron with α-spin at r_1 and another electron with α-spin at r_2. The other terms have analogous meaning. The expression $P_2^{\alpha\alpha}(r_1, r_2) = P_1^\alpha(r_1) P_1^\alpha(r_2) - P_1^{\alpha\alpha}(r_1, r_2)$ is usually referred to as Fermi's correlation probability function for two particles. In the limit that $r_1 = r_2 = r$, we have $P_2(r_1, r_2) = 2 P_1^\alpha(r) P_1^\beta(r)$.

This implies that it may be possible to have two electrons at a point in space with opposite spins—antiferromagnetic state. Hence, it is no surprise that the singlet state represents the state with higher energy.

HEISENBERG REPRESENTATION OF EXCHANGE COUPLING

We may represent the splitting between the triplet and singlet states in terms of spin variables only, as in the Heisenberg representation of the exchange coupling:

$$H = -J \vec{S}_1 \cdot \vec{S}_2, \ S_1 = S_2 = S = 1.$$

For $\vec{S}_1$ parallel to $\vec{S}_2$ (triplet or ferromagnetic state),

$$E_{//} = H = -J.$$

For $\vec{S}_1$ opposite to $\vec{S}_2$ (antiferromagnetic or singlet state),

$$E_\perp = J.$$

Thus, the splitting between the two states is $2J$ as calculated above.

MULTIPLET STATES

Let's now consider a realistic or practical situation where we have two electrons in the d-shell ($3d^2$). A 3d electron has five possible orbits to choose from in order to orbit around the nucleus. There are

$$\frac{10 \times 9}{2!} = 45$$

45 independent distributions or permutations of two electrons distributed over 10 possible orbital and spin motions. However, whatever the assignment of distributing two electrons, it must be consistent with Pauli's exclusion principle. For a given distribution, there are associated new quantum numbers or equivalent representations much like the Heisenberg representation, L and m_L, S, m_S.

$$m_L = \sum m_l; \ m_S = \sum m_s,$$

where m_L are components of L. Symmetry in the number of distributions for $\pm m_L$ and for $\pm m_S$ reduces the number of distributions from 45. A simple distribution would be, for example, that one electron with spin $m_s = +1/2$ is assigned to the $m_l = 2$ orbital and the other electron, $m_s = +1/2$, is assigned to $m_l = 1$. The Pauli exclusion principle is obeyed and the new quantum numbers are, for example,

$$m_L = 3 \text{ and } m_S = 1.$$

Since m_L can range from $+3$ to -3, $L=3$ and $S=1$. This state is designated as ^3F. It corresponds to seven distributions or it is sevenfold degenerate. The rest of the distributions are the following: $^3P(L=1, S=1)$, $^1D(L=2, S=0)$, $^1D(L=4, S=0)$, and $^1S(L=0, S=0)$. These states are referred to in the literature as term states or multiplet states. Let's describe exactly the functional dependence on r, θ, and φ of one of these states, for example, ^3F. It consists of two wave functions:

$$\left|3,2,2,+\frac{1}{2}\right\rangle \text{ and } \left|3,2,1,+\frac{1}{2}\right\rangle.$$

The reader should refer to the exact dependences on r, θ, and φ of these states in the previous section. Thus,

$$\left|^3F\right\rangle = \frac{1}{\sqrt{2}} \det \begin{pmatrix} \Phi_a(1)|+\rangle & \Phi_b(1)|+\rangle \\ \Phi_a(2)|+\rangle & \Phi_b(2)|+\rangle \end{pmatrix},$$

where

$$\Phi_a(i) = R_{3,2}(r_i) Y_{2,2}(\theta_i, \varphi_i); \ i = 1,2,$$

$$\Phi_b(i) = R_{3,2}(r_i) Y_{2,1}(\theta_i, \varphi_i); \ i = 1,2, \text{ and}$$

$$|+\rangle = |S = 1/2, m_s = 1/2\rangle.$$

The above determinant is referred to as the Slater determinant and it ensures that the ^3F wave function is antisymmetric. The other four terms or multiplets may also be represented by a 2×2 Slater determinant similar but with different orbitals, since the combination of orbitals is different.

The five terms represent 25 ways to distribute two electrons over 10 possible atomic motions or degrees of freedom. The term ^3F represents 7 ways ($2L+1=7$), and therefore, it is sevenfold degenerate. Now that we have the wave function as expressed above, let's calculate the average electrostatic interaction for this combination or distribution of electrons. This has been tabulated in terms of the Racah parameters, A, B, and C. For $3d^2$, the average energies of each term are

$$\left\langle {}^3F \left| \frac{e^2}{r_{12}} \right| {}^3F \right\rangle = A - 8B,$$

$$\left\langle {}^{3}P \left| \frac{e^2}{r_{12}} \right| {}^{3}P \right\rangle = A - 7B,$$

$$\left\langle {}^{1}G \left| \frac{e^2}{r_{12}} \right| {}^{1}G \right\rangle = A + 4B + 2C,$$

$$\left\langle {}^{1}D \left| \frac{e^2}{r_{12}} \right| {}^{1}D \right\rangle = A - 3B + 2C,$$

and

$$\left\langle {}^{1}S \left| \frac{e^2}{r_{12}} \right| {}^{1}S \right\rangle = A + 14B + 7C.$$

The Racah parameters have been deduced from ionization experiments (see Griffith, 1961), and they are for the 3d electrons: $A \cong 3,000$, $B = 670$, and $C = 2,500\,\text{cm}^{-1}$. According to Griffith, the core energy of a single d electron is $\sim -200,000\,\text{cm}^{-1}$ $(-U)$. If we use the formula that $U = -Z/n^2(13.6\,\text{eV})$ and convert into cm^{-1}, we obtain $U \sim -300,000\,\text{cm}^{-1}$, which is well above the quoted value here. This may be an indication that the nucleus charge is being shielded somewhat by the core electrons. Slater (1937, 1961) approximated the screening by assuming the radial wave functions to be of this form

$$R_{3,2} = \frac{4}{3\sqrt{10}} \alpha^{\frac{7}{2}} r^2 e^{-\alpha r}, \ \alpha = \frac{Z - \gamma}{3 a_0}, \ \gamma > 0.$$

where γ is an adjustable parameter.

In Figure 3.4, energy diagram for $3d^2$ configuration including the multiplet or term energies is shown.

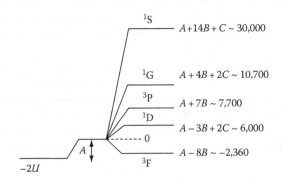

FIGURE 3.4 Multiplet energy splittings in units of cm^{-1} for $3d^2$ electronic configuration.

For $3d^2$ configuration, we may calculate the average energy of e^2/r_{12} weighted according to the degeneracy of each term $(2L+1)(2S+1)$ or

$$\frac{e^2}{r_{12}}_{3d^2} = (A-8B)\frac{21}{45} + (A+7B)\frac{9}{45} + (A+4B+2C)\frac{9}{45}$$

$$+ (A-3B+2C)\frac{5}{45} + (A+14B+7C)\frac{1}{45}$$

$$= A - \frac{14}{9}B + \frac{7}{9}C = 3,900\,\text{cm}^{-1}.$$

We estimate $\langle r_{12}\rangle \approx 2.6\,\text{Å}$. The implication is that two particles in the $3d^2$ configuration are on the average 2.6Å apart! According to Orgel (1970), the size of a transition metal ion is in the order of 0.9 Å. This implies that 3d electrons range well outside of the core electrons.

HUND RULES

As we have seen from the previous sections, the primary effect of the exchange interaction is to couple spins to form a resultant or representative $\vec{S}$ to give rise to a lowest energy state or multiplet ground state. The orbital angular momentum is then coupled together to form a resultant L consistent with the exclusion principle. These two rules are known as Hund's rules. For example, consider the electronic configuration

$$1s^2, 2s^2, 2p^6, 3s^2, 3p^6, 3d^3$$

Hund's rule says that the lowest energy is obtained when all spins within the 3d shell are parallel to each other or the total spin of the atomic system is

$$S = \frac{1}{2} + \frac{1}{2} + \frac{1}{2} = \frac{3}{2}.$$

Since $n=3$ and $l=2$, $m_l=2, 1, 0, -1, -2$. The second Hund's rule states that the three electrons in the d-shell must take on three different m_l values (consistent with the Pauli exclusion principle) in order to maximize L or

$$L = 2 + 1 + 0 = 3.$$

By Hund's rules, the lowest energy state or the lowest multiplet state of the three-electron state is $L=3$ and $S=3/2$ which is designated as the ${}^4F_{3/2}$ multiplet energy state. The superscript 4 is obtained from $2S+1$, the letter F is a symbolic representation for $L=3$, and the subscript 3/2 is to denote $|J|=|L \pm S|$. The minus sign is applicable, if the d-shell is half occupied and the plus sign more than half full. There are $2L+1=7$ values of m_L (angular momentum quantum number) 3, 2, 1, 0, −1, −2, −3, which make up seven of possible permutation states in the 3d shell. The other multiplet states belong to the $L=1$ manifold or excited states ($m_L=1$, 0, -1). This excited state is designated as $^4P_{1/2}$. Notice that the subscript is still $|J|=|L-S|$, but for this case, $L=1$ and $S=3/2$. For most microwave magnetic materials where the frequency of excitations is relatively low (~GHz), ground-state multiplets are excited. It would take optical frequencies' excitations to induce transitions to higher energy-level multiplets.

Let's apply Hund's rule to the 3d and 4f shell electrons (see Tables 3.1 and 3.2).

TABLE 3.1
Ground-State Multiplet for Transition Metal Ions

3d Electrons	Ion	S	L	J	Ground State	g_J
0	Ca^{2+}, Ti^{4+}, Sc^{3+}	0	0	0	$^1S_0$	—
1	Ti^{3+}, V^{2+}	1/2	2	3/2	$^2D_{3/2}$	0.800
2	V^{3+}	1	3	2	$^3F_2$	0.667
3	V^{4+}, Cr^{3+}	3/2	3	3/2	$^4F_{3/2}$	0.400
4	Mn^{3+}, Cr^{3+}	2	2	0	$^5D_0$	—
5	Mn^{2+}, Fe^{3+}	5/2	0	5/2	$^6S_{5/2}$	2.000
6	Fe^{2+}	2	2	4	$^5D_4$	1.500
7	Co^{2+}	3/2	3	9/2	$^4F_{9/2}$	1.333
8	Ni^{2+}, Co^{3+}	1	3	4	$^3F_4$	1.25
9	Co^{2+}	½	2	5/2	$^2D_{5/2}$	1.20
10	Zn^{2+}, Cu^{1+}	0	0	0	$^1S_0$	—

TABLE 3.2
Ground-State Multiplet for Rare Earth Ions

4f Electrons	Ion (3+)	S	L	J	Ground State	g_J
0	Ce	0	0	0	$^1S_0$	—
1	Ce	1/2	3	5/2	$^2F_{5/2}$	0.857
2	Pr	1	5	4	$^3H_4$	0.800
3	Nd	3/2	6	9/2	$^4I_{9/2}$	0.727
4	Pm	2	6	4	$^5I_4$	0.60
5	Sm	5/2	5	5/2	$^6H_{5/2}$	0.286
6	Eu	3	3	0	$^7F_0$	—
7	Gd	7/2	0	7/2	$^8S_{7/2}$	2.000
8	Th	3	3	6	$^7F_6$	1.500
9	Dy	5/2	5	15/2	$^6H_{15/2}$	1.333
10	Ho	2	6	8	$^5I_8$	1.250
11	Er	3/2	6	15/2	$^4I_{15/2}$	1.200
12	Tm	1	5	6	$^3H_6$	1.167
13	Yb	1/2	3	7/2	$^2F_{7/12}$	1.143
14	Ln	0	0	0	$^1S_0$	—

SPIN–ORBIT INTERACTION

Consider a single electron orbiting around the nucleus. Imagine you are sitting on the electron and orbiting the nucleus. How do you see the nucleus? If the electron is the motional reference point, then the nucleus is orbiting around the electron with orbital angular momentum $\hbar \vec{L}$, where L is the quantum mechanical operator for angular momentum. The magnetic moment associated with $\vec{L}$ is then

$$\vec{m} = |\gamma| \hbar Z \vec{L} = \frac{|e| \hbar}{2 m_e} Z \vec{L},$$

where
   Z is the atomic number
   m_e is the mass of the electron

The angular momentum of the nucleus is $Z\vec{L}$, where $\vec{L}$ is the orbital angular momentum of the electron. In this frame of reference, the electron is fixed in space! The $\vec{B}$ field at the electron site due to the nucleus is then (see Chapter 2)

$$\vec{B} = \frac{\mu_0 \vec{m}}{2 \pi r^3}$$

or

$$\vec{B} = \frac{\hbar Z |\gamma| \vec{L} \mu_0}{2 \pi r^3}.$$

The magnetic potential energy due to this $\vec{B}$ field is then

$$U = -\vec{m}_S \cdot \vec{B},$$

$$\vec{m}_S = g_s \gamma \hbar \vec{S}, \ g_s = 2,$$

and

$$\gamma = -|\gamma|.$$

where $\vec{S}$ is the spin angular momentum operator. Since the electron is fixed in space in this frame of reference, there is only spin angular momentum at the electron site or G_S. As such, there is no contribution to the potential energy from the orbital angular momentum of the electron. We can write U as

$$U = \lambda \vec{L} \cdot \vec{S},$$

where

$$\lambda = \frac{\mu_0}{2\pi r^3} g_s \beta^2 Z\left(\frac{1}{2}\right), \; \beta = |\gamma|^{th}.$$

The factor (1/2) is a relativistic correction introduced by Thomas. A formal treatment including relativistic corrections may be found in Dirac's book.

LANDE g_J-FACTOR

As implied by the spin–orbit interaction, $\vec{L}$ and $\vec{S}$ are coupled to each other. The coupling Hamiltonian is of the form $H = \lambda \vec{L} \cdot \vec{S}$, where L and S may be taken from Table 3.1, for example, and λ is the spin–orbit interaction parameter. We can also write H as

$$H = -\vec{m}_S \cdot \vec{B}_L,$$

where

$$\vec{B}_L = -\lambda \vec{L} / \gamma_S \hbar \text{ and}$$

$$\vec{m}_S = \gamma_S \hbar S.$$

Equivalently, we may write

$$H = -\vec{m}_L \cdot \vec{B}_S,$$

where

$$\vec{B}_S = -\lambda \vec{S} / \gamma_L \hbar \text{ and}$$

$$\vec{m}_L = \lambda_L \hbar \vec{L}.$$

From the equation of motion, it is simple to show that (see Chapter 2)

$$\frac{d\vec{L}}{dt} = \frac{(-\lambda)}{\hbar}\left(\vec{L} \times \vec{S}\right)$$

and

$$\frac{d\vec{S}}{dt} = \frac{(-\lambda)}{\hbar}\left(\vec{S} \times \vec{L}\right).$$

Introduce a resultant angular momentum $\vec{J} = \vec{L} + \vec{S}$ and substitute it into the equation of motion. Thus, we have

$$\frac{d\vec{L}}{dt} = \frac{(-\lambda)}{\hbar}\left(\vec{L} \times \vec{J}\right) \text{ and}$$

$$\frac{d\vec{S}}{dt} = \frac{(-\lambda)}{\hbar}\left(\vec{S} \times \vec{J}\right).$$

This means that both S and L precess about J with radial angular velocity of $\lambda/\hbar$. Thus, $\vec{J}$ is the reference direction by which everything else is quantized, $\vec{L}$, $\vec{S}$, m_L, and m_S, as illustrated graphically in Figure 3.5.

The projection of $\vec{m}_S$ and $\vec{m}_L$ onto this reference direction is simply m_J:

$$m_J = m_L \cos\alpha_1 + m_S \cos\alpha_2, \tag{3.14}$$

where

$$\cos\alpha_1 = \frac{L(L+1) + J(J+1) - S(S+1)}{2\sqrt{L(L+1)J(J+1)}}$$

and

$$\cos\alpha_2 = \frac{-L(L+1) + J(J+1) + S(S+1)}{2\sqrt{J(J+1)S(S+1)}}.$$

Substituting the above into Equation 3. 14 and using $m_J = g_J\beta\sqrt{J(J+1)}$, we obtain

$$g_J = \frac{3}{2} + \frac{S(S+1) - L(L+1)}{2J(J+1)}.$$

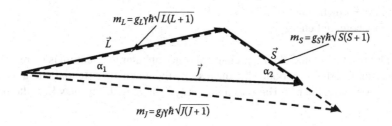

FIGURE 3.5 Vectorial representation of angular momentum and magnetic moment.

The above relation is usually referred to as the *Lande's g-formula*. Let's examine it for two limits of interest:

For $S=0$, $g_J=1$ (orbit only).
For $L=0$, $g_J=2$ (spin only).

Finally, in a coupled situation, the potential magnetic energy may be written as

$$H = -\left(\vec{m}_S + \vec{m}_L\right)\cdot\vec{B} = -\vec{m}_J\cdot\vec{B}.$$

In summary, the introduction of a nucleus as in an atom is to allow for many closed path orbits or closed loops, since the motion is three-dimensional. For the most part, these orbits are degenerate in energy and don't give rise to magnetism. The other consequence of the nucleus is that it gives rise to the so-called spin–orbit interaction. The magnetic fields associated with this interaction are enormous, but only within the free atom. Outside of the free atom, this field is zero. Finally, the existence of many electrons with various orbits gives rise to repulsion between them. The effect of this is the so-called exchange interaction. We will see that this interaction is the main or most important interaction in the field of magnetism, especially when electrons come from different sites.

With this chapter, it ends the process of taking a macroscopic body (a current loop) and reducing its size to a microscopic body. In the ensuing chapters, we will reverse the process and eventually discuss the microwave properties of practical magnetic materials.

EFFECTS OF MAGNETIC FIELD ON A FREE ATOM

With no magnetic field, the total energy of a free atom may be given as

$$H_0 = \frac{p_0^2}{2m} + V(\vec{r}),$$

where
$p_0^2/2m$ = kinetic energy
$V(\vec{r})$ = potential energy

In applying a magnetic field, H_0 is changed, since angular momentum is increased for an electron in orbital motion. This means that the magnetic potential energy is also increased. Hence, the new Hamiltonian energy is increased. The new Hamiltonian is

$$H = \frac{p^2}{2m} + V(\vec{r}).$$

Clearly, $p \neq p_0$, but we have assumed that $V(\vec{r})$ is unaffected. Thus, increase in magnetic potential energy is included in the kinetic energy. Let's now calculate this increase. Starting with

$$\nabla \times \vec{E} = -\frac{\partial \vec{B}}{\partial t} = -\frac{\partial}{\partial t} \nabla \times \vec{A},$$

we obtain

$$\vec{E} = -\frac{\partial \vec{A}}{\partial t} + \text{constant}.$$

The constant of integration is arbitrary and chosen to be equal to $-\vec{\nabla}\Phi$. Thus, $\vec{E} = -\left(\partial \vec{A}/\partial t\right) - \vec{\nabla}\Phi$, where Φ can represent an external applied voltage. Since no external electric field is applied, $\vec{E} = -\partial \vec{A}/\partial t$. For a particle of charge q, the rate of change of momentum is by Lorentz's law

$$\frac{\partial \vec{p}}{\partial t} = q\vec{E} = -q\frac{\partial \vec{A}}{\partial t},$$

where q is the electrical charge. Integrating the above equation, we get

$$\vec{p} = -q\vec{A} + \vec{p}_0.$$

Again, we have chosen the constant of integration to be arbitrary and equal to $\vec{p}_0$. Note that at $t=0$, $\vec{p} \rightarrow \vec{p}_0$ (no B field). Now if $q - e$, $\vec{p} = \vec{p}_0 + e\vec{A}$. The term $e\vec{A}$ represents the added linear momentum upon the application of a $\vec{B}$ field. The Hamiltonian may now be rewritten as

$$\vec{H} = \frac{\left(\vec{p}_0 + e\vec{A}\right)^2}{2m} + V(\vec{r}).$$

We can write the new Hamiltonian in terms of H_0 (the Hamiltonian with no external magnetic field):

$$H = H_0 + H',$$

where

$$H_0 = \frac{p_0^2}{2m} + V(\vec{r})$$

and

$$H' = \frac{e}{m}\vec{A}\cdot\vec{p}_0 + \frac{e^2}{2m}A^2.$$

H' represents the added kinetic energy upon the application of the $\vec{B}$ field. For a steady magnetic field, we have

$$\vec{A} = \frac{1}{2}\vec{B} \times \vec{r} = -\frac{1}{2}\vec{r} \times \vec{B}.$$

This results in

$$\vec{A} \cdot \vec{p}_0 = -\frac{1}{2}(\vec{r} \times \vec{B}) \cdot \vec{p}_0 = -\frac{1}{2}\vec{B} \cdot (\vec{r} \times \vec{p}_0).$$

Finally, we have

$$\vec{A} \cdot \vec{p}_0 = -\frac{1}{2}\vec{B} \cdot \vec{G}_0,$$

where $\vec{G}_0$ is the orbital angular momentum before the application of the $\vec{B}$ field. In addition, we have

$$A^2 = \frac{1}{4}\left(r^2 B^2 - (\vec{r} \cdot \vec{B})^2\right).$$

Assuming $\vec{B} = B\vec{a}_z$, then

$$H' = -\frac{e}{2m}BG_{oz} + \frac{e^2 B^2}{8m}x^2 + y^2.$$

Let's identify G_0 with the angular momentum of atomic orbitals in which L designates the particular orbital, as in the scheme developed before. Thus, write

$$\vec{G}_0 = \vec{G}_L \text{ or } \vec{G}_0 = \vec{G}_S.$$

Let's now apply a magnetic field and consider the equation of motion for the angular momentum. In general,

$$\dot{\vec{G}} = \vec{m} \times \vec{B} = \gamma\vec{G} \times \vec{B}, \text{ where } \vec{B} = B\vec{a}_z.$$

In components form, the equation of motion becomes

$$\frac{dG_x}{dt} = \gamma BG_y,$$

$$\frac{dG_y}{dt} = -\gamma BG_x,$$

$$\frac{dG_z}{dt} = 0.$$

The solutions to the coupled differential equations are of the form

$$G_x = A\cos(\gamma Bt + \alpha),$$

$$G_y = A\sin(\gamma Bt + \alpha),$$

and

$$G_z = \text{constant}.$$

The angle α depends on the initial condition. The basic difference is that the angle between the total angular momentum, G and G_z, takes on discrete values in quantum mechanics but is continuous in classical mechanics (Chapter 2). Let's denote this angle as $^c\vartheta$.

Let's now calculate $^c\vartheta$, when G is discrete. For example, for $n=2$ and $l=1$. The magnitude of the orbital angular momentum is

$$\left|\vec{G}_l\right| = \hbar\sqrt{l(l+1)} = \sqrt{2}\hbar. \tag{3.15}$$

In a magnetic field, the angular momentum along the z-direction can take on three values and they are $\hbar$, 0, $-\hbar(m_l = 1, 0, -1)$. Hence, the angle $^c\vartheta$ can be $45°$, $90°$, and $135°$ with respect to the z-axis depending on whether or not the orbitals are occupied by the electrons. Certainly, if all three orbitals are occupied, then the net angular momentum is zero. Similarly, the magnitude of the spin angular momentum is $\left|\vec{G}_s\right| = \hbar/\sqrt{2}$. But the projections of the spin angular momentum along the z-direction are $\pm\hbar/2$. In this case, $^c\vartheta = 54°$ and $126°$. Again, the net angular momentum along the B or z-direction depends on the occupation number. This means that the corresponding magnetic moments are $\vec{m}_L = \gamma_L\vec{G}_L$ or $\vec{m}_S = \gamma_S\vec{G}_S$. $\vec{m}_L$ is the magnetic moment due to orbital motion and $\vec{m}_S$ due to spin motion. Finally, the increase in kinetic energy may be written as follows:

$$H' = -\vec{m}_S\cdot\vec{B} - \vec{m}_L\cdot\vec{B} + \frac{e^2 B^2}{8m}x^2 + y^2. \tag{3.16}$$

Remarkably, the above form for the energy is exactly of the same form as the magnetic potential energy derived from classical arguments in Chapter 2. Let's now pay attention to the third term in H'. Let

$$x^2 = r^2\cos^2\alpha_1 = \frac{1}{3}r^2 \text{ and } y^2 = r^2\cos^2\alpha_2 = \frac{1}{3}r^2,$$

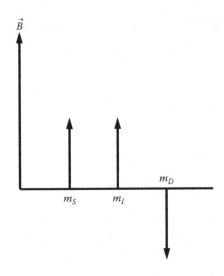

FIGURE 3.6   Magnetic moments in a B field.

where α_1 and α_2 are angles of the particle's position with respect to the x- and y-axis, respectively. We can then estimate the net magnetic moment either parallel to or along the z-direction.

$$m = m_S^z + m_L^z - \frac{e^2 B}{6m} r^2.$$

Thus, the magnetic moment due to orbital and spin motions is polarized along the same direction as $\vec{B}$, but the third term contributes to a magnetic moment in the opposite direction to $\vec{B}$. This last term is called the diamagnetic contribution to the total magnetic moment and is usually identified as m_D (Figure 3.6).

We can write m_D in terms of the change in current due to the application of a $\vec{B}$ field:

$$m_D = I_D \pi r^2.$$

This implies that $I_D = -e^2 B/6\pi m$. Thus, the $\vec{B}$ field induces a change in current (I_D) such that the resultant flux opposes the applied $\vec{B}$ field. This effect is also known in the literature as Lenz law. I_D can also be obtained from the following equation:

$$I_D = \frac{-dE_D}{d\Phi_B},$$

where the diamagnetic energy may be expressed simply as

$$E_D = \frac{e^2 \Phi_B^2}{12m} r^2,$$

the flux as

$$\Phi_B = BA$$

and

$$A = \pi r^2.$$

It is noted that the induced magnetic field due to changes in spin and orbital angular momentum is in the same direction as the applied field. Let's define susceptibility, χ_D, as $n m_D / B$. We see that $\chi_D = -(n e^2 / 6m)\langle r^2 \rangle$, where n is the number of atoms per cm³. The susceptibility of diamagnetic materials is always less than 1, but the susceptibility due to m_S and m_L can be much greater than 1, which is very useful for microwave applications, as we will see later.

What is the total magnetic moment, if all the electronic states are occupied in a free atom? Since the net angular momentum is zero, the magnetic moment is zero. Even in an unfilled shell, the magnetic moment is zero in the absence of an external magnetic field. This is due to the fact that the motion is $(2l + 1)$ degenerate for given l. There are $(2l + 1)$ ways by which an electron orbits around the nucleus and still has the same energy. On the average, the angular momentum is zero, and the electron spends equal time in each type of orbit designated by

$$m_l = l, l - 1, \ldots, -l.$$

However, in the presence of an external magnetic field, H, the motion of the electron is confined to the lowest energy for which a specific orbit is identified by m_l, i.e., the degeneracy is removed. As such, for unfilled shells and in the presence of an H field, free atoms may be polarizable. Clearly, one needs atoms with unfilled shells in order for an atom to contribute to the magnetic moment of a material. So far, we have assumed no interaction between electrons. We will show that these interactions play a major role in affecting the magnetic moment in an atom. This is due to the fact that, besides the external field, the atom generates internal fields which compete with the external field in magnetically polarizing the atom. These internal fields arise from two basic interactions: (1) spin–orbit interaction and (2) electrostatic interactions. An example of electrostatic interactions is the exchange coupling, which we have discussed above, and another example is the crystal field in ions.

CRYSTAL-FIELD EFFECTS ON MAGNETIC IONS

We are now in a position to "assemble" a magnetic material utilizing atoms or ions that are intrinsically magnetic. The fundamental building "blocks" or magnetic cells or magnetic sublattices of a magnetic insulating material are the octahedron and tetrahedron unit cells whereby the magnetic ion on an atom is located at the center of the unit cell. There are other unit cells, such as the dodecahedron, but they are uncommon or play a secondary role in affecting the magnetic properties of an

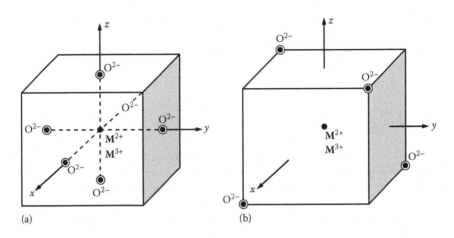

FIGURE 3.7   Magnetic ions in (a) octahedral and (b) tetrahedral sites.

insulating magnetic material. For magnetic metals, usually there is only one unit cell to consider, but the mechanism to explain magnetism is somewhat different from the insulating ones. For example, magnetic metals are usually ferromagnetics and the insulating ones ferrimagnetics. There are purely antiferromagnetic insulators, but for microwave applications, these materials are not useful. However, they are useful at optical frequencies. Our interest is only on microwave properties of ordered magnetic materials. For now, let's limit our discussions to magnetic insulators, such as spinel, garnet, and hexagonal ferrites, and focus firstly on the electrostatic interaction between the oxygen anions and magnetic cations located at the center of an octahedral or tetrahedral coordination of ions, see Figure 3.7. The tetrahedral and octahedral sites are two fundamental magnetic sublattices. In our first discussion, let's consider a magnetic 3d cation at the center of an octahedron.

The interaction between the magnetic ion and O^{2-} is purely electrostatic. However, it reveals itself magnetically via the measurement of magnetic anisotropy, for example. The distance between the magnetic cation and the oxygen anion is about 2.1–2.2 Å (Orgel). The radius of an oxygen anion is in the order of 1.0 Å. There are two point of views in calculating the electrostatic interaction between the two ions. One point is to say that the overlap of wave functions between ions is small or negligible so that one can represent the oxygen anion as a point charge of $-2|e|$. The electrostatic interaction involves a potential calculation at a point in space $P(x, y, z)$. The interaction energy is simply the product of the potential due to the six anions (octahedral coordination) at P times the electron charge $-|e|$. Hutchings provides a review of this approach (see References). The other approach allows for overlapping of wave functions and is sometimes referred to as the covalent approach. In view of the above dimensionalities of distance and radius of the ions, one may conclude that the point change approach may be appropriate. One advantage of this approach is that it lends itself to straightforward estimation of the crystal-field effect on the magnetic ion. However, the reader should not put too much credence to this approach. It is, after all, an approximate approach. If indeed there is no overlap of wave functions, then a magnetic ion in one octahedral site would not "see" the existence of another magnetic ion in an adjacent octahedral site. It is well known that superexchange interactions—interaction between sites—can only exist, if there is some covalency between the cations and anions. We will address this later.

As we learned from the calculations of multiplets or terms, the ground state for $3d^2$ is the ^3F term or multiplet and it is sevenfold degenerate—see energy diagram in Figure 3.4. The crystal-field interaction removes this degeneracy. We are going to present a simple argument for the splitting of the ^3F term, for example. Let's introduce a new set of wave functions in terms of the hydrogenic wave functions Y_{l,m_l}. Specifically,

$$d_{x^2-y^2} = \sqrt{\frac{4\pi}{5}} \times \frac{1}{\sqrt{2}} \left(Y_{2,+2} + Y_{2,-2} \right) = \frac{\sqrt{3}}{2} \left(\frac{x^2 - y^2}{r^2} \right),$$

$$d_{z^2} = \sqrt{\frac{4\pi}{5}} \times Y_{2,0} = \frac{1}{2} \left(\frac{3z^2 - r^2}{r^2} \right),$$

$$d_{xy} = \sqrt{\frac{4\pi}{5}} \times \frac{1}{\sqrt{2}} j \left(-Y_{2,+2} + Y_{2,-2} \right) = \sqrt{3} \frac{xy}{r^2},$$

$$d_{yz} = \sqrt{\frac{4\pi}{5}} \times \frac{1}{\sqrt{2}} j \left(Y_{2,1} + Y_{2,-1} \right) = \sqrt{3} \frac{yz}{r^2},$$

and

$$d_{zx} = \sqrt{\frac{4\pi}{5}} \times \frac{1}{\sqrt{2}} \left(-Y_{2,1} + Y_{2,-1} \right) = \sqrt{3} \frac{zx}{r^2}.$$

The factor $\sqrt{4\pi/5}$ is a normalizing factor. The r-dependence is still the same for all wave functions and is equal to $R_{3,2}(r)$. The wave functions $d_{x^2-y^2}$ and d_{z^2} belong to the so-called e_g set of orbitals and d_{xy}, d_{yz}, and d_{zx} to the t_{2g} set of orbitals. In a similar fashion, we do the same for the oxygen ion wave functions and construct the following set:

$$P_x = \frac{x}{r} = \sqrt{\frac{4\pi}{3}} \times \frac{1}{\sqrt{2}} \left(-Y_{1,1} + Y_{1,-1} \right),$$

$$P_y = \frac{y}{r} = \sqrt{\frac{4\pi}{3}} \times \frac{1}{\sqrt{2}} j \left(Y_{1,1} + Y_{1,-1} \right),$$

and

$$P_z = \frac{z}{r} = \sqrt{\frac{4\pi}{3}} Y_{1,0}.$$

The factor $\sqrt{4\pi/3}$ is a normalizing factor. The r-dependence is contained in $R_{2,1}(r)$:

$$R_{2,1}(r) = \frac{1}{2\sqrt{6}}\left(\frac{Z}{a_0}\right)^{5/2} re^{-\frac{Zr}{2a_0}}.$$

In Figure 3.8, we sketch the wave functions at both the cation and anion sites for both the octahedral and tetrahedral coordination of ions.

In the octahedral coordination, we see that the p-orbitals are closest to the e_g set of functions relative to the t_{2g} set. The reader is reminded that the so-called p_x, wave function of the oxygen ion, means that the amplitude of the wave function is maximum along the x-axis or the electron is likely to be along the x-axis or the bonding axis of the crystal (see Figure 3.8). Thus, the repulsive electrostatic energy would be greater for the e_g set (E_g), since their wave functions are also maximum along the bonding axes. The energy level corresponding to the $t_{2g}(E_t)$ set would be relatively lower. The splitting in energy between the two levels, $E_g - E_t$, is designated in the literature as Δ or $10Dg$, see Figure 3.9.

The value of Δ ranges from 8,500 to 14,000cm⁻¹. This value of Δ should be compared with the multiplet splittings (Figure 3.4). Ideally, it is simpler to assume Δ to be weak compared to the multiplet splittings. In some cases where Δ is "strong,"

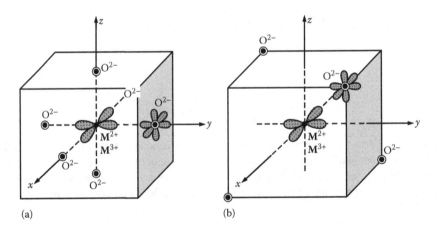

(a)   (b)

FIGURE 3.8   3d wave functions of magnetic ions distributed in (a) octahedral and (b) tetrahedral sites.

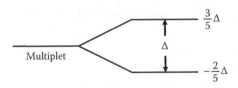

FIGURE 3.9   Splitting of multiplet energy by crystal field.

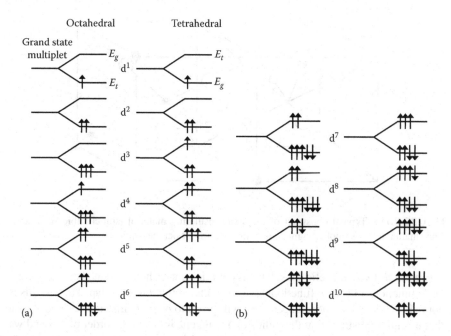

FIGURE 3.10 3d electrons in octahedral and tetrahedral complexes.

calculations of the crystal-field splittings may precede the multiplet splittings and calculations are rather complex. For simplicity, we will assume weak Δ splittings compared to the multiplet splittings. As such, we can assume that Hund's rule still applies and there is an orderly way of populating E_g and E_t energy levels depending on the 3d electronic configuration (see Figure 3.10).

For most temperatures of interest, only the ground-state multiplet is usually occupied by 3d electrons.

SUPEREXCHANGE COUPLING BETWEEN MAGNETIC IONS

In putting together a magnetic material, we envision a tetrahedral or octahedral site next to a similar magnetic unit cell or sublattice. These sites are usually connected via an oxygen ion, as shown in Figure 3.11.

The black dots represent the tetrahedral magnetic sites and the white octahedral sites, as shown in Figure 3.11. The distance between two nearest-neighbor magnetic cation sites of similar coordination (octahedral or tetrahedral) is about 4.3 Å. The distance for dissimilar sites (between octahedral and tetrahedral) is about 3.85 Å. Without going much into the theory of superexchange, one may argue that the exchange coupling between magnetic sites is proportional to the distance separating the ions. Let's consider a very simple theoretical situation whereby one single 3d electron is located in one site and another 3d electron in another site and ask the question: Is it ferromagnetic or antiferromagnetic coupling between the two magnetic sites? In order to simplify the analysis further, we remove the oxygen ion site

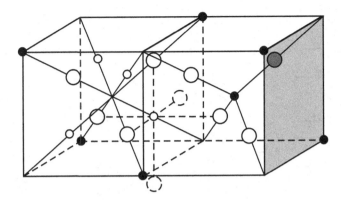

FIGURE 3.11 Typical construct of magnetic insulating material (spinel cubic structure). The smaller spheres imply magnetic ions (tetrahedral or octahedral sites).

between the two magnetic ions. The assumption made here is that their respective wave functions (3d wave functions) are so extended that they overlap. Though this is not reasonable, let's proceed because it will prove to be quite instructive. The reason that it is not realistic is that the radius of a magnetic ion is in the order of 1 Å and we are asking the ions to "see" each other at about 4 Å away. It is a little bit of a stretch. Clearly, the role of the oxygen ion is well defined. It mediates the coupling between localized electrons even though they are well separated. For now, let's ignore the oxygen site and consider a two-body problem depicted in Figure 3.12. Geometry of two-body problem (3d ions) is shown in Figure 3.12.

The Hamiltonian of the system is

$$H = H_1 + H_2 + \frac{Z_A Z_B e * e}{r_{AB}} + \frac{e^2}{r_{12}},$$

where
$Z_A |e| =$ charge of nucleus at site A,
$Z_B |e| =$ charge of nucleus at site B,

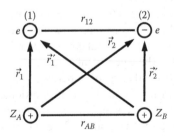

FIGURE 3.12 Geometry of two interacting electrons.

$$r_{12} = |\vec{r}_1 - \vec{r}_2|$$

$$\vec{r}_1 = \vec{r}_{1'} + \vec{r}_{AB}$$

$$\vec{r}_2 = \vec{r}_{2'} - \vec{r}_{AB}$$

$$H_1 = \frac{P_1^2}{2m_1} - \frac{Z_A e^2}{r_1} - \frac{Z_B e^2}{r_{1'}}$$

$$H_2 = \frac{P_2^2}{2m_2} - \frac{Z_A e^2}{r_2} - \frac{Z_B e^2}{r_{2'}}$$

$$m_1 = m_2 = m_e$$

Since both electronic configurations are 3d', there is no multiplet to be concerned about. However, crystal-field corrections to the energy are needed to be included for a realistic calculation, but for now, let's omit them. There is one more correction we need to be aware of: the transfer of one electron from one site to the other would indeed introduce a ground-state energy term to the site with one extra electron. Let's now address our question. Let's assume some overlap of the 3d orbitals so that the orbitals are modified as follows:

$$\Phi_A = u_A - \frac{\theta}{2} u_B,$$

$$\Phi_B = u_B - \frac{\theta}{2} u_A,$$

and

$$\theta = \int u_A(r_1) u_B(r_1') dV \equiv u_A \mid u_B.$$

There are three singlet states [$\psi_s(AA)$, $\psi_s(BB)$, and $\psi_s(AB)$] and one triplet state $\psi_t(AB)$. The singlet state $\psi_s(AA)$ means that, at site A, there are two electrons whose spins oppose each other. The same applies for $\psi_s(BB)$. The singlet state $\psi_s(AB)$ means that there is one electron at site A and the other at B with their spins in opposite direction. This represents the antiferromagnetic state. The triplet state $\psi_t(AB)$ implies that the two spins are parallel to each other at the two sites. This represents the ferromagnetic state. The wave functions may for the singlet states be expressed as follows:

$$\psi_s(AB) = \frac{1}{2} \left[\left(\phi_A(1)\phi_B(2) + \phi_B(1)\phi_A(2) \right)\left(\alpha(1)\beta(2) - \beta(1)\alpha(2) \right) \right] = |S_{AB},$$

where

α's $\equiv$ up spin
β's $\equiv$ down spin
$\phi_A(1) \equiv \phi_A(r_1)$

$$(2) = \phi_B(r_2')$$

$$\psi_s(AA) = \frac{1}{\sqrt{2}} \phi_A(1)\phi_A(2)[\alpha(1)\beta(2) - \beta(1)\alpha(2)] = |S_{AA}, \text{ and}$$

$$\psi_s(BB) = \frac{1}{\sqrt{2}} \phi_B(1)\phi_B(2)[\alpha(1)\beta(2) - \beta(1)\alpha(2)] = |S_{BB}$$

Finally, the triplet state may be expressed as

$$\psi_t(AB) = \frac{1}{\sqrt{2}} \alpha(1)\alpha(2)[\phi_A(1)\phi_B(2) - \phi_A(2)\phi_B(1)] = |t.$$

It is noted that all of the wave functions are antisymmetric with respect to 1 and 2.

Care must be exercised between primed and unprimed systems. The reader should refer to Figure 3.12 for clarification. We are setting up an eigenvalue problem by operating H over the set of wave functions postulated here. Since we have four wave functions, the H matrix is necessarily a 4×4 matrix.

It is noted that matrix elements of the following form are zero:

$$\langle t|H|S \rangle = 0,$$

since they contain terms like $\langle \alpha|\beta \rangle = \langle \beta|\alpha \rangle = 0$. Let's consider the matrix element

$$\langle t|H|t \rangle = \langle t \left| H_1 + H_2 + \frac{Z_A Z_B e^2}{r_{AB}} + \frac{e^2}{r_{12}} \right| t \rangle$$

or

$$\langle t|H|t \rangle = \frac{1}{2} \iint [\phi_A(1)\phi_B(2) + \phi_A(2)\phi_B(1)] \left[H_1 + H_2 + \frac{Z_A Z_B e^2}{r_{AB}} + \frac{e^2}{r_{12}} \right]$$

$$\times [\phi_A^*(1)\phi_B^*(2) - \phi_A^*(2)\phi_B^*(1)] dV_1 dV_2$$

$$= \frac{1}{2} \left\{ \begin{array}{l} \overbrace{\int \phi_A(1) H_1 \phi_A^*(1) dv_1}^{} \overbrace{\int \phi_B(2) \phi_B^*(2) dV_2}^{1} + \overbrace{\int \phi_B(1) H_1 \phi_B^*(1) dv_1}^{} \overbrace{\int \phi_A(2) \phi_A^*(2) dV_2}^{1} \\[6pt] -\underbrace{\int \phi_A(1) H_1 \phi_B^*(1') dv_1}_{0} \int \phi_B(2') \phi_A^*(2) dV_2 - \underbrace{\int \phi_B(1) H_1 \phi_A^*(1) dv_1}_{0} \int \phi_A(2) \phi_B^*(2') dV_2 \\[6pt] +\int \phi_B(2') H_2 \phi_B^*(2') dv_2 \underbrace{\int \phi_A(1) \phi_A^*(1) dV_1}_{1} + \int \phi_A(2) H_2 \phi_A^*(2) dv_2 \underbrace{\int \phi_B(1') \phi_B^*(1') dV_1}_{1} \end{array} \right\}$$

$$+ \langle t \left| \frac{e^2}{r_{12}} \right| t + t \left| \frac{Z_A Z_B e^2}{r_{AB}} \right| t \rangle .$$

In summary, we have that

$$\langle t | H | t \rangle = \frac{1}{2} \left\{ (E_1)_A + (E_1)_B + (E_2)_B + (E_2)_A \right\} t \left| \frac{e^2}{r_{12}} \right| t + E_S$$

or

$$\langle t | H | t \rangle = (E_1)_A + (E_2)_B t \left| \frac{e^2}{r_{12}} \right| t + E_S ,$$

where

$$E_S = \left\langle \frac{Z_A Z_B e^2}{r_{AB}} \right\rangle \cong \frac{Z_A Z_B}{r_{AB}} e^2 ,$$

and E_1, E_2 are the core energies plus crystal-field energies. Finally, we have the exchange integral:

$$\langle t \left| \frac{e^2}{r_{12}} \right| t \rangle = \frac{1}{2} \underbrace{\iint \phi_A(1) \phi_B(2') \frac{e^2}{r_{12}} \phi_A^*(1) \phi_B^*(2') dV_1 dV_2}_{K_{AB}}$$

$$+ \frac{1}{2} \underbrace{\iint \phi_B(1') \phi_A(2) \frac{e^2}{r_{12}} \phi_B^*(1') \phi_A^*(2) dV_1 dV_2}_{K_{BA}}$$

$$- \frac{1}{2} \underbrace{\iint \phi_A(1) \phi_B(2') \frac{e^2}{r_{12}} \phi_B^*(1') \phi_A^*(2) dV_1 dV_2}_{J_{AB}}$$

$$-\frac{1}{2}\iint \phi_B(1')\phi_A(2)\underbrace{\frac{e^2}{r_{12}}\phi_A^*(1)\phi_B^*(2')}_{J_{BA}}dV_1\,dV_2.$$

Thus,

$$\left\langle t\left|\frac{e^2}{r_{12}}\right|t\right\rangle = K_{AB} - J_{AB}$$

and

$$\boxed{\therefore \langle t|H|t\rangle = (E_1)_A + (E_2)_B + K_{AB} - J_{AB} + E_S}.$$

In a similar manner, we obtain

$$_{AB}\langle S|H|S\rangle_{AB} = (E_1)_A + (E_2)_B + K_{AB} + E_S,$$

$$_{AA}\langle S|H|S\rangle_{AA} = 2(E_1)_A + K_{AA} + E_S, \text{ and}$$

$$_{BB}\langle S|H|S\rangle_{BB} = 2(E_2)_B + K_{BB} + E_S.$$

K_{AA} and K_{BB} are recognized as the repulsive Coulomb energy within sites A and B, respectively. Let's now consider the off-diagonal elements $_{AB}\langle S|H|S\rangle_{AA}$ representing transition probability of transferring one electron from B to the A site such that the B site is empty and the A site contains two electrons.

It is written explicitly as

$$_{AB}\langle S|H|S\rangle_{AA} = \sqrt{2}t_{B\to A},$$

where

$$_{AB}\langle S|H|S\rangle_{AA} = \frac{1}{\sqrt{2}}\iint dV_1\,dV_2\left[\phi_A(1)\phi_B^*(2') + \phi_B(1')\phi_A(2)\right]$$

$$\times\left[H_1 + H_2 + \frac{e^2}{r_{12}} + \frac{Z_A Z_B e^2}{r_{AB}}\right]\left[\phi_A^*(1)\phi_A^*(2)\right].$$

The transition of one electron from site A to site B requires a "transfer" energy, as calculated in the above matrix element. The transfer energy is represented by $t_{B\to A}$. Expanding each term, we have

$$= \frac{1}{\sqrt{2}} \iint \phi_A(1)\phi_B(2') \underbrace{H_2 \phi_A^*(1)\phi_A^*(2)}_{\frac{1}{\sqrt{2}}(E_2)_{BA}} dV_1 dV_2$$

$$+ \frac{1}{\sqrt{2}} \iint \underbrace{\phi_B(1')\phi_A(2) H_1}_{\frac{1}{\sqrt{2}}(E_1)_{BA}} \phi_A^*(1)\phi_A^*(2) dV_1 dV_2$$

$$+ \frac{1}{\sqrt{2}} \left\{ \iint \phi_A(1)\phi_B(2') \frac{e^2}{r_{12}} \phi_A^*(1)\phi_A^*(2) dV_1 dV_2 \right.$$

$$\left. + \iint \phi_B(1')\phi_A(2) \frac{e^2}{r_{12}} \phi_A^*(1)\phi_A^*(2) dV_1 dV_2 \right\}$$

$$+ \frac{1}{\sqrt{2}} \left[\underbrace{\iint dV_1 dV_2 |\phi_1(A)|^2 \phi_B(2') \frac{e^2}{r_{12}} \phi_A^*(2) \iint dV_1 dV_2 |\phi_2(A)|^2 \phi_B(1') \frac{e^2}{r_{12}} \phi_A^*(1)}_{\sqrt{2}K_{AB,AB}} \right]$$

$$+ \frac{Z_A Z_B e^2}{\sqrt{2} r_{AB}} \iint dv_1 dv_2 \left[\phi_A(1)\phi_B(2')\phi_A^*(1)\phi_A^*(2) \atop 0 \right].$$

Thus, $t_{B \to A} = (E_1)_{BA} + K_{AB,AA}$.

By symmetry, we have that

$$_{AB}\langle S|H|S \rangle_{BB} = \sqrt{2} t_{A \to B} = \sqrt{2} \left((E_1)_{AB} + K_{AB,BB} \right).$$

Also

$$_{AA}\langle S|H|S \rangle_{BB} = J_{AB^*}.$$

$$J_{AB^*} = \iint dV_1 dV_2 \phi_A(1)\phi_A(2) \frac{e^2}{r_{12}} \phi_B^*(1')\phi_B^*(2).$$

Normally, J_{AB} is defined as follows:

$$J_{AB} = \iint dV_1 dV_2 \phi_A(1)\phi_B(2') \frac{e^2}{r_{12}} \phi_B^*(1')\phi_A^*(2).$$

The *H* matrix in the basis representation proposed here is then

$$
[H]=
\begin{array}{c}
 \\
{}_{AB}\langle s| \\
{}_{AA}\langle s| \\
{}_{BB}\langle s|
\end{array}
\begin{bmatrix}
\langle t| & |t\rangle & |s\rangle_{AB} & |s\rangle_{AA} & |s\rangle_{BB} \\
 & 2E + K_{AB} - J_{AB} + E_s & 0 & 0 & 0 \\
 & 0 & 2E + K_{AB} + E_S & \sqrt{2}t_{B\to A} & \sqrt{2}t_{A\to B} \\
 & 0 & \sqrt{2}t_{B\to A} & 2E + K_{AA} + E_S & J_{AB^*} \\
 & 0 & \sqrt{2}t_{A\to B} & J_{AB^*} & 2E + K_{BB} + E_S
\end{bmatrix}.
$$

Clearly, we need to only solve for the eigenvalues of the 3×3 matrix. The eigen energies may be solved simply as follows. Assume this linearized representation of the eigenvalue problem:

$$
\det
\begin{pmatrix}
E_1 - \lambda & a & b \\
a & E_2 - \lambda & c \\
b & c & E_3 - \lambda
\end{pmatrix}
= \det
\begin{pmatrix}
W_1 - \lambda & 0 & 0 \\
0 & W_2 - \lambda & 0 \\
0 & 0 & W_3 - \lambda
\end{pmatrix}.
$$

with the assumptions that

$$
W_1 + \Delta_1 = E_1,
$$

$$
W_2 + \Delta_2 = E_2,
$$

and

$$
W_3 + \Delta_3 = E_3.
$$

By equating the two determinants and linearizing solutions for Δ_1, Δ_2, and Δ_3, we obtain the following solutions for the Δ's:

$$
\Delta_1 = -\frac{a^2}{E_1 - E_2} - \frac{b^2}{E_1 - E_3} - \frac{2abc}{(E_1 - E_2)(E_1 - E_3)},
$$

$$
\Delta_2 = -\frac{a^2}{E_2 - E_1} - \frac{b^2}{E_2 - E_3} - \frac{2abc}{(E_2 - E_1)(E_2 - E_3)},
$$

and

$$
\Delta_3 = -\frac{a^2}{E_3 - E_1} - \frac{b^2}{E_3 - E_2} - \frac{2abc}{(E_3 - E_1)(E_3 - E_2)}.
$$

This result is recognized as a result obtained from perturbation theory. The first two terms are first-order corrections and the last term is the second-order correction to the perturbation. Thus, the eigenvalues are approximately

$$W_1 = \lambda_1 = E_1 - \Delta_1,$$

$$W_2 = \lambda_2 = E_2 - \Delta_2,$$

and

$$W_3 = \lambda_3 = E_3 - \Delta_3,$$

where E_1, E_2, and E_3 are the diagonal terms in the H matrix.

The eigen energy λ_1 corresponds to the energy state of the antiferromagnetic state, $E_S(AB)$:

$$E_S(AB) \cong 2E + K_{AB} + E_S - \frac{2t_{A \to B}^2}{K_{AB} - K_{AA}} - \frac{2t_{A \to B}^2}{K_{AB} - K_{BB}}.$$

The energy of the ferromagnetic state is

$$E_t(AB) \cong 2E + K_{AB} - J_{AB} + E_S.$$

Clearly, the stable magnetic state is the one with the lower energy or eigenvalue. Thus, write

$$E_t(AB) - E_s(AB) \cong -2J_{AB} + \frac{2t_{B \to A}^2}{K_{AB} - K_{AA}} + \frac{2t_{A \to B}^2}{K_{AB} - K_{BB}} + \dots. \qquad (3.17)$$

The magnetic state is antiferromagnetic if the right-hand side of the above equation is positive and ferromagnetic if it is negative. In the literature, the above relation is written as follows:

$$E_t(AB) - E_s(AB) = -2J_{A,B} + \frac{2t_{B \to A}^2}{U_{B \to A}} + \frac{2t_{A \to B}^2}{U_{A \to B}} + \dots.$$

$U_{B \to A}$ is defined as the difference in Coulomb repulsion energies as an electron is transferred from site B to A. Similar definition applies to $U_{A \to B}$. In summary, we have that for

$$2J_{AB} > \frac{2t^2}{U} \text{'s},$$

the ferromagnetic state is more stable, since the triplet state is lower than the singlet state or $E_t < E_s$.

For

$$2J_{AB} < \frac{2t^2}{U}\text{'s},$$

the antiferromagnetic state is more stable, since $E_s < E_t$. It is very interesting to note that antiferromagnetism or ferrimagnetism results from first-order perturbation theory which accounts for an electron "migrating" from one site to another. This is exactly what the oxygen provides to neighboring magnetic sites. It provides that conduit whereby electrons can easily transfer from one magnetic site to another. In essence, the oxygen anion is a mediator or facilitator for ferrimagnetism.

Our mathematical formulation is not exactly correct. It did not include the effect of the mediator—the oxygen ion. Let's do that now. The inclusion of the oxygen ion complicates the algebra, but not the methodology introduced in this section. It now becomes a three-body problem (four electrons) and three nuclei (see Figure 3.13).

The Hamiltonian operator for this system becomes the following:

$$H = \sum_{i=1}^{4}\left(-\frac{\hbar^2}{2m}\nabla_i^2 - \frac{Z_A e^2}{4\pi\varepsilon_0 r_i} - \frac{Z_B e^2}{4\pi\varepsilon_0 r_i'} - \frac{Z_C e^2}{4\pi\varepsilon_0 r_i''}\right)$$

$$+ \sum_{\substack{i,j=1 \\ i<j}}^{3}\frac{Z_i Z_j e^2}{4\pi\varepsilon_0 R_{ij}} + \sum_{\substack{i,j=1 \\ i<j}}^{4}\frac{e^2}{4\pi\varepsilon_0 r_{ij}},$$

(3.18)

where the unprimed system applies to one of the magnetic ions, the single primed to the other magnetic ion, and the double primed to the oxygen ion.

The terms inside the parenthesis include the kinetic and potential energies of the four electrons whose nuclei are located at sites A, B, and C. The other two terms

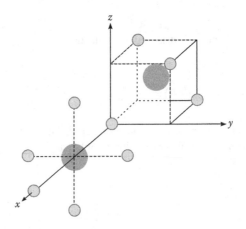

FIGURE 3.13 The magnetic ion (larger sphere) in the octahedral site is located at $(a/4,0,0)$ and in the tetrahedral site at $(-a/8,a/8,a/8)$, where a is the lattice constant (~8.4 Å).

represent electrostatic energy between nuclei which plays a minor role here and repulsion energy between electrons.

The four basis functions now become

$$\psi_t\left(A\uparrow O\downarrow O\uparrow B\uparrow\right)=|t\rangle,$$

$$\psi_s\left(A\uparrow O\downarrow O\uparrow B\downarrow\right)=|s\rangle_{AB},$$

$$\psi_s\left(A\uparrow O\downarrow O\uparrow A\downarrow\right)=|s\rangle_{AA},$$

and

$$\psi_s\left(B\uparrow O\downarrow O\uparrow B\downarrow\right)=|s\rangle_{BB},$$

where A and B represent the 3d wave functions shown in the previous sections and O represents the oxygen p wave functions. Note that we have assumed that the two spins on the oxygen ion have opposite directions for all the basis functions chosen. A typical singlet wave function is shown below:

$$|s_{AB}=\psi\left(A\uparrow O\uparrow \underline{O}\downarrow B\downarrow\right)$$

$$= N \det \begin{vmatrix} \phi_A(1)\alpha_A(1) & \phi_A(2)\alpha_A(2) & \phi_A(3)\alpha_A(3) & \phi_A(4)\alpha_A(4) \\ \phi_O(1)\alpha_O(1) & \phi_O(2)\alpha_O(2) & \phi_O(3)\alpha_O(3) & \phi_O(4)\alpha_O(4) \\ \phi_{\underline{O}}(1)\beta_{\underline{O}}(1) & \phi_{\underline{O}}(2)\beta_{\underline{O}}(2) & \phi_{\underline{O}}(3)\beta_{\underline{O}}(3) & \phi_{\underline{O}}(4)\beta_{\underline{O}}(4) \\ \phi_B(1)\beta_B(1) & \phi_B(2)\beta_B(2) & \phi_B(3)\beta_B(3) & \phi_B(4)\beta_B(4) \end{vmatrix}.$$

where ϕ_O and $\phi_{A,B}$ are wave functions corresponding to the oxygen ion and magnetic ions A and B. The mathematical methodology follows the same as in the two-body problem outlined above.

DOUBLE SUPEREXCHANGE COUPLING

The fundamental question arises: when the possibility exists that $E_1=E_2$ or $E_1=E_3$ (see previous section), we have degeneracy in the system. How do we modify the previous approach in determining magnetism for the ensemble of spins? That is, for example, the situation arises when transferring an electron from one site to another does not change E_1 or E_2. This implies that both states are in the ground state. This situation is often referred to as the double-exchange phenomena; exchanging an electron from one site to another or vice versa does not alter the energy states. Let's cite an example of this situation. Titanium ion may exist into two valence states, Ti^{2+} and Ti^{3+}. However, it may also exist as Ti^{4+}. In this example, we are not considering this valence state, since there are no 3d electrons in this valence state. Let's imagine a hypothetical situation where an oxygen ion is located to the left of Ti^{2+}, and of course, Ti^{3+} is to the right. Let's label this electronic configuration as $\psi_1 = |1\rangle$ and the opposite electronic configuration as $|2\rangle$. Clearly, the mathematical descriptions of $|1\rangle$ or $|2\rangle$ must necessarily be the same, because it is the same configuration or state. Hence, we have twofold degeneracy. How does nature go about in removing this degeneracy? Let's construct a new set of basis functions, namely,

$$\Psi_+ = \Psi_1 + \Psi_2 = |-\rangle = |1\rangle + |2\rangle$$

and

$$|-\rangle = |1\rangle - |2\rangle.$$

Let's operate this set on the Hamiltonian, H, and obtain

$$E_+ = \langle + | H | + \rangle = 2E_0 + 2\Delta,$$

$$E_- = \langle - | H | - \rangle = 2E_0 - 2\Delta,$$

where

$$E_0 = \langle 1 | H | 1 \rangle = \langle 2 | H | 2 \rangle$$

and

$$\Delta = \langle 1 | H | 2 \rangle.$$

Thus,

$$E_+ - E_- = 4\Delta.$$

The basic question to ask now is: What must the nature of $|1\rangle$ and $|2\rangle$ be to remove the degeneracy? There are only two possibilities: states $|1\rangle$ and $|2\rangle$ are either ferromagnetic or ferrimagnetic. For simplicity, the ferromagnetic states would take this form assuming site A contains two electrons (Ti^{2+}) and site B one electron (Ti^{3+}).

$$|1\rangle = \frac{1}{\sqrt{6}} \det \begin{pmatrix} \phi_A(1)\alpha(1) & \phi_A(2)\alpha(2) & \phi_A(3)\alpha(3) \\ \phi_{A'}(1)\alpha(1) & \phi_{A'}(2)\alpha(2) & \phi_{A'}(3)\alpha(3) \\ \phi_B(1)\alpha(1) & \phi_B(2)\alpha(2) & \phi_B(3)\alpha(3) \end{pmatrix},$$

where orbitals ϕ_A and $\phi_{A'}$ are associated with the two electrons at site A (Ti^{2+}). The prime is to indicate that the two electrons must take on different orbitals, since the Pauli exclusion principle requires it to be so. Orbital ϕ_B is associated with the single electron state at site B (Ti^{3+}). The spin at site B must be aligned with the other two spins to give rise to a ferromagnetic state. The other ferromagnetic state, $|2\rangle$, is the opposite of $|1\rangle$. Namely, at site B, we now have two orbitals, ϕ_B and $\phi_{B'}$, associated with Ti^{2+} and one electron at site A, ϕ_A, associated with Ti^{3+}.

The ferrimagnetic state will contain the same orbitals, but the spin arrangements at sites A and B must necessarily oppose each other. However, the exact functional dependences are still contained in expanding a 3×3 states Slater determinant. It is a ferrimagnetic state, since there is now a "net" moment when the spins oppose each other at the two sites. Clearly, for the ferrimagnetic state,

$$\langle 1 | H | 2 \rangle = 0,$$

since products of the form $\langle \alpha | \beta \rangle = 0$. This means that the ferrimagnetic state does not remove the degeneracy. However, the ferromagnetic state does remove the degeneracy as a result of

$$\langle 1 | H | 2 \rangle \neq 0.$$

Products like $\langle \alpha | \alpha \rangle \neq 0$.

FERROMAGNETISM IN MAGNETIC METALS

As implied in the previous section, conductivity implies ferromagnetism. Let's examine magnetism in magnetic metals and adopt the algebraic formalism developed for the double-exchange modeling. We model carriers in magnetic metals as being free to conduct or be part of the free electron gas. In the absence of repulsion between carriers, they occupy one common ground state, much like the double-exchange situation. Let's apply the same ideas as we have done in the above section. The ground-state free electron wave functions taking two electrons at a time may be defined as follows:

$$\psi_k(r_i)\psi_{k'}(r_j) = |1,$$

$$\psi_{k'}(r_i)\psi_k(r_j) = |2,$$

and

$$\psi_k(r_i) = \frac{1}{\sqrt{V}} e^{j\vec{k} \cdot \vec{r_i}}.$$

With properties

$$\langle 1|1\rangle = \langle 2|2\rangle = 1$$

and

$$\langle 1\,\|\,2\rangle = 0.$$

By including the spin motion contribution, we must choose between the ferromagnetic and the antiferromagnetic state. The antiferromagnetic state is composed as follows:

$$|+\rangle = \frac{1}{2}\big[|1\rangle + |2\rangle\big]\big(\alpha_i\beta_j - \alpha_j\beta_i\big),$$

and the ferromagnetic state as

$$|-\rangle = \frac{1}{2}\big[|1\rangle - |2\rangle\big]\alpha_i\alpha_j. \tag{3.19}$$

The subscripts i and j are to indicate pair of electrons in the formulation.

The Hamiltonian is of the following form:

$$H = \sum_i \frac{p_i^2}{2m} + V(r) + \sum_{i \neq j} \frac{e^2}{r_{ij}},$$

where the first term is the kinetic energy, the second term is the electrostatic energy due to positive charge background at lattice sites, and the last term is the repulsion energy between pair electrons.

As in the previous arguments, we construct matrix elements based on the above representation of wave functions. Specifically,

$$\langle -|H|-\rangle = \sum_k \frac{\hbar^2 k^2}{2m} - \frac{1}{V} \sum_{k'}^{k_F} \int \frac{e^2}{|\vec{r} - \vec{r}'|} e^{-j(k-k')(\vec{r}-\vec{r}')} dr' = E, \tag{3.20}$$

$$\langle +|H|-\rangle = 0,$$

and

$$\langle +|H|+\rangle = \sum_k \frac{\hbar^2 k^2}{2m}.$$

It is noted that operating H with the ferromagnetic state results in lowering the total energy from the ground-state energy consisting only of the kinetic energy (first term in Equation 3.20). The minus sign is a result of the "cross" terms in the execution of the matrix element in Equation 3.20, which is manifestation of the exchange coupling. The cross term for the antiferromagnetic state is zero, since $\langle\alpha||\beta\rangle=0$. The total electrostatic energy is zero as the electron gas is electrically neutral. For a given electron charge, it "sees" electrostatically the same amount of positive (lattice sites) and negative (other electrons) charges in the free electron gas ensemble.

We sum over all possible values of k' due to the second particle. Now let's obtain the total energy by summing over k. Kubler shows that the average energy per electron (E/N) of free electron gas is

$$\frac{E}{N}=\sum_{k}^{k_F}E_i(k,k')=\frac{3}{5}\left(\frac{\hbar^2 k_F^2}{2m}\right)-\frac{3}{2}\left(\frac{e^2}{\lambda_F}\right),$$

where $k_F\equiv$ Fermi's radius $(2\pi/\lambda_F)$.

The above expression says that the average energy of an electron in a gas is made up of 3/5 of its kinetic energy evaluated at the Fermi energy and contains an exchange energy term. The factor of 3 is understood as having 3 degrees of freedom. The factor of 1/2 is the result of averaging. Finally, e^2/λ_F is the usual potential energy of two particles separated by λ_F on the average. Dimensionally, we need the second term to be multiplied by $1/4\pi\varepsilon_0$ to put it in units of Joules. Further simplification in the above equation may be written as

$$\frac{E}{N}=\frac{1}{N}\sum_{k}\left(\frac{\hbar^2 k^2}{2m}-\frac{e^2 k_F}{2\pi}\right)=\frac{3}{5}E_F-\frac{3e^2}{4\pi}k_F,$$

where $E_F=$ Fermi energy $=\dfrac{\hbar^2 k_F^2}{2m}$.

The ferromagnetic state is stable for $E/N<0$. Let's equate the second term representing the exchange coupled energy (ferromagnetically) as an "internal" molecular exchange field aligning the spins of the electron along this internal field. Heisenberg model for this coupling is of the form

$$E_{exc}=-2\sum_{j\neq i}^{N}J_{ij}\vec{S}_i\cdot\vec{S}_j.$$

In a molecular field approximation, we write E as

$$E_{exc}\sim-JNz\vec{S}_i\cdot\vec{S}_j,$$

where z is the number of nearest neighbors to site i, for example. The factor of 2 is omitted in order to avoid double counting. Assuming $\vec{S}_i$ parallel to $\vec{S}_j$ and $S_i=S_j=S$, we have also assumed $J_{ij}\cong J$ (nearest neighbors only). Writing E_{exc} in the following form:

$$E_{exc}=-JNz\frac{Ng\mu_B SNg\mu_B S}{N^2\left(g\mu_B\right)^2}=-\frac{1}{2}\lambda M^2,$$

where $M = Ng\mu_B S$.

The exchange self-energy may be calculated from

$$E_{exc} = -\int \vec{H}_{exc} \cdot d\vec{M}, \text{ where}$$

$$\vec{H}_{exc} \cdot \lambda \vec{M}, \text{ and}$$

$$E_{exc} = -\frac{1}{2}\lambda M^2.$$

Thus,

$$\lambda = \frac{2Jz}{N(g\mu_B)^2}.$$

E_{exc} can be thought of as the exchange "self"-energy or spontaneous self-energy. Let's equate the above to the second term of the total energy obtained for the gas model, i.e.,

$$\frac{1}{2}\lambda(g\mu_B)^2 = \frac{3e^2 k_F}{4\pi n} \frac{1}{4\pi\varepsilon_0}.$$

We have inserted the factor $1/4\pi\varepsilon_0$ on the right-hand side to balance the units of both sides (J m³) of the equation. Solving for λ, we obtain

$$\lambda = \frac{2 \times 3e^2 k_F}{2\pi n} \frac{1}{4\pi\varepsilon_0} \frac{1}{g^2\mu_B^2},$$

$$k_F = \frac{2\pi}{\lambda_F},$$

where $k_F = (3\pi^2 n)^{\frac{1}{3}}$,

$$n = \frac{2}{a^3} \text{ (BCC)},$$

$$n = \frac{2}{(2.86)^3 \times 10^{-30}} = 8.56 \times 10^{28} \text{ m}^{-3} \text{ (iron)},$$

$$k_F = 1.37 \times 10^{10} \text{ m}^{-1},$$

$$\lambda = 2 \times \frac{3 \times 2.56 \times 10^{-38} \times 1.37 \times 10^{10}}{\underbrace{\frac{1}{9} \times 10^{-9}}_{4\pi\varepsilon_0} \times 4\pi \times 8.56 \times 10^{28} \times g^2 \times \underbrace{1.36 \times 10^{-58}}_{\mu_B^2}},$$

$$\lambda = 2 \times 14.82 \times 10^9 \text{ Am/Wb, MKS,}$$

$$\lambda = 2 \times 14.82 \times 10^9 \left(\frac{(1/10) \times 100}{10^8} \right), \text{ Gaussian units,}$$

$$\lambda = 2,960.$$

The internal molecular field is in the order of

$$H_{exc} = \lambda M = 1,482 \times 1,672 \text{ Oe,}$$

$$H_{exc} = 2.48 \times 10^6 \text{ Oe!}$$

Let's estimate the value of J:

$$J = \frac{\lambda n g^2 \mu_B^2}{z},$$

$$J = 2 \times 1,482 \times \frac{8.56}{z} \times 10^{22} \times (9.27)^2 \times 10^{-42} g^2 \text{ erg,}$$

$$J = 2 \times 1.1 \times 10^{-14} \frac{g^2}{z} \text{ erg,}$$

$$J = 1.2 \times 10^{-14} \text{ erg.}$$

PROBLEMS

3.1

    a. Calculate the resonant frequency of a TE_{101} rectangular cavity with the following dimensions (in Å): $1 \times 1 \times 1/2$.

    b. Do the same for a TE_{011} cylindrical cavity with the following dimensions (in Å): $r = 1/2$ and $L = 1$.

    c. Assuming that an electron beam excites the cavity to resonance, calculate the electron velocity in (a) and (b) assuming the De Broglie relation.

    d. Calculate the energy of the electrons using the relationships

$$E = \frac{p^2}{2m} \text{ and } E = \frac{hc}{\lambda},$$

    where $c = 3 \times 10^{10}$ cm/s.

3.2   Assume $r = (1/2)$ Å, $m = 9.1 \times 10^{-28}$ g, calculate v for $n = 1$, and compare with Problem 3.1.

3.3   Calculate β in CGS and MKS systems of units.

3.4   Show that $2 \sum_{\ell}^{n-1} (2\ell + 1) = 2n^2$.

3.5

    a. Let $f = S_x = \dfrac{1}{2} \begin{bmatrix} 0 & 1 \\ 1 & 0 \end{bmatrix}$. Construct the $[f]$ matrix using the spin functions

$$\psi_1 = \begin{bmatrix} 1 \\ 0 \end{bmatrix} \text{ and } \psi_2 = \begin{bmatrix} 0 \\ 1 \end{bmatrix}.$$

    b. Let $f = S_y = \dfrac{1}{2} \begin{bmatrix} 0 & -i \\ i & 0 \end{bmatrix}$. Construct the $[f]$ matrix.

    c. Let $f = S_z = \dfrac{1}{2} \begin{bmatrix} 1 & 0 \\ 0 & -1 \end{bmatrix}$. Construct the $[f]$ matrix.

    d. Are the $[f]$ matrices in (a), (b), and (c) Hermitian?

3.6

    a. Assume a microwave cavity with dimensions of (in Å): $1 \times 1 \times 1/2$. Calculate the eigenvalue of a single electron in the cavity. Assume

$$\psi_n \propto \sin k_{1x} \sin k_{2x}.$$

    b. Compare this to the previous calculations of the resonant frequency of a TE_{101} cavity with same dimensions.

3.7 Calculate χ_D.

Assume $\langle r^2 \rangle = (1/4) \times 10^{-16} cm^2$, $n = 6 \times 10^{23}$ particles/cm³, and $B = 1,000$ G. Calculate χ_S assuming $S = 1/2$.

3.8 Assume $Z = 1$, $r = 1$ Å, and $\ell = 1$. Calculate the magnetic field induced by the spin–orbit interaction.

3.9 Consider the following differential equation:

$$\frac{d^2\psi}{dx^2} = -\lambda(1+x)\psi.$$

Set up an eigenvalue problem by choosing

$$\psi_1 = \sin k_{1x}, \ k_1 = \frac{\pi}{t}$$

and

$$\psi_2 = \sin k_{2x}, \ k_2 = \frac{\pi}{t}.$$

For $0 \le x \le t$, find eigenvalues and eigen functions.

3.10 Calculate g_J for ions which assume 3d and 4f electronic configurations.

3.11 Assume an infinitely deep well with width a. The first two excited state wave functions are

$$\psi_1 = N_1 \sin \frac{\pi x}{a}$$

and

$$\psi_2 = N_2 \sin \frac{2\pi x}{a}$$

a. Calculate N_1, N_2, E_1, and E_2.
b. Place two electrons in the well and calculate the matrix $[H]$ based upon the above function representation in which

$$H = \frac{e^2}{|x_1 - x_2|}.$$

c. Calculate the new eigenvalues.
d. Calculate new wave functions including spin representations.

APPENDIX 3.A: MATRIX REPRESENTATION OF QUANTUM MECHANICS

Electronic states in which the energies have definite discrete values are called stationary states. Examples of stationary states are orbital states, electromagnetic standing modes in finite geometries, acoustic standing modes, and solitons. The energy value or resonant frequency remains constant with time. These states are described by a wave function ψ_n which is an eigen function of the Hamilton operator, H:

$$H\psi_n = E_n\psi_n, \tag{3.A.1}$$

where

E_n = energy eigenvalue
ψ_n = eigen wave function
H = Hamiltonian operator

H represents the total energy of the system. In a quantum mechanical system, it represents the kinetic $(p^2/2m)$ and potential $(V(\vec{r}))$ energies:

$$H = \frac{p^2}{2m} + V(\vec{r}).$$

It is convenient to write H in differential operators or

$$H = -\frac{\hbar^2}{2m}\nabla^2 + V(\vec{r}). \tag{3.A.2}$$

The wave function ψ completely determines the state of the physical system in quantum mechanics. This means that ψ is solved in terms of $\vec{r}$ and t. If ψ is an eigen function of the system, we may write

$$H\psi_n = i\hbar\frac{\partial}{\partial t}\psi_n$$

or

$$i\hbar\frac{\partial}{\partial t}\psi_n = E_n\psi_n.$$

The time dependence of ψ_n is then of the form

$$\psi_n(\vec{r},t) = \exp\left(-\frac{i}{\hbar}E_nt\right)\psi_n(\vec{r}).$$

$\psi_n(\vec{r})$ may be calculated from

$$\left(-\frac{\hbar^2}{2m}\nabla^2 + V\right)\psi_n(\vec{r}) = E_n\psi_n(\vec{r}).$$

Sometimes, it is mathematically convenient to expand the wave function as a linear combination of all stationary states or

$$\psi = \sum_{n=1}^{N} a_n \psi_n(\vec{r},t) = \sum_{n=1}^{N} a_n \exp\left(-\frac{i}{\hbar}E_n t\right)\psi_n(\vec{r}). \tag{3.A.3}$$

The squared modulus $|a_n|^2$ determines the probability of a particle of having energy E_n or occupying state ψ_n. The probability function of locating a particle is given as

$$P(\vec{r},t) = \left|\psi_n(\vec{r},t)\right|^2 = \left|\psi_n(\vec{r})\right|^2.$$

We see that P is independent of time (for stationary states). The same is true of the mean values of an operator f defined as

$$f = \int Pf dv = \int \psi_n^*(\vec{r}) f \psi_n(\vec{r}) dv. \tag{3.A.4}$$

$\langle f \rangle$ is independent of time. f can represent the operators x, y, z, s, p, etc. In general, we may write

$$f = \int \psi^* f \psi dv = \int \sum a_n^* \psi_n^*(\vec{r},t) f \sum a_m \psi_m(\vec{r},t) dv, \tag{3.A.5}$$

$$f = \sum_{n,m} a_n^* a_m f_{nm}(\vec{r},t),$$

where

$$f_{nm}(\vec{r},t) = \int \psi_n^*(\vec{r},t) f \psi_m(\vec{r},t) dv.$$

Now f_{nm} means more than the average value of a physical quantity. Clearly, f_{nn} or f_{mm} implies some sort of averaging, but what about f_{nm}? It is the transition probability of a particle to be excited from one stationary state to another. f_{nm} is also referred to simply as a matrix element of f. Time dependence of f_{nm} may be simply included by writing

$$f_{nm}(\vec{r},t) = f_{nm}(\vec{r}) e^{i w_{nm} t},$$

where $w_{nm} = (E_n - E_m)/\hbar$ = transition frequency between states n and m.
    For real physical observables, we require

$$f_{nm} = f_{nm}^*$$

Such matrices whose off-diagonal matrix elements obey the above property are said to be Hermitian. Examples of observable quantities are $\vec{G}$, $\vec{x}$, $\vec{p}$, H, $P^2/2m$, S_x, S_y, and S_z.

BIBLIOGRAPHY

P.W. Anderson, *Phys. Rev.* 79, 350, 1950.

P.W. Anderson and H. Hasegawa, *Phys. Rev.* 100, 675, 1955.

C.J. Ballhausen, *Introduction Ligand Field Theory*, McGraw-Hill Book Co., Inc., New York, 1962.

C.J. Ballhausen and H.B. Gray, *Molecular Orbital Theory*, W.A. Benjamin, Inc., New York, 1964.

A.D. Beclee, *Phys. Rev. Brief Rep.* **38**, 3098, 1988.

I.B. Bersuken, *Electronic Structure and Properties of Transition Metal Compounds*, John Wiley & Sons, New York, 1970.

B. Bleaney and K.W.H. Stevens, *Rep. Prog. Phys.* **16**, 108, 1953.

B.I. Bleaney and B. Bleaney, *Electricity and Magnetism*, Oxford Press, Amen House, London, UK, 1962.

B.I. Bleaney and B. Bleaney, *Electricity and Magnetism*, Clarendon Press, Oxford, UK, 1965.

R.M. Bozorth, *Ferromagnetism*, D. Van Nostrand Co., Inc., New York, 1964.

P.G. De Gennes, *Phys. Rev.* **118**, 141, 1960.

R.H. Dicke and J.P. Witke, *Introduction to Quantum Mechanics*, Addison-Wesley, Reading, MA, 1961.

J.B. Goodenough, *Magnetism and the Chemical Bond*, R.E. Krieger Co., New York, 1976.

E.W. Gorter, *Philips Res. Rep.* **9**, 295, 1954.

J.S. Griffith, *Theory of Transition-Metal Ions*, Cambridge Press, New York, 1961.

M.T. Hutchings, *Solid State Phys.* **16**, 227, 1964.

C. Kittel, *Introduction to Solid State Physics*, John Wiley & Sons, Inc., New York, 1963.

C. Kittel, *Introduction to Solid State Physics*, John Wiley & Sons, Inc., New York, 1966.

H.A. Kramers, *Physica* **1**, 191, 1934.

J. Kubler, *Theory of Itinerant Electron Magnetism*, Oxford Science Publications, Clarendon Press, Oxford, UK, 2000.

L.D. Landau and E.M. Lifshitz, *Quantum Mechanics*, Pergamon Press, Oxford, UK, 1958.

L.D. Landau and E.M. Lifshitz, *Mechanics*, Pergamon Press, Oxford, UK, 1962.

E.M. Larsen, *Transitional Elements*, W.A. Benjamin, Inc., New York, 1965.

A.H. Morrish, *The Physical Principles of Magnetism*, John Wiley & Sons, New York, 1965.

L. Neel, *Ann. Phys.* **18**, 5, 1932.

L.E. Orgel, *An Introduction to Transition-Metal Chemistry Ligand-Field Theory*, Redwood Press Ltd., Trowbridge, UK, 1970.

J.C. Slater, *Phys. Rev.* **52**, 198, 1937.

J.C. Slater, *Quantum Theory of Atomic Structure*, McGraw-Hill, New York, 1960.

K.W.H. Stevens, *Magnetic Ions in Crystals*, Princeton University Press, Princeton, NJ, 1997.

S. Sugano, *Multiplets of Transition-Metal Ions in Crystals*, Academic Press, New York, 1970.

L.H. Thomas, *Nature* **117**, 514, 1926.

J.H. Van Vleck, *The Theory of Electric and Magnetic Susceptibility*, Oxford University Press, Oxford, UK, 1932.

M. Weissbluth, *Atoms and Molecules*, Academic Press, New York, 1978.

W.P. Wolf, *Phys. Rev.* **108**, 1152, 1957.

H.J. Zeiger and G.W. Pratt, *Magnetic Interactions in Solids*, Clarendon Press, Oxford, U. K., 1973.

C. Zener, *Phys. Rev.* 82, 403, 1951.

X. Zuo and C. Vittoria, *Phys. Rev. B* 66, 184420, 2002.

SOLUTIONS

3.1

a. The cavity (see Figure S3.1a) dimensions are given as follows: $a=b=1\,\text{Å}$ and $d=1/2\,\text{Å}$. Resonance frequency of this cavity is

$$\frac{\omega_{101}}{c^2} = \left(\frac{\pi}{a}\right)^2 + \left(\frac{\pi}{d}\right)^2,$$

we have the following:

$$f_{101} = \frac{c}{2}\sqrt{\frac{1}{a^2} + \frac{1}{d^2}},$$

where $c = 3 \times 10^8\,\text{m/s}$.

$$f_{101} = 3.35 \times 10^9 \text{ GHz}.$$

b. The cylindrical cavity (see Figure S3.1b) dimensions are given as follows: $R = 1/2\,\text{Å}$ and $l = 1\,\text{Å}$. The resonance frequency for this cavity is

$$\frac{\omega_{011}}{c^2} = \left(\frac{J_{01}}{R}\right)^2 + \left(\frac{\pi}{l}\right)^2,$$

we have the following:

$$f_{011} = \frac{c}{2\pi}\sqrt{\left(\frac{J_{01}}{R}\right)^2 + \left(\frac{\pi}{l}\right)^2},$$

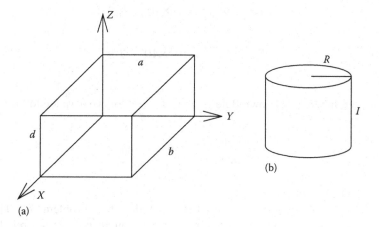

FIGURE S3.1   Shapes of microwave cavities.

where $J_{01} = 2.4$ (see the table of Bessel function).

$$f_{011} = 2.73 \times 10^9 \text{ GHz.}$$

c. From the well-known De Broglie relation

$$m\upsilon = \frac{h}{\lambda},$$

and if you have an assumption that those electrons move nearly at the speed of light

$$\lambda f = \upsilon,$$

then you can have velocity formula as follows:

$$\upsilon = \sqrt{\frac{hf}{m_e}},$$

where $h = 6.626 \times 10^{-34}$ J s and $m = 9.1 \times 10^{-31}$ kg.
Velocity is given as

$$\upsilon_e^{101} = \sqrt{\frac{hf_{101}}{m_e}} = 0.494 \times 10^8 \text{ m/s}$$

and

$$\upsilon_e^{011} = \sqrt{\frac{hf_{011}}{m_e}} = 0.445 \times 10^8 \text{ m/s.}$$

d. Energy of electrons beams may be written as follows:

$$E_{101} = hf_{101} = 3.27 \times 10^{-15} \text{ J,}$$

$$E_{011} = hf_{0111} = 2.94 \times 10^{-15} \text{ J.}$$

3.2 From the well-known Bohr relation $\oint p \, dq = nh$, we have the following:

$$\upsilon = \frac{nh}{2\pi r m_e} = 0.219 \times 10^7 \text{ m/s,}$$

where $n = 1$.

This velocity is 10 times smaller than Problem 3.1. That means that the assumption for electron speed was not reliable in Problem 3.1.

3.3 MKS $\Rightarrow \beta = \dfrac{e\hbar}{2m_e} = 9.274 \times 10^{-24}$ J/T.

$$CGS \Rightarrow \beta = \frac{e\hbar}{2m_e c} = 9.27 \times 10^{-21} \text{ erg/G}.$$

3.4 Simple way to solve this problem using formula of summing is as follows:

$$2 \sum_{0}^{n-1} (2l+1) = 2S_n = 2n^2,$$

where

$$S_n = \frac{n(a_1 + a_n)}{2} d$$

$d = 1$
$a_1 = 1$
$a_n = 2n - 1$

3.5 The given states are

$$|\psi_1\rangle = \begin{pmatrix} 1 \\ 0 \end{pmatrix}, |\psi_2\rangle = \begin{pmatrix} 0 \\ 1 \end{pmatrix}.$$

We can construct the following matrix expression:

$$[f]_i = \langle \psi_n | S_i | \psi_m \rangle,$$

where i (x, y, and z), $n(1,2)$, and $m(1,2)$. By using simple matrix multiplication, we have matrices as follows:

a. $[f]_x = \dfrac{1}{2} \begin{pmatrix} 0 & 1 \\ 1 & 0 \end{pmatrix}.$

b. $[f]_y = \dfrac{1}{2} \begin{pmatrix} 0 & -i \\ i & 0 \end{pmatrix}.$

c. $[f]_z = \dfrac{1}{2} \begin{pmatrix} 1 & 0 \\ 0 & -1 \end{pmatrix}.$

d. The above matrices are Hermitian because $[f]_i = ([f]_i)^+$.

3.6

a. The eigenvalue can be derived by using the well-known Schrodinger equation. One can easily have eigenvalue of cubic box

$$E_{lmn} = \frac{\hbar^2}{2m_e}\left[\left(\frac{l\pi}{a}\right)^2 + \left(\frac{m\pi}{b}\right)^2 + \left(\frac{n\pi}{d}\right)^2\right].$$

b. $E_{101} = 3.01 \times 10^{-17}$ J. This energy is smaller than Problem 3.1. That means that the calculated energy of electron beam was not reliable in Problem 3.1.

3.7   Consider an atom in the uniform magnetic field. We have the Hamiltonian as follows:

$$\hat{H} = \frac{p^2}{2m_e} - \frac{Ze^2}{r} - \frac{eB}{2m_e}G_j + \frac{e^2B^2}{8m_e}\langle x^2 + y^2 \rangle,$$

where G_j stands for angular and spin momentum. The magnetic moment can be estimated by averaging the following expression:

$$m = -\frac{\partial \hat{H}}{\partial B}.$$

One may pay attention that the averaging needs the following relationship:

$$\langle x^2 + y^2 \rangle = \frac{2}{3}\langle r^2 \rangle.$$

Then, we have the total magnetic moment as follows:

$$m = \langle m_l \rangle + \langle m_s \rangle - \frac{e^2B}{6m_e}\langle r^2 \rangle,$$

where $\langle m_s \rangle = (e/2m_e)G_s$, $\langle m_l \rangle = (e/2m_e)G_l$, and we let $m_d = (e^2B/6m_e)\langle r^2 \rangle$, then we have

$$\chi_d = \frac{m_d}{B} = \frac{ne^2}{6m_e}\langle r^2 \rangle = 0.7,$$

$$\chi_s = \frac{g_s\beta S}{B} = 9.27 \times 10^{-18}.$$

3.8   Magnetic moment due to spin–orbit interaction can be written as follows: $\vec{m} = b\beta \vec{L}$, where z is the atomic number, β is the Bohr magneton, and L is the orbital angular momentum.

$$\vec{B} = \frac{\mu_0|\vec{m}|}{2\pi r^3}.$$

From the above two equations, we know that

$$\vec{B} = \frac{\mu_0 \left| z\beta\vec{L} \right|}{2\pi r^3},$$

then, $B = 20.96$ T.

3.9 We have the equation

$$\frac{d^2\psi_i}{dx^2} = -\lambda(1+x)\psi_i,$$

where $\psi_i = \sin(k_i x)$, $k_i = \pi/t$, multiply ψ by ψ_j^* and integrate the previous equation as follows:

$$\int_0^t \psi_j^* \frac{d^2\psi_i}{dx^2} dx = -\lambda \int_0^t (1+x)\psi_j^* \psi_i dx,$$

if i, j wave functions are orthogonal to each other, then integration is trivial. Either $i=j=1$ or $i=j=2$, we have the following:

$$\lambda_{1,2} - \frac{\pi^2}{t(t+\pi)}.$$

3.10

$$g_J = \frac{3}{2} + \frac{S(S+1)-L(L+1)}{2J(J+1)}.$$

Table for 3d shell

Ion	Ca²⁺	Mn²⁺	Fe²⁺	Co²⁺	Ni²⁺	Zn²⁺	V²⁺	Cr²⁺	Cu²⁺	Fe³⁺	Ti³⁺
$^{2S+1}X_J$	$^1S_0$	$^6S_{5/2}$	$^5D_4$	$^4F_{9/2}$	$^3F_4$	$^1S_0$	$^4F_{3/2}$	$^5D_0$	$^2D_{5/2}$	$^6S_{5/2}$	$^2D_{3/2}$
g_J	—	2	1.5	1.33	1.25	—	1.31	—	1.2	2.0	1.0
g_{Exp}	—	1.97	1.19	0.98	0.69	—	1.96	—	0.63	1.97	0.87

Table for 4f shell

Ion	Ce³⁺	Pr³⁺	Pm³⁺	Yb³⁺	Tb³⁺	Ho³⁺	Gd³⁺	Nd³⁺	La³⁺	Sm³⁺	Lu³⁺
$^{2S+1}X_J$	$^2F_{5/2}$	$^3H_4$	$^5I_4$	$^4F_{9/2}$	$^7F_6$	$^5I_8$	$^8S_{7/2}$	$^4I_{9/2}$	$^1S_0$	$^6H_{7/2}$	$^1S_0$
g_J	0.85	0.8	0.6	1.14	1.5	1.25	2	0.73	—	0.28	—
g_{Exp}	0.34	0.77	—	1.09	1.48	1.23	2	0.69	—	0.52	—

Source: B.I. Bleaney, and B. Bleaney, *Electricity and Magnetism*, Clarendon Press, Oxford, UK, 1965.

3.11

a. The wave functions are $\psi_1 = N_1 \sin(\pi x/a)$ and $\psi_2 = N_2 \sin(2\pi x/a)$. Normalization requires that

$$\int_0^a |\psi_i|^2 \, dx = 1,$$

$$N_1 = \sqrt{\frac{2}{a}} \text{ and } N_2 = \sqrt{\frac{2}{a}}.$$

E_1 and E_2 can be derived by eigenvalue equation

$$-\frac{\hbar^2}{2m_e} \frac{\partial}{\partial x^2} \psi_i = E_i \psi_i,$$

$$E_1 = \frac{\hbar^2 \pi^2}{2ma^2} \text{ and } E_2 = \frac{22\hbar^2 \pi^2}{ma^2}.$$

b, c, d. We have the normalized wave functions

$$u(x_1) = \sqrt{\frac{2}{a}} \sin\left(\frac{\pi}{a} x_1\right) \text{ and } v(x_2) = \sqrt{\frac{2}{a}} \sin\left(\frac{2\pi}{a} x_2\right).$$

The singlet state is

$$\psi_s = \frac{1}{\sqrt{2}} [u(x_1)v(x_2) + u(x_2)v(x_1)] \frac{1}{\sqrt{2}} [\alpha(1)\beta(2) - \beta(1)\alpha(2)].$$

The triplet states are

$$\psi_t = \frac{1}{\sqrt{2}} [u(x_1)v(x_2) - u(x_2)v(x_1)] \left| \begin{array}{c} \alpha(1)\beta(2) \\ \alpha(1)\beta(2) \\ \frac{1}{\sqrt{2}} [\alpha(1)\beta(2) - \alpha(2)\beta(1)] \end{array} \right..$$

The energy of each state can be determined from $\langle s|H|s \rangle$ and $\langle t|H|t \rangle$, where

$$H = \frac{e^2}{|x_1 - x_2|} = \begin{cases} \dfrac{e^2}{x_2} \displaystyle\sum_{n=0}^{\infty} \left(\dfrac{x_1}{x_2}\right)^n, & \text{if } x_1 < x_2 \\[4mm] \dfrac{e^2}{x_1} \displaystyle\sum_{n=0}^{\infty} \left(\dfrac{x_2}{x_1}\right)^n, & \text{if } x_2 < x_1 \end{cases}.$$

We would expect the total energy to be of this form (Figure S3.11a)

$$u + \langle s \mid H \mid s \rangle$$

$$u + \langle t \mid H \mid t \rangle$$

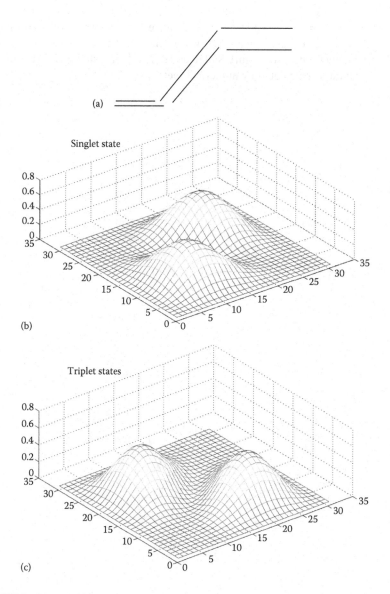

(a)

Singlet state

(b)

Triplet states

(c)

FIGURE S3.11 (a) Displacement of energy of two electron system. Density wave functions of (b) singlet and (c) triplet states.

where u is the Coulomb repulsion energy. If both electrons occupy the same n state, only the singlet state is possible as Pauli would want. If we have lattice constant $a = 3\,\text{Å}$, then we can estimate Coulomb repulsion energy as follows:

$$u = \frac{e^2}{4\pi\varepsilon_0 \bar{x}} = 11.87 \text{ eV},$$

where $\bar{x} = 1.2$ Å is the average distance between two electrons. The parameter $\bar{x}$ was estimated by wave function density for either singlet or triplet states, see Figure S3.11b and c. The $\langle s|H|s\rangle$ and $\langle t|H|t\rangle$ integrals should be calculated by numerical methods.

4 Deposition of Ferrite Films at the Atomic Scale by the ATLAD Technique

HISTORICAL BACKGROUND TO THE DEVELOPMENT OF THE ATLAD TECHNIQUE

As the interest in ferrite materials came to a crescendo in the early 1960s throughout the world, the laser was invented. By the 1970s, the interest in ferrites began to shift more and more toward garnet materials for bubble memory device applications. YIG ($Y_3Fe_5O_{12}$) was a sought-after ferrite to address MMIC (Monolithic microwave integrated circuit) devices, since YIG exhibited the narrowest FMR (Ferromagnetic resonance) linewidth ever measured at X-band (8–12 GHz). Much progress was also evident in other technologies, such as semiconductors, laser, composites, nanomaterial structures, and computer. The emphasis by the 1980s was on how to combine these technologies, and, most importantly, make them compatible to produce the ultimate composite or invention or application.

There was a concerted effort in the 1980s to deposit oxide films, ferrite films included, in an efficient and inexpensive means. Conventional and LPE (Liquid phase epitaxy) depositions were rather involved and expensive. It was clear that the conventional way of depositing metal films was not going to be able to deposit oxide materials, since the melting temperature of oxides was fairly high. Besides, after melting the oxide in a vacuum chamber, how would the free metal element chemically combine with oxygen in the vacuum chamber. This presented more problems than solutions. Nevertheless, E-beam (electron beam) deposition techniques were deployed to deposit oxides. Usually, this technique required extremely high vacuum (~10^{-9} torr) to deposit high-quality metal films. Filling the chamber with oxygen only produced low-quality oxide films.

About the same time, laser applications expanded enormously as well as medical, optical, and material preparations. The focus in research laboratories in the 1980s was to produce magnetic materials compatible with materials utilized in the MMIC technology. Addressing the requirements of the microwave technology, YIG was deposited by laser ablation on single crystal silicon, Corning glass, single crystal MgO, and quartz substrates. The composites exhibited practical magnetic and microwave properties. FMR linewidth of 55 Oe was measured on films deposited on glass substrates, for example. Partial success was obtained sufficiently to issue the first patent on ferrite films production by laser deposition on various substrates (C. Vittoria).

This success marked the turning point toward the deposition of ferrite films at the atomic scale, because, besides the raw data quoted above, it revealed two aspects of

DOI: 10.1201/9781003431244-4

depositing YIG on various substrates: (1) Each laser pulse impinging upon the YIG target caused the deposition of ~1 A of oxide material. (2) The deposition did not separate Fe from the oxygen but deposited both as ions. The obvious conclusion or the derivative of these two points was that it may be possible to construct a crystal unit lattice with proper number of laser pulses, since the lattice constant of YIG is 12.65 A, a spinel ferrite 8.5–9.0 A, M-type hexaferrite 23.05 A, etc. Clearly, it became a question of how many laser pulses would it take to construct a unit crystal lattice. However, there are many other variables that needed to be controlled besides the number of pulses: sticking coefficient of various oxides, vacuum pressure in the growth chamber, substrate temperature, pulse rate, etc. To some extent, it was empirical but achievable goal. The resulting ferrite film was compared with a standard samples and X-ray analyzed. As always, trial and errors were never reported! Only good results. At least, there was a slim hope.

DEPOSITION OF FERRITE FILMS BY THE LASER ABLATION TECHNIQUE

As implied in this chapter, most ferrites are ferrimagnetic with lower magnetization than the ferromagnetic metals. Ferrites of interest are either cubic (spinels and garnets) or hexagonal (M, Y, W, Z-type, etc.) structure. The chemical composition within a given plane of the unit crystal lattice structure is chemically uniform with magnetic moments either parallel to each other or antiparallel. On adjacent planes (either side of the plane of interest), the opposite may be expected. The distance between successive planes is usually greater than 1 A lending an opportunity to deposit one type of magnetic layer in one plane and something else on the adjacent one. Hence, at least two targets would be needed to deposit a film whose crystal structure is identical to the standard one. Hence, an artificial ferrite whose crystal structure identical to a spinel or a hexagonal one was feasible. The same conclusion applied to the garnet structure. There was much confusion in distinguishing this ATLAD (alternate target laser ablation deposition) technique from the so-called superlattice composite structures. The superlattice composite consisted of many layers. Each layer is being different from others in terms of thickness and chemical composition. The intent was to generate physical properties of the composite different from the single layers that make up the composite. The thickness of the single layer itself was usually much greater than its lattice constant. The crystal structure of the composite was rather complicated and never intended to simulate a cubic or a hexagonal structure.

In order to simulate the unit crystal lattice and the magnetic properties of a standard ferrite and, then, "play" the devil's advocate by placing cations in the unit cell not allowed by the natural growth process. Usually, the constituents are in a molten state whereby site occupations is governed by temperature, rate of temperature gradients to the molten flux, time, oxygen exposure, etc. Clearly, there is no external control as to where the cations may migrate in the crystal unit cell. It was decided early on to concentrate on the spinel and hexaferrite materials. There was not much to add to the garnet story, since every imaginable garnet structured was covered with the advent of "bubble" garnet materials. However, YIG could have been considered initially, but the initial thrust was on the spinel structure being the simplest—only two crystal sublattices.

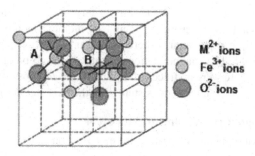

FIGURE 4.1 Relative positions of each ion in the spinel structure. Courtesy of Semantic Scholar.

DEPOSITION OF SPINEL FERRITE FILMS AT THE ATOMIC SCALE—ATLAD TECHNIQUE

There are five different types of chemical composition of spinel ferrites (Oxides, Sulfides, Selenides, Tellurides, and Nitrites) that exhibit cubic crystal structures (Figure 4.1). We are concentrating only on oxide spinels in which the main magnetic cation is Fe^{3+}. Typically, the chemical formula of spinel ferrites is MFe_2O_4, where M refers to a divalent metal ion. Ferrites consist of magnetic sublattices in which a transition metal cation, typically Fe^{3+}, is at the center of a tetrahedral or octahedral site surrounded by oxygen ions. The magnetic ions within the crystal lattice form a plane. For the spinel structure, magnetic ions in tetrahedral sites are referred to as (A) sites to form the (A) sublattice and there are 8 of them in the crystal lattice. Magnetic ions that occupy octahedral sites are referred to as the (B) sublattice and there are 16 of them in the crystal lattice. Altogether there are 56 Oxygen ions surrounding the 24 sublattices. The divalent metal ions (Co, Fe, Ni) prefer B sites and (Mn, Zn) the A sites. In the normal spinel structure, the eight divalent metal ions occupy the A sites and Fe^{3+} ions occupy the B sites. However, if the 8 divalent metal ions occupy the B sites, displacing eight Fe^{3+} ions into the A sites, then we have an inverted spinel. However, in most cases, there are partial displacements.

In the simplest case, the spins on either site (A or B) are ferromagnetically ordered, but are coupled to each other antiferromagnetically (see Chapter 3).

Films of Lithium Ferrite ($Li_{0.5}Fe_{2.5}O_4$) Doped with Al_2O_3

The object was to impinge a laser pulse upon two separate targets alternatively such that the thickness of each deposit was in the order of 1–2 A. The two targets ($Li_{0.5}Fe_{2.5}O_4$ and Al_2O_3) were mounted on a rotating wheel controlled by a trigger signal from a computer. Calibration runs were made separately singly from each target and it was determined experimentally that 3 laser pulses impinged on each target induced about 2.5 A deposition on a substrate. Single crystal (100) MgO was used as substrate material. The lattice constant of 4.2 A is compatible with the lattice constant of spinel structure (8.5–9.0 A). From a magnetism perspective, the main objective was to eliminate any exchange coupling between sequential layers deposited, since Al ions are non-magnetic. Hence, this was a method to localize any exchange interactions within Fe layer and induce an antiferromagnetic interaction between magnetic ions. In such situations, the net magnetic moment should vanish.

As in any first-time experiments, there were more questions than answers. Just the opposite occurred. The magnetization was measured to be about 2,000 Gauss compared to zero. There were no traces of lithium in the composite film based on our chemical analysis. X-ray data was consistent with the calibration runs.

Deposition of Single Crystal Films of MnFe$_2$O4

It is well known that manganese ferrite (MnFe$_2$O$_4$) is a partially inverse spinel ferrite in which 20% of the Mn^{2+} ions migrate from A to B sites. The degree of migration between sites is governed by the natural growth process in this spinel as in any other spinel ferrites. The fundamental question addressed in utilizing the ATLAD technique is the following: Can one alter the degree of inverse ion migration in spinel ferrites by artificial, such as the ATLAD technique, means to be different from that provided by the natural or conventional growth process described above. If so, then it may be possible to affect systematically the antiferromagnetic coupling between A and B sites by artificially depositing non-magnetic ions in A or B site. This was the main motivation in depositing the films by the ATLAD technique. As a follow-up and from a magnetic perspective, is it possible to rearrange the position of the magnetic ions artificially to be different from their natural arrangement while maintaining the integrity of the chemical composition as well as the spinel crystal structure? Indeed, it was proven that it was indeed feasible.

Artificial films were deposited by alternating layers of MnO and Fe$_2$O$_3$ on (111) and (100) MgO substrates consistent with the well-known spinel chemical formula of MnFe$_2$O$_4$. The MnO and Fe$_2$O$_3$ targets were prepared by pressing and sintering the powders at 1,200°C in amphibian oxygen pressure, respectively. The targets were mounted on a revolver-like target rotator driven by a servomotor and synchronized with the trigger of the pulse laser.

As a simple rule, the number of targets equals the number of sublattices to simulate. For example, the ATLAD deposition of single crystal hexaferrites required three targets consistent with three sublattices at most. However, the advantage of the ATLAD is that one can always subdivide a sublattice with more targets than normal (see Figure 4.2).

In each cycle of deposition, there were four laser pulses at 400 mJ impinging on the MnO target and eight on Fe$_2$O$_3$. It is noted that the ratio of pulses impinging on ferric oxide target versus pulses aimed at manganese oxide was 2:1, corresponding to the chemical formula MnFe$_2$O$_4$ or the number of B sites versus A sites. Thus, the aim of this deposition trial was to place ferric ions in the B sites and Mn ions in the A sites, a normal spinel ferrite. This represented a violation of the natural growth process in which 20% of Mn ions migrate to the B sites. However, the number of pulses (4) impinging on MnO target was arrived on an empirical basis. Specifically, the ratio of MnO to Fe$_2$O$_3$ laser pulse shots was varied from 1:10 to 6:10, with a ratio of 4:10 resulting in magnetic properties matching the bulk values. In addition, X-ray θ −2θ scans of MnFe$_2$O$_4$ films deposited on a (100) MgO substrate at an oxygen pressure of 1 mTorr were consistent with the epitaxial growth on MgO. Twice the lattice constant of MgO (a=4.216 A) nearly matched that of MnFe$_2$O$_4$ (a=8.511 A). Only the

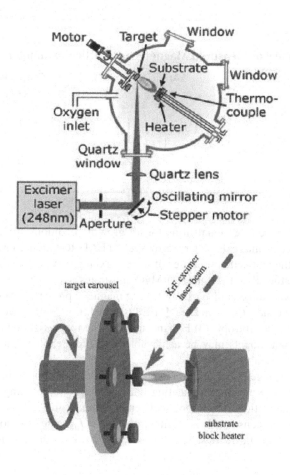

FIGURE 4.2 ATLAD apparatus minus computer controller. Potentially, many targets may be mounted on the wheel.

{0,0,4n} peaks of the spinel phase were observed. The temperature of the substrate during deposition was maintained at 700°C.

Studies of manganese ferrite films deposited by ATLAD on (111) MgO substrates as a function of oxygen partial pressure have shown the inversion parameter of the films increasing from the near bulk value of 17%±4% at 1 mTorr to over 35% at 50 mTorr oxygen pressure. For ATLAD films deposited on (100) MgO substrate, the inversion parameters were 30% increasing to more than 50% at high pressures, respectively. The saturation magnetization was determined to decrease linearly with increasing inversion parameter of the films. The value of the saturation magnetization of 4.5 kG was measured for the ATLAD films deposited at 1 mTorr. For manganese ferrite thin films deposited by conventional PLD (pulse laser deposition), using $MnFe_2O_4$ targets, a value of 3.3 kG was obtained under the same oxygen partial pressure (7). The saturation magnetization for the ATLAD films was consistently

TABLE 4.1

Summary of Measured Magnetic and Lattice Parameters

	Artificial	Conventional
a (A)	8.5133	8.5064
$4\pi M_S$ (KG)	4.5	3.3
T_N (°K)	475	543
J_{AB} (°K)	−14.82	16.07
J_{BB} (°K)	11.25	11.70
X (%)	35	20

higher than that of conventional films for all growth conditions studied. Thus, by depositing artificial manganese ferrite by the ATLAD technique, much control is exercised over the magnetic properties of the epitaxial growth of the films. Also, by depositing on either the (100) or (111) MgO substrates and controlling the oxygen pressure in the vacuum chamber, the following magnetic properties were controlled: (1) Magnetic inversion factor in the $MnFE_2O_4$ spinel. (2) Control over the induced uniaxial magnetic anisotropy. (3) Enhancement of the magnetization. It is believed that further enhancements may be achievable with more exploration with various growth conditions (Table 4.1).

J_{AB} is the exchange coupling parameter between magnetic ions in the A and B sites, and the sign is negative indicating antiferromagnetic coupling; $4\pi M_S$ is the saturation magnetization. T_N is the Neel temperature, x the inversion parameter, and $a(A)$ the lattice constant. It is assumed that $J_{AA} < J_{BB}$, since in most cases non-magnetic ions occupy the A sites. The sign of J_{BB} is positive indicating ferromagnetic exchange coupling.

Deposition of Single Crystal Film of CuFe₂O₄

Placing a non-magnetic ion (Cu^{2+}) in the A site purportedly may affect the antiferromagnetic exchange coupling, J_{AB}. As it is well known in spinel ferrites, $J_{AB} > J_{AA}$ or J_{BB}. However, in the absence of J_{AB}, the 16 magnetic spins in the B sites will couple antiferromagnetically, substantially lowering the magnetization. Furthermore, inversion will certainly occur, as in the previous trials. Bulk copper ferrite is 85% inverse with a fairly low saturation magnetization of 1.7 kG.

Thin films of copper ferrite were deposited on (100) MgO substrate by sequentially ablating CuO and Fe_2O_3 targets under a variety of growth conditions. The crystal phase and the texture of the films were verified by XRD measurements confirming crystallinity. Highest saturation magnetization values were obtained for films grown at 650°C and oxygen partial pressure of 90 mTorr. As evident from the equilibrium cation distribution in bulk copper ferrite, Cu^{2+} cations have a strong preference for the octahedral sublattice (B sites), since the inversion parameter is 85% (~7 copper ions have migrated from the tetrahedral to the octahedral site), thwarting in lowering J_{AB}.

An important hint was gathered from the previous trial. The magnetization in the ATLAD films was enhanced in comparison to films deposited from a single target.

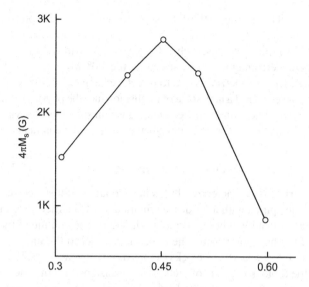

FIGURE 4.3 Saturation magnetization plotted as a function of the ratio of laser pulse shots impinging upon targets of CuO and Fe_2O_3, respectively (A. I. Geiler).

Clearly, the magnetization depends strongly on the inversion parameter. A simple way to affect this parameter is to slightly vary the number of pulses near the critical ratio of 4/10 (ratio used in the previous runs) while still maintaining crystallinity. In this trial, the number of laser shots on the Fe_2O_3 was kept fixed at 10, but the number of shots on the CuO target was varied from 3 to 6 (ratio varying from 0.3 to 0.6). The maximum value of the magnetization was reached for the ratio of 0.45 (5/9).

The maximum saturation magnetization of 2.8 kG was measured, approximately 65% enhancement over the bulk value of 1.7 KG (Figure 4.3). Lowering the inversion factor increases the saturation magnetization. The intended value of x was zero to maximize the magnetization. EXAFS analysis (9) showed that the highest value of the saturation magnetization was reached for $x = 52\%$ inverse, compared to an 85% inversion parameter for bulk copper ferrite.

As in the manganese ferrite system, the saturation magnetization was observed to increase with the lowering of the inversion parameter x. In the limit that the inversion parameter is 0%, the ferric ions in the B site couple antiferromagnetically. On the other hand, in the limit that the inversion parameter is 100%, 8 ferric ions migrate to the A sites. Hence, the net magnetic moment either way is zero. For 50% inversion, simple argument would say that 12 ferric ions are on the B sites and 4 on the A sites with 8 net magnetic moments. As such, the resulting magnetization should have been as much as that of the ATLAD $MnFe_2O_4$ or any other bulk spinel with eight net magnetic moments. Thus, the ATLAD technique allows for the possibility of controlling the inversion parameter systematically.

In general, spinel ferrites are ferrimagnetic, if both A and B sites are occupied by magnetic cations, and antiferromagnetic if only one of the sublattices is occupied by magnetic ions. This is due to the 180° superexchange (see Chapter 3) interaction which is the dominating factor in transition metal oxides. Ferromagnetism was only found in Mn^{4+} ($3d^3$) spinel ferrites, such as $Cu^+[(Mg^{2+})_{0.5}(Mn^{4+})_{1.5}O_4]$, $LiZn_{0.5}Mn_{1.5}O_4$, and $CuRhMnO_4$. Ferromagnetism arises from the 90° Mn^{4+}–O^{2-}–Mn^{4+} exchange coupling. Hence, ferromagnetism is due to direct exchange coupling between the two Mn cations. We have entered a new stage of inquisition into the physics of spinel ferrites, entertaining the thought of enhancing ferromagnetism in spinel ferrites via the ATLAD technique with the purpose of affecting magnetism in a very fundamental way.

Deposition of Hexaferrite Films at the Atomic Scale—ATLAD Technique

This class of ferrites is characterized by its hexagonal crystal structure. The crystal unit cell is parallelepiped with a 60-degree rhombus with a lattice constant (a) (hexagon side dimension) and height (c) normal to the base. The volume of the unit cell is equal to 1/3 of the hexagon prism. The c-axis is parallel to the uniaxial axis of the hexagon. The unit cell consists of blocks of ions designated as the S, R, and T blocks. For example, the S block consists of metal ions consistent with the chemical formula MFe_2O_4, the composition of a spinel ferrite. Typically, M is a transition metal ion, including Fe^{3+}. The S block is the smallest size of the blocks, because the size of the transition metal ions is smaller (diameter of ~1.2 A) than the barium ion (~4.4 A) contained in the R and T blocks. The dimension of the S block along the c-axis is ~4.3 A and that of the R block is ~7.3 A.

Interestingly that in magnetic garnets, octahedral sites (O_h) are referred to as (a) sites, tetrahedral sites (T_d) as (d) sites, and dodecahedral as (c) sites. In spinel ferrites, (T_d) sites are referred to as A sites and (O_h) as B sites. However, in hexaferrites, the so-called Wyckoff positions of (O_h) and (T_d) sites take on a totally different designation, depending on where these sites are located, S, R, or T block. There are four O_h sites whose magnetic moment contributes to the majority of magnetic sublattices and two T_d sites contributing to the minority moment. Thus, there are two net moments in the majority (2α). In the R block, there are no T_d sites with spins opposing the majority of spin directions. In the R block, five O_h sites, three spins aligned with the majority and two opposing (3α–2β) and one trigonal site (α), sometimes referred to as bi-pyramidal 5-fold co-ordination. The net magnetic moment is 2α. There is no trigonal site in the T block, but there are 2 T_d sites (2β) and six O_h sites (4α–2β). Thus, the net magnetic moment is zero.

Let's now designate the Wyckoff O_h and T_d sites only in the S and R blocks, since the focus here is on the growth of artificial M-type hexaferrite utilizing the ATLAD technique for one of many types of hexaferrites. Nevertheless, a summary of the properties of the most often investigated types is reported. Barium ferrite, also known by the trade name ferroxdure, consists of alternating series of S and R blocks. The unit cell comprises of SRS*R* blocks in sequence. The chemical content of the S block is Fe_6O_8 and the R block is $BaFe_6O_{11}$. Together they form the chemical formula $BaFe_{12}O_{19}$. The O_h sites bordering the S and R blocks are designated as the (12k) sites, Table 4.2. Within the S block, the two T_d sites are designated as (4f_1) and one O_h at the edge of the unit cell designated as (2a). The trigonal site in the R block is designated as (2b). In a unit cell, there is a net majority moment equivalent to 8 ferric moments.

TABLE 4.2

Dimensions of Lattice Constant (a); S, R, and T Blocks (W); and Spin Orientations in Each Block

Block	a (A)	W (A)	T_d	O_h	Trigonal ($2b$)	Net Spins
S	5.89	~4.3	2β	4α	–	2
R	5.89	~7.3	–	3α-2β	1α	2
T	5.89	~10.2	2β	4α-2β	–	0

α—spin aligned along the majority spins (parallel to the c-axis).
β—spin along the opposite direction.

TABLE 4.3

Block Makeup of Common Hexaferrites, Chemical Formulas, and Lattice Constants

Type	Blocks	Formula	a (A)	c (A)
M	SRS*R*	$BaFe_{12}O_{19}$	5.88	23.2
W	SSRS*S*R*	$BaM_2Fe_{16}O_{27}$	5.88	32.8
Y	STSTST	$Ba_2M_2Fe_{12}O_{22}$	5.88	43.6
Z	RSTSR*S*T*S*	$Ba_3M_2Fe_{24}O_{41}$	5.89	52.3

Physical parameters of the chemical blocks that make up Hexaferrites Types M, W, Y, and Z are listed in Table 4.3. There are more types of hexaferrites (X, U, etc.) which will not be covered here. As the size of the heavy ion is increased, the crystal unit lattice also is increased. Empirically, as the unit lattice increases, it requires higher temperatures during growth to stabilize the crystal structure. For example, it requires temperatures as high as 1,150°C to grow Z-type hexaferrites. Thus, it becomes problematical to deposit films at those high temperatures, since growth chambers are usually designed to sustain temperatures only up to ~1,000°C.

The crystal structure of the unit cell in Strontium hexaferrite is shown in Figure 4.4. Sr^{2+} can be replaced by Ba^{2+} or Pb^{2+} (Figure 4.5).

Deposition of Single Crystal Films of Barium Hexaferrite—$BaFe_{12}O_{19}$ by ATLAD

Since WWII, this class of ferrites warranted much basic research, since practical implication were immense. After 75 years, hexaferrites are still of great research interest in terms of understanding the physics of magnetism and applications thereof. The ATLAD technique is an important tool toward exploring the possibilities of re-arranging the basic ionic structure in the unit cell. The purpose in this section is two-fold: (1) to ascertain that the unit lattice of barium ferrite can indeed be reproduced with the ATLAD technique–standardization and (2) to place transition metal ions

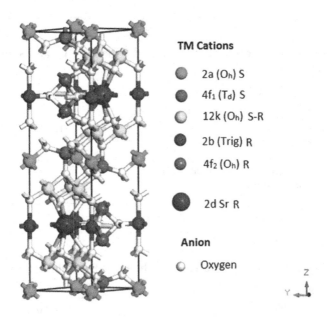

FIGURE 4.4   Crystal structure of a unit cell of an M-type hexaferrite with ionic distributions.

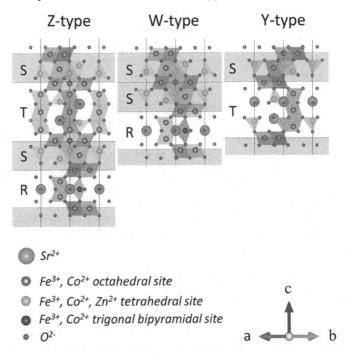

FIGURE 4.5   Two dimensional images of the Z, W and Y-Type hexaferrites (not inclusive of the inverse blocks). The Z-Type hexaferrite structure is symbolized as consisting of RSTSR*S*T*S* blocks, where the asterik indicates 180 degrees – rotation about the C-axis. The W-Type is symbolized as SSRS*S*R* and Y-Type as STST (no inverse blocks).

at any arbitrary sites within the unit cell to study the ramifications in the change of the ionic structure from the standard. The ultimate goal is to place these ions in sites that normally "don't belong there," according to the natural growth process, ushering a new era of hexaferrites and ferrites research in general, similar to the times when barium ferrite first appeared on the scene in the 1950s.

Based on the successful growth of single crystals spinel films by the ATLAD technique (see the previous section), the two targets chosen to deposit barium ferrite films by the ATLAD technique were orthorhombic $BaFe_2O_4$ and Fe_2O_3, hematite. The first target was used to simulate the R block, since that block contains the barium ion. The other target was used to simulate the S block. A single crystal substrate of rhombohedral sapphire (Al_2O_3) was used. The lattice constants of sapphire are 4.76 and 12.99 A (along the c-axis). Deposition was on the basal plane. The basal plane dimension of barium ferrite is 5.89 A. However, there are other substrates which may be suitable for deposition, such as (111) MgO (Cubic) whose lattice constant in the basal plane is 5.95 A. The ideal substrate would have been another hexagonal structured substrate in which the ferric ions are replaced by a non-magnetic transition metal ion. However, these substrates are not easily accessible. One possibility is to transition from sapphire or MgO substrate to a thin non-magnetic hexagonal substrate. The crystal quality of a deposited magnetic film is very much dependent on the uniformity of the magnetic properties at the interface between the substrate and the magnetic film.

The targets were prepared by conventional ceramic processing. Pressed Fe_2O_3 powder was sintered at 1,150°C for 4 hours. After milling, the powder was pressed again at 2,000 psi. After sintering at 1,350°C for 4 hours, a high density (>90%) was reached. Powders of $BaFe_2O_4$ were obtained by admixing powders of $BaCO_3$ and Fe_2O_3 and grinded. The same procedure, as in the other target, was adopted in producing a high-density target of single phase $BaFe_2O_4$. These are necessary precautions to be able to deposit single crystals of barium ferrite films by the ATLAD technique. Since the mismatch between substrate and film was 7%, it implied biaxial stress in the film after cooling from 925° C to room temperature after deposition. Furthermore, care was exercised not to induce diffusion at the interface by rapid (20 minutes) annealing at 1,050°C.

The ATLAD films were deposited in a high-purity oxygen atmosphere of 300 mTorr. A KrF excimer laser with a wavelength of 248 nm, energy of 400 mJ/pulse, and 25 ns full width at half maximum pulse width was focused to an energy density of 10 J/cm^2 on the target surface. The distance between target and substrate was set to 5 cm. The rotation of the target carousel was synchronized with the laser trigger signal to allow the targets to be alternated during film deposition and the laser beam rastered to maximize target surface usage. Three laser shots were aimed at $BaFe_2O_4$ and 33 at the Fe_2O_3 targets. The growth rates for Fe_2O_3 and $BaFe_2O_4$ were approximately 1.4 and 0.6 A/pulse shot, respectively. Altogether, 300 repetitions of the deposition routine resulted in a thickness of 6,500 A, as measured by a scanning surface profilometer. Upon completion of the run, the films were annealed in flowing oxygen at 1,050°C and rapidly removed from the furnace.

From XRD measurements, only <$002n$> barium ferrite peaks were visible, suggesting that the films were single phase and possess a high degree of c-axis

TABLE 4.4

Summary of XRD Measurements of Barium Ferrite

Technique	a (A)	c (A)	Δω (degrees)
ATLAD	5.89	23.17	0.25
BULK	5.88	23.19	–
LAD (PLD)	5.89	23. 18	0.15
LPE	5.88	23.17	0.08

TABLE 4.5

Summary of Static and Dynamic Magnetic Field Measurements of Barium Ferrite

	H_A (kOe)	$4\pi M_S$ (kG)	T_N (°K)	ΔH (Oe)	G
ATLAD	16.5	4.6	738	42	1.996
BULK	17.0	4.54	724	52	1.99
PLD	16.4	4.2	–	23	2
LPE	16.4	4.4	–	27	–

PLD, pulse laser deposition; LPE, liquid phase epitaxy; Bulk, conventional flux grown ferrites; H_A, uniaxial magnetic anisotropy; $4\pi M_S$, saturation magnetization; T_N, Neel temperature; ΔH, ferrimagnetic resonance at ~55 GHz; g, factor.

orientation. Rocking curve measurement of the highest intensity <008> diffraction peak exhibited c-axis deviation at different points on the film surface of only $\Delta\omega = 0.25$ degrees. Pole figure measurement of the <107> diffraction peak illustrated the sixfold symmetry of the ionic arrangement in the basal plane of barium ferrite. In summary, epitaxial deposition of single crystal barium ferrite films was achieved by the ATLAD technique (Table 4.4).

The slight variations in $\Delta\omega$ may be due to small curvatures of the films upon different methods of annealing the films. In order to complement the XRD results with static and dynamic field measurements of the ATLAD films, other films deposited by different techniques are reported in Table 4.5.

In Tables 4.4 and 4.5, there is some slight variations in the measured parameters, but the variations are smaller than the experimental errors. In summary, the ATLAD technique has succeeded in depositing single crystals of the spinel and hexagonal crystal structures. However, the aim of the ATLAD technique is not to simply reproduce what has been achieved before (standardization), but to be the "devil's advocate" in placing ions in sites within the crystal unit cell that the natural growth process does not allow! The purpose is to address the new and artificial ferrites for novel applications. The quantum "jump" from productions of artificial ferrite materials to novel applications requires fundamental research into understanding the physics of magnetism in re-organizing a unit cell, much like when ferrites appeared on the scene in Europe after WWII.

There are a number of magnetic and non-magnetic ions that can substitute for ferric ions in barium ferrite, as long as the size of the substituted ion is compatible in size (0.5–0.6 A radius) to the ferric ion. Size matters because the distance between oxygen anions and cations (center to center) is only about 2 A. Much data has been accumulated in the literature on manganese-doped barium ferrite, more so that other magnetic ions substituted in barium ferrite. The database is sufficiently extensive to allow quantitative comparison of physical and magnetic measurements of both ATLAD films with bulk ones, even if the data on the bulk materials may need to be linearly extrapolated. However, the focus is not on the comparison between measurements, but whether the ATLAD technique has succeeded in placing Mn^{2+}, $BaFe_{12-x}Mn_xO_{19}$, at interstitial sites different from the bulk materials. As it is well known, the magnetic properties of ferrites are governed by site occupancies of cations.

In the manganese-doped bulk barium ferrite, Mn^{3+} cations occupy $(2a)$ and $(12k)$ sites in the S block and $(2b)$ and $(4f_2)$ in the R block. Since manganese cations were dispersed throughout the unit cell even for small substitutions ($BaFe_{12-x}Mn_xO_{19}$, $x < 0.5$). While in the previous sections the ATLAD technique was utilized to place specific magnetic or non-magnetic cations in A sites in the spinel crystal unit cell, in the section below, it is utilized to omit substitutions of manganese cations in the R block of the barium ferrite crystal unit cell while maintaining epitaxial growth and integrity of the hexagonal crystal structure. The effects of this omission are examined in terms of the measured physical and magnetic properties. Theoretical implications or ramifications of the measurements are explored.

Deposition of Single Crystal Films of $BaFe_{12-x}Mn_xO_{19}$

Epitaxial $BaFe_{10.5}Mn_{1.5}O19$ thin films were deposited by the ATLAD technique on c-plane sapphire (Al_2O_3) substrates. The deposition chamber was evacuated to a base pressure of 10^{-6} Torr. High-purity oxygen gas was introduced and a stable partial pressure of 300 mTorr was maintained throughout the deposition. During the deposition, the substrates were heated to 925°C by a resistive block heater. Laser pulses from a KrF excimer laser with a wavelength of 248 nm, energy of 410 ± 10 mJ/pulse, and pulse width of 25 ns full width at half maximum were optically focused on the target surface to an energy density of 10 ± 1 J/cm². The ATLAD system is shown schematically in Figure 4.6.

Targets were prepared by conventional ceramics processing techniques. Barium ferrite thin films were deposited by sequential ablation of $BaFe_2O_4$, Fe_2O_3, $MnFe_2O_4$, and Fe_2O_3 targets. The targets were mounted on a carrousel synchronized with the laser trigger signal via a computer. To maximize surface usage, all three targets were rastered and rotated throughout the deposition. The deposition routine consisted of three pulses on the $BaFe_2O_4$. Instead of impinging 33 pulses onto the Fe_2O_3 target only, 11 pulses were impinged on the $MnFe_2O_4$ target, preceded and superseded by 11 pulses on the Fe_2O_3 target. As a result, the manganese cations were placed only within either the interstitial site of the S or S* block as opposed to being distributed randomly throughout the unit cell. Thus, the Mn cations were purposely omitted from the R block. Normally (bulk samples), Mn cations are spread in the S and

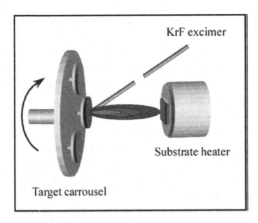

FIGURE 4.6   Targets of $BaFe_2O_4$, Fe_2O_3, and $MnFe_2O_4$ mounted sequentially on the rotating carousel.

R blocks. Therefore, the distribution of cations within the unit cell is controlled at the atomic scale. The placement of cations is determined by the order in which the targets were ablated, while the substitution amount is controlled by the total number of laser pulses impingent on each target. Thus, based on the number of pulses on each target, the nominal composition was $BaFe_{10.5}Mn_{0.5}O_{19}$. Growth rates from $BaFe_2O_4$, Fe_2O_3, and $MnFe_2O_4$ targets were determined to be 0.14, 0.05, and 0.04 nm/pulse, respectively. Approximately 24 Å of the films were deposited per routine, commensurate to the height of the barium ferrite unit cell ($c = 23.17$ Å). About 250 repetitions of the deposition routine resulted in a film thickness of 0.5 μm, measured by a scanning stylus profilometer. Laser trigger frequency was set to 1 Hz for the first 10 repetitions of the deposition routine, 5 Hz for the subsequent 10 repetitions, and finally, 10 Hz for the remaining 230 repetitions.

After the deposition, the films were "flash" annealed in flowing high-purity oxygen gas. A 2″ tube furnace was preheated to 1,050°C. Thin films were inserted into the furnace in a high-purity alumina crucible and annealed for 20 minutes. Upon completion, the films were quickly removed from the furnace and cooled in air. Annealing beyond the 20-minute interval resulted in the deterioration of microwave properties of the films, which was interpreted as evidence of interdiffusion at the film–substrate interface. The chemical composition of the film was confirmed by the energy dispersive X-ray spectroscopy.

As in the previous section, X-ray diffraction measurements, indeed, revealed that the films were of very high-quality single crystals, $c = 23.19$ A, dispersion of the c-axis from point to point on the film surface was 0.25 degrees, and the sixfold symmetry in the plane of the film intact. The static and dynamic magnetic measurements of manganese-doped ATLAD films of M-type hexaferrite are compared with $x = 0$ (no Mn^{3+}) measurements (Table 4.6).

The sites occupancy and valence states of manganese-doped barium ferrite prepared by the ATLAD technique were determined by the EXAFS (extended X-ray Absorption Fine Structure) technique which is a widely used experimental technique

TABLE 4.6
Measured Dynamic and Static Field Parameters of ATLAD Hexaferrite Films

	G	ΔH (Oe)	D (erg/cm)	$4\pi M_S$ (kG)		H_A (kOe)
	300°K	300°K	300°K	300°K	T_N (°K)	300°K
$BaFe_{12}O_{19}$ $(x=0)$	1.99	39	6.3×10^{-7}	4.6	738	16.5
$BaFe_{10.5}Mn_{0.5}O_{19}$	1.96	81	4.1×10^{-7}	4.3	683	17.0

D is defined as the Exchange stiffness constant.
H_A is the uniaxial magnetic anisotropy field.

TABLE 4.7
Sites Occupancy and Valence States of Manganese-Doped Barium Ferrite

	2a (S)	$4f_1$ (S)	12k (SR)	$4f_2$ (R)	2b (R)	Valence	Ref.
Bulk	Yes	No	Yes	Yes	Yes	2+	Von Aulock
ATLAD	Yes	Yes	Yes	No	No	2.82+(av)	Geiler

Yes—Manganese cations occupy that particular site.
No—Manganese cations do not occupy the site.
Av, average; S, S block; R, R block; SR, located in between S and R blocks.

for determining the local geometric and/or electronic structure of matter. The data is obtained by tuning the photon energy of an X-ray to a range where core electrons can be excited, from the quantum number of an atom, $n = 1$, 2, and 3, corresponding to K, L, and M edges to unoccupied states in the transition metal ions of ferrites, for example. The most intense are the electric-dipole allowed transitions to unoccupied final states. In Table 4.7, the results of the EXAFS analysis are summarized.

It is noticed that there are no manganese cations in the R block for the ATLAD films. That was by design! The point is that the ATLAD technique was able to place cations into sites in the unit cell either accidentally or on purpose. Thus, cation distributions at particular sites at the atomic scale in ferrites have been realized by the ATLAD technique opening new possibilities for a new class of ferrites, applications, and re-assessment of past theories of magnetism.

Let's examine the ramifications of tampering with the unit cell as in Table 4.7. The effect of omitting Mn cations in the R block was to release almost three electrons from the manganese atom to form an ionic state of approximately 3+ (+2.82). However, in order to electrically neutralize the unit cell, the valence state of ferric cations must reduce from 3+ to 2+ which implies ferrous ions must co-exist. This explains why the FMR linewidth increased from 39 to 81 Oe. The presence of 3+ in Mn cations promotes a free electron "floating" between ferric and ferrous sites causing eddy currents to increase in the ferrite.

Clearly, it is difficult to compare theoretically changes in measured magnetic parameters, when the valence state of the manganese cation in bulk grown ferrites is not precisely known. The valence state affects all magnetic parameters. For example, in the ATLAD films, lack of manganese cations in the R block appears to reduce the exchange coupling between S and R blocks as reflected in the measurements of T_N and D. Also, increasing the valence state of manganese cations from 2+ to 2.82+ reduced the saturation magnetization from 4.6 to 4.3 kG and the net moment of the cation. To what extent the omission of manganese cations in the R block affected its valence state is not clear. This is further to be investigated.

The ATLAD technique can readily be applied to the Y- and Z-type hexaferrites, since the number of pulses required to simulate the S and R blocks has been established in depositing the M-type films by the ATLAD technique. For the Y-type, it requires the S and T blocks combination (see Table 4.3). For the Z-type, it requires the S, R, and T blocks (see Table 4.3). Thus, only the T block needs to be simulated to deposit both types. We estimated that it requires about six laser pulses on a target of $Bafe_2O_4$ to simulate the T block, since there are two Ba ions in that site. The unit crystal cell of the Y- and Z-type is significantly larger than the M-type (see Table 4.3). As such, the larger unit cells require very high substrate temperatures in excess of 1,150°C in order to stabilize growth through the "bulk" process or depositing films from a single target. Most, if not all, high vacuum chambers where ferrite films are deposited cannot easily sustain high temperatures beyond 1,000°C. They are not designed to be high temperature chambers. It has been shown that Y-type hexaferrite films deposited by the ATLAD technique only require a heating temperature of the substrate of 910°C. This implies that high temperatures are required for bulk growth or PLD or LPE but not for the ATLAD technique.

CONCLUDING REMARKS

Deposition by the ATLAD technique does indeed place cations in ferrite materials at any site within the unit crystal cell by proper selection of targets and by sequential order of targets to be laser pulsed. It is no longer an experimental technique. At least for the spinels and hexaferrites, it has become a routine technique. As for the garnets, a simple extension of the spinel work should be applicable. It is a question of measuring the deposition rate from each single targets. However, the nemesis, ever since the 1950s, is the high temperature deposition of magnetic films which is still haunting. Interdiffusion at the substrate–film interface and excitations of Mn^{3+} and Fe^{2+} have deleterious effects on the quality of the films. In ATLAD films, these are minor concerns as measurements of the FMR linewidths indicate.

In Table 4.8, a summary of targets and number of pulse shots is proposed to deposit Y- and Z-type hexaferrites by the ATLAD technique based on the deposition of barium hexaferrite (see the above sections).

TABLE 4.8

Estimated Number of Laser Pulse Shots on Targets Fe_2O_3 (Simulating S Block), $BaFe_2O_4$ (R Block), and $Ba_{0.5}M_{0.5}Fe_2O_4$ (T Block) to Utilize the ATLAD Technique

	Chemical Formula/Blocks	S	R	T
Y-type hexaferrite	$Ba_2M_2Fe_{12}O_{22}$ (STSTST)	22		8
Z-type hexaferrite	$Ba_3M_2Fe_{24}O_{41}$ (RSTSR*S*T*S*)	33	6	6

Nevertheless, the above estimates should be compared with the calibrated deposition rates of each target. The above estimates are just extrapolations!

M signifies any transition metal ion.

PROBLEMS

4.1

a. Assume that four Mg^{2+} ions in $MgFe_2O_4$ have migrated from the A sites to the B sites. Usually, Mg ions prefer A sites. What is the inversion factor x?

b. Calculate the saturation magnetization at $0°K$.

c. Would you expect T_N to be lower and why?

d. Assume a normal spinel, can you expect ferromagnetism or antiferromagnetism? Explain.

4.2 Applying the ATLAD technique to deposit M-type hexaferrite film, what two targets should be used such that Mn^{2+} ions are placed only in the R block?

4.3 Calculate the saturation magnetization of Y-type and Z-type hexaferrite.

4.4 Applying the ATLAD technique to deposit YIG ($Y_3Fe_5O_8$) films,

a. What single crystal material should be considered for substrate?

b. What target materials should be used?

c. How many laser pulse to be used to impinge on each target?

REFERENCES

ATLAD REFERENCES

A.L. Geiler, Ph.D. Thesis Dissertation, Atomic Scale Design and Control of Cation Distribution in Hexagonal Ferrites for Passive and Tunable Microwave Magnetic Device Applications, Northeastern University, 2009.

R. Karim, C. Vittoria, et al., *IEEE Trans. Mag.*, **31**, 3485, 1995.

R. Karim and C. Vittoria, *J. Magn. Magn. Mater.*, **167**, 27, 1997.

C. Vittoria, Fabrication of Ferrite Films Using Laser Deposition, U.S. Patent 5,227,204, 1993.

C. Vittoria, Method of Engineering Single Phase Magnetoelectric Hexaferrite Films (at Atomic Scale), U.S. Patent 10,767,256 B2, 2020.

Xu Zuo, C. Vittoria, et al., *J. Magn. Magn. Mater.*, **272-276**, Supplement, E1795, 2004.

Xu Zuo, C. Vittoria, et al., *Appl. Phys. Letters,* **87**, 152505, 2005.

SUPPLEMENTAL REFERENCES

B.I. Bleaney and B. Bleaney, *Electricity and Magnetism*, Oxford University Press, London, 1965.

S. Chikazumi, *Physics of Magnetism*, John Wiley & Sons, Inc., New York, 1964.

H. Kojima, *Ferromagnetic Materials* **3**, North-Holland Press, Amsterdam, 1982.

Landolt-Bornstein, *Numerical Data and Functional Relationships in Science and Technology* **4**, Editor K. H. Hellwege, Springer Press, Berlin, 1970.

D.E. Sayers and B.A. Bunker, *X-Ray Absorption: Principles, Applications, Techniques of EXAFS, SEXAFS and XANE*, John Wiley & Sons, Inc., New York, 1988.

J. Smit and H.P.J. Wijn, *Ferrites*, John Wiley & Sons, Inc., New York, 1959.

W.H. Von Aulock, *Handbook of Microwave Ferrite Materials*, Academic Press, New York, 1965.

W.P. Wolf, *Phys. Rev.*, **108**, 1152, 1957.

Xu Zuo and C. Vittoria, et al., *Appl. Phys. Letters*, **87**, 152505, 2005.

SOLUTIONS

4.1

    a. $x = 4/8 = 0.50$ (50%).

    b. T_N would be lower, because the superexchange coupling between A and B sites is lowered by the introduction of non-magnetic ions in the A sites.

    c. The only reason spins in either the A or B sites are parallel to each other is because in (b) magnetic ions transferred over to the A site. Removing them from the A sites would only introduce antiferromagnetism in the B sites.

4.2 Fe_2O_3 and $BaFe_{2-x}Mn_{3x/2}O_4$.

4.3 The saturation magnetization at $0°K$ temperature is calculated as follows:

$$M_S = (N/V)g\mu_\beta S,$$

where N is the net number of spins pointing along the majority spin direction. For Y-type, $N = 6$ and for Z-type, $N = 12$. V is the volume of the crystal unit lattice, $V = \sqrt{3}a^2c/2$. For Y- and Z-type, $a = 5.89$ A; $c = 43.56$ A for Y-type and 52.30 A for Z-type. The g-factor (g) is assumed to be equal to 2.0. For Fe^{3+} ions $S = 5/2$. $\mu_\beta = 9.27 \times 10^{-21}$ emu $=$ Bohr magneton. Thus,

$$M_S = 215.7 \text{ emu/cm}^3, \text{ Y-type and}$$

$$M_S = 359.3 \text{ emu/cm}^3, \text{ Z-type.}$$

4.4

    a. MgO ($a = 4.2$ A) and $Gd_3Ga_5O_{12}$ ($a = 12.36$ A). Lattice constant of YIG is 12.365 A.

    b. Y_2O_3 and Fe_2O_3.

    c. Extrapolating from the spinel work 5 pulse shots impinge on the yttrium target and 11–12 on Fe_2O_3. However, little is known about the sticking coefficient of Y_2O_3 deposition. Thus, there is a need to calibrate its deposition rate compared to Fe_2O_3.

5 Free Magnetic Energy

THERMODYNAMICS OF NONINTERACTING SPINS: PARAMAGNETS

The total magnetic moment in a solid is the sum of all the moments at each site:

$$M = Nm. \tag{5.1}$$

The brackets denote the thermal average value of m, the magnetic moment, at each site and N is the number of ions or atoms/cm^3.

For simple cubic (SC), $N = 1/a^3$, where a is the lattice constant, $N = 2/a^3$ and $N = 4/a^3$ for body center cubic (BCC) and face center cubic (FCC), respectively. We may write $\langle m \rangle$ more explicitly as follows:

$$\langle m \rangle = \vec{g}\beta\langle m_J \rangle,$$

$$m_J = J, J-1, J-2, \ldots, -J,$$

$$J = L \pm S,$$

$$\beta = \text{Bohr magneton} = \gamma\hbar,$$

$$\gamma = \frac{e}{2mc} \cong 1.4 \times 2\pi \times 10^6 \, \text{Hz/Oe. CGS.}$$

The thermal average of m may be obtained by invoking Bernoulli's equation or theorem:

$$\langle m \rangle = \frac{\displaystyle\sum_{-J}^{J} m_J P_J}{\displaystyle\sum_{-J}^{J} P_J},$$

where P_J is the probability of a magnetic moment with energy $E_J = -m_J g_J \beta H$ at temperature T and m_J is the total angular momentum quantum number. H is the magnetic field. We invoke Boltzmann statistics to calculate P_J:

$$P_J = e^{-E_J/kT},$$

$$P_J = e^{+m_J g_J \beta H/kT},$$

DOI: 10.1201/9781003431244-5

and, therefore,

$$\langle m \rangle = g_J \beta \frac{\sum m_J e^{m_J g_J \beta H/kT}}{\sum e^{m_J g_J \beta H/kT}}. \tag{5.2}$$

We note that as $H \to 0$, $\langle m \rangle \to 0$ which implies random orientation of spins from site to site. Substituting Equation 5.2 into Equation 5.1, we obtain the following:

$$M(T) = Ng_J J \beta \left[\left(\frac{2J+1}{2J} \right) \coth \left(\frac{2J+1}{2J} \right) y - \frac{1}{2J} \coth \left(\frac{y}{2J} \right) \right],$$

where $y = Jg_J \beta H/kT$. The quantity in square parenthesis is usually abbreviated as $B(y)$, where $B(y)$ denotes the Brillouin function. Thus, M may be written as $M(T) = Ng_J J \beta B(y)$, and

$$B(y) = \left(\frac{2J+1}{2J} \right) \coth \left(\frac{2J+1}{2J} \right) y - \frac{1}{2J} \coth \left(\frac{y}{2J} \right).$$

For $J = 1/2$, we have a simple expression for $M(T)$.

$$M(T) = \frac{N}{2} g_J \beta \tanh \left(\frac{g\mu_\beta H}{2kT} \right).$$

We may write $M(T)$ as $M(T) = M(0) \tanh(g\mu_\beta H/2kT)$. In Figure 5.1, we plot $M(T)$ versus T.

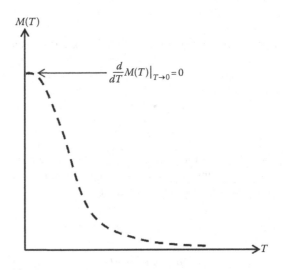

FIGURE 5.1 $M(T)$ for a paramagnetic material is plotted as a function of temperature.

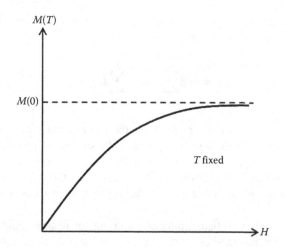

FIGURE 5.2 Magnetization as a function of H for paramagnetic state.

As $T \to 0$, $(dM(T)/dT) \to 0$. This can be demonstrated from

$$\frac{\partial}{\partial x}\tanh(x)|_{x \to \infty} = 0, \; since \; x \propto \frac{1}{T}.$$

In Figure 5.2, $M(T)$ is plotted as function of H at fixed temperature, T. For small H, we may write $M(T)$ as

$$M(T) \cong M(0)g_J\frac{\beta H}{2kT} = \frac{Ng^2\beta^2 H}{4kT}.$$

Let's introduce a susceptibility, χ, defined as

$$\chi = \frac{M(T)}{H} = \frac{Ng^2\beta^2}{4kT} = \frac{\lambda}{T},$$

where $\lambda = Ng^2\beta^2/4k$.

This is referred to as the Curie's law for a paramagnetic state.

FERROMAGNETIC INTERACTION IN SOLIDS

So far, we have assumed that the position of an atom is represented by a point in a solid far removed from any other atom. In fact, electronic orbitals do overlap with neighboring atoms in a solid. This, of course, implies repulsion between electronic charges, and hence, the exchange interaction between sites in a solid. In the literature, this is referred to as the localized model for magnetism. We may represent this type of interaction as follows. Assume for simplicity $J=S$ and $L=0$ and write the magnetic potential energy of a solid as follows:

$$E = -g\beta \sum_{i=1}^{N} \vec{S}_i \cdot \vec{H} - \sum_{i \neq j}^{N} J_{ij} \vec{S}_i \cdot \vec{S}_j. \qquad (5.3)$$

The first term is the Zeeman energy term and the second term is the exchange energy term between spins $\vec{S}_i$ and $\vec{S}_j$ at sites i and j. J_{ij} is the strength of the exchange interaction or the overlap integral between two sites. The Zeeman energy is recognized also as the magnetic potential energy of spins $\vec{S}_i$ at site i under the influence of a magnetic field. The reader is reminded that we will exclusively use the CGS system of units in this book.

Usually, the overlap exchange integral is strong for nearest neighbors for electronic orbitals centered at sites in close proximity or nearest neighbors. Thus, we may write

$$E = -g\beta \sum_{i=1} \vec{S}_i \cdot \vec{H} - J \sum_{i \neq j} \vec{S}_i \cdot \vec{S}_j. \qquad (5.4)$$

where we assume $J_{ij}=J$ for nearest neighbors and $J_{ij}=0$ for next nearest neighbors. We introduce a molecular field of the form $\vec{H}_{MF} = Jz\vec{S}_{ij}/g\beta$, where z is the number of nearest-neighbor ions to site i. For a simple cubic solid, $z=6$, for example. We may now write the magnetic potential energy as follows:

$$E = -g\beta \sum_{i}^{N} \vec{S}_i \cdot \left(\vec{H} + \vec{H}_{MF} \right).$$

Usually, $\vec{H}_{MF}$ may be written also as $\vec{H}_{MF} = \lambda \vec{M}$ or

$$\vec{H}_{MF} = Jz \frac{Ng\beta\vec{S}_j}{Ng^2\beta^2}.$$

H_{MF} is in units of Oe. Thus, we identify λ as equal to

$$\lambda = \frac{Jz}{Ng^2\beta^2}.$$

For iron, $J=2.16 \times 10^{-14}$ erg, $g=2.09$, $z=8$, and lattice constant $a=2.86 \times 10^{-8}$ cm. Since iron has a BCC structure,

$$N = \frac{2}{a^3} = 8.54 \times 10^{22} \text{ iron atoms/cm}^3.$$

We calculate λ to be at low temperatures:

$$\lambda \cong 5,400.$$

This value is an experimental value based on a measurement of the exchange constant, J, made by others; see Chikazumi, 1964. This value is to be compared with our estimated value of ~3,000, in Chapter 3, based on the simple free gas model.

The internal spontaneous or exchange field is approximately

$$H_{MF} = \lambda M,$$

where for iron $M \approx 1,672$ emu/cm^3 resulting in

$$H_{MF} \cong 9 \times 10^6 \text{ Oe}!$$

Clearly, this is a significant field to polarize the spins spontaneously!

The bracket indicates a thermal average. Let $\vec{H}_T = \vec{H} + \vec{H}_{MF}$, where $\vec{H}_T$ is the total field acting at site i. For $S = 1/2$, the average thermal magnetization is

$$M(T) = \frac{Ng\beta}{2} \tanh\left(\frac{g\beta}{2kT}(H_T) \right).$$

If all the spins were totally aligned, $M(0) = Ng\beta S$. As $T \to 0$ K, $\tanh(y) \to 1$, where $y = g\beta H_T/2kT$. The magnetization reaches the maximum value, $M(0)$. Thus,

$$M(T)\big|_{T \to 0} \to Ng\beta S - M(0), \, T \to 0K.$$

Let's now examine the rate of change of $M(T)$ with temperature as $T \to 0$ K. This means we examine the following derivative:

$$\frac{d}{dT}\left[\tanh(y) \right]_{T \to 0} = \frac{C}{T^2 e^{2y}}\bigg|_{T \to 0} \to 0.$$

The constant C in the numerator is a constant not dependent on temperature. Since the exponential e^{2y} contains terms of the form

$$\frac{1}{T^n},$$

the derivative of $\tanh(y)$ will approach zero as $T \to 0$ K. In summary, $M(T) \to M(0)$ and $(d/dT)M(T) \to 0$ as $T \to 0$ K. This conclusion is certainly true for all ordered magnetic materials as well as paramagnetic materials.

At high temperatures, we would expect H_{MF} to vanish, since neighboring spins to a given site would be randomly oriented as in a paramagnetic state. For small arguments of H_T, we have for $S = 1/2$ that

$$M(T) \approx Ng\beta S\left[\frac{(S+1)}{3kT}g\beta H_T\right] = M(0)\left[\frac{(S+1)}{3kT}g\beta H_T\right].$$

Let's rewrite the above equation as

$$M(T) = \left[\frac{Ng^2\beta^2 S(S+1)}{3kT}\right]H_T$$

or

$$M(T) = \chi_0 H_T = \chi_0\left(H + \lambda M(T)\right), \qquad (5.5)$$

where $\chi_0 \equiv$ susceptibility $= Ng^2\beta^2 S(S+1)/3kT$.

From Equation 5.5, one may derive the susceptibility ratio M/H as

$$\chi = \frac{M}{H} = \frac{\chi_0}{1-\chi_0\lambda} = \frac{Ng^2\beta^2 S(S+1)3K}{T - J_z S(S+1)/3k} = \frac{C}{T-T_c},$$

where

$$T_c = \frac{JzS(S+1)}{3k}.$$

For iron, $T_c \sim 1,000$ K. T_c is referred to as the Curie temperature. It is the temperature at which χ is maximum and below which there is spontaneous magnetization in the absence of an external magnetic field. Thus, an ordered magnetic state results from strong exchange coupling between spins. Let's now demonstrate how the magnetization varies with temperature below T_c. From Equation 5.5, we have that

$$\frac{M(T)}{M(0)} = \tanh\left[\frac{g\beta S(H + \lambda M(T))}{kT}\right], \qquad (5.6)$$

where $M(0) = Ng\beta S$. Let's now plot $M(T)$ versus T for $H = 0$. Basically, one needs to solve a transcendental equation in order to solve for $M(T)$. A scheme introduced in the literature is to let

$$y = \frac{M(T)}{M(0)}$$

and

$$x = \frac{g\beta S(H + \lambda M(T))}{kT} = \frac{g\beta S}{kT}(H + \lambda M_0 y).$$

For $H = 0$,

$$x = \frac{g\beta S}{kT} \times \frac{Jz}{Ng^2\mu_\beta^2} \times Ng\beta S \times y = \frac{JzS^2 y}{kT} = \frac{JzS(S+1)y}{3kT}, \quad S = \frac{1}{2}.$$

Hence, we have the simple result that

$$y = \frac{T}{T_c} x. \tag{5.7a}$$

The other functional relationship between y and x is that

$$y = \tanh(x). \tag{5.7b}$$

We plot both functional relationships, Equations 5.7a and 5.7b, between y and x in Figure 5.3. The point where the two plots intersect is indeed a solution for the transcendental equation.

We now plot a family of lines in Figure 5.3 in which the ratio T/T_c is varied between 0 and 1. The intersection labeled by point (1) is a solution to the transcendental equation, Equation 5.7, y_1 and x_1. This is repeated however many times until a plot of M versus T is obtained, as in Figure 5.4.

Let's now estimate H_{MF} for a typical alloy material like permalloy. Typically, $\lambda \sim 3,000$, $M \approx 1,000$ G. Therefore, $H_{MF} = \lambda M = 1 \sim 3 \times 10^6$ Oe.

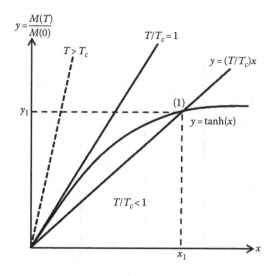

FIGURE 5.3 Normalized magnetization is plotted as a function of x for various values of T/T_c graphical solution.

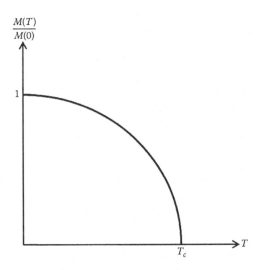

FIGURE 5.4   A plot of $y = \tanh(x) = M(T)/M(0)$ versus T.

FERRIMAGNETIC ORDERING

Besides ferromagnetic ordering, there are other types of magnetic ordering in which spontaneous magnetization is different at different sites in a solid. For example, let's consider ferrimagnetic ordering in which part of the solid contains spontaneous magnetization $\vec{M}_1$ and another, $\vec{M}_2$. We may refer to these regions as sublattice magnetizations. They are exchange coupled. Let's now ask the question what is the molecular field at region (1) or sublattice (1) (see Figure 5.5).

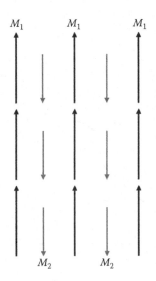

FIGURE 5.5   Two collinear magnetic sublattices.

$$\vec{H}_{MF}^{(1)} = \lambda_{11}\vec{M}_1 - \lambda_{12}\vec{M}_2; \; \lambda_{11}, \lambda_{12} > 0. \tag{5.8}$$

The molecular field at site or sublattice (2) is of the same form:

$$\vec{H}_{MF}^{(2)} = \lambda_{22}\vec{M}_2 - \lambda_{12}\vec{M}_1; \; \lambda_{22} > 0. \tag{5.9}$$

λ_{ij} are the antiferromagnetic exchange parameters. The ferrimagnetic coupling spontaneously aligns the magnetization in the opposite direction in a given sublattice. However, in most ferrites, the exchange coupling parameters are such that $\lambda_{12} \gg \lambda_{11}$ and λ_{22}. This means that $\vec{M}_1$ is in the opposite direction to $\vec{M}_2$, although the natural tendency within a sublattice is to align spins in the opposite direction. Thus, all the λ's have the same sign, and for simplicity, we have taken them to be positive.

Thus, magnetization at sublattice (1) is

$$M_1(T) = M_1(0)B\left[\frac{g\beta S_1}{kT}\left(H_{MF}^{(1)} + H\right)\right]$$

$$\approx \frac{C_1}{T}\left(\lambda_{11}M_1(T) - \lambda_{12}M_2(T) + H\right), \tag{5.10}$$

and at sublattice (2)

$$M_2(T) = M_2(0)B\left[\frac{g\beta S_2}{kT}\left(H_{MF}^{(1)} + H\right)\right]$$

$$\approx \frac{C_2}{T}\left(\lambda_{22}M_2(T) - \lambda_{12}M_1(T) + H\right). \tag{5.11}$$

Let's assume antiferromagnetic alignment so that

$$C_1 = C_2 = C = g\beta\frac{M(0)(S+1)}{3k}$$

and

$$M_1(T) = -M_2(T).$$

Also, we assume $\lambda_{11} = \lambda_{22} \ll \lambda_{12}$ so that $M_1(T) \approx (C/T)(\lambda_{12}M_1(T) + H)$ and $M_2(T) \approx (C/T)$ $(\lambda_{12}M_2(T) + H)$. We divide by H and obtain

$$\frac{M_1(T)}{H} = \frac{M_2(T)}{H} = \frac{C}{(T - \lambda_{12}C)}.$$

We now define a new ordering parameter and it is called the Neel temperature, T_N, where $T_N = C\lambda_{12} = Ng^2\beta^2(S(S+1)\lambda_{12}/3k)$. For $|M_1| \neq |M_2|$ and $\lambda_{11}, \lambda_{22} \ll \lambda_{12}$, the same result for T_N may be used. However, S needs to be reinterpreted as an effective S or a "net" S. One can show that mathematically the ferrimagnetic state ($|M_1| \neq |M_2|$) may

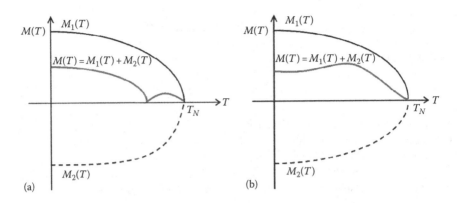

FIGURE 5.6 Examples of some interesting temperature dependences of the magnetization for ferrimagnetic materials.

be treated as a "ferromagnetic state" dealing only with the net magnetization, M_{net}, and net spin S, which is the difference in the sublattice magnetizations and respective spins. We will illustrate this using a numerical example. Of course, at optical frequencies' excitations, one may not make such drastic approximation. Contrary to pure ferromagnetic materials (iron, nickel, etc.), the temperature dependence of the "net" magnetization can be rather complex. In some cases, there may be a compensation temperature whereby the "net" magnetization is zero for temperatures well below T_N, as shown in Figure 5.6. The measured magnetization can never be less than zero, since one measures the component of the magnetization along an external magnetic field. Hence, if the net magnetization is calculated to be less than zero, one simply takes its magnitude in order to compare with experiments. However, as $T \to 0$, $dM(T)/dT$ approaches zero, since all the sublattice magnetizations behave like Brillouin functions.

As an example, let's calculate λ_{12} and T_N for yttrium iron garnet (YIG). In the literature, λ_{12} is often referred to λ_{ad}, where a refers to tetrahedral sites of iron ions and d to octahedral sites. Iron ions are in an S-state ($L=0$) with $S=5/2$. The quoted value for J_{ad} is $10\,\mathrm{cm^{-1}}$ which corresponds to (see appendix in Chapter 1) $J_{ad} \sim 2 \times 10^{-15}$ ergs. Again, we utilize the expression for λ_{ad} or λ_{12} as

$$\lambda_{ad} \approx \frac{J_{ad}}{Ng^2\beta^2}.$$

Note that the expression for λ_{ad} is very similar to a ferromagnetic λ's. The key to estimating λ_{ad} is the interpretation of N. In YIG, there are 24 iron ions (sublattice) in d-sites and 16 in a-sites. This means that there are a "net" of 8 ions which give rise to the "net" magnetization of YIG. Hence,

$$N = \frac{8}{a^3},$$

where a is the lattice constant $= 12.65 \times 10^{-8}\,\mathrm{cm}$, $g \approx 2$, and $\beta = 9.27 \times 10^{-21}$ emu.

Putting the above values into λ_{ad}, we obtain

$$\lambda_{ad} \sim 2,000.$$

Now, let's calculate T_N from

$$T_N = Ng^2\beta^2 \frac{S(S+1)\lambda_{ad}}{3k}.$$

There are two ways to interpret the expression for λ_{ad}. Either N incorporates the factor of 8 or the value of S, but it cannot be included twice! Anyway, we have

$$T_N \approx 560 \text{ K.}$$

The experimental value is 550 K.

SPIN WAVE ENERGY

Clearly, as T approaches 0 K, $M(T)|_{T\to 0} = M(0)$ and the potential energy becomes $E = -\lambda M(0)^2$. As the temperature increases from $T=0$ K, thermal energy is imparted into dynamic motion of the magnetization. This spin dynamic motion is often referred to as the spin wave motion. Thermal fluctuations or excitations induce motion of the magnetization. For example, consider two regions in a solid characterized by $\vec{M}_1$ and $\vec{M}_2$ (they may not be the same).

We assume that both $M_{1,2}$ consist of a static component and a time-varying component. For simplicity, both static components are equal to each other and equal to M_0. At an instantaneous point in time, the physical picture looks like the figure below.

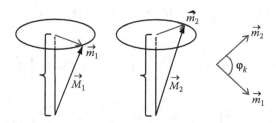

These two regions are coupled by exchange interaction, and the potential energy may be written as follows in a phenomenological way as

$$E = -\vec{M}_1 \cdot \lambda \vec{M}_2,$$

where

$$\vec{M}_1 = \vec{M}_0 + \vec{m}_1(t) \text{ and } \vec{M}_2 = \vec{M}_0 + \vec{m}_2(t).$$

For now, let's drop the explicit dependence on time and write

$$E = -\lambda M_0^2 - \lambda m_1 m_2 \cos \varphi_k.$$

Note that

$$\vec{M}_0 \cdot \vec{m}_1 (E) = 0 \text{ and also } \vec{M}_0 \cdot \vec{m}_2 = 0.$$

At $T=0$ K, $m_1 = m_2 = 0$, since there are no thermal vibrations of the magnetization and $M_0 = M(0)$. For $T > 0$, one needs to account for the extra energy, the spin wave energy (the second term in the expression for E) again, be mindful that for $T > 0$ K, M_0 will be temperature dependent. M_0 may be viewed as the saturation magnetization at T or $M_0(T) = M_S(T)$. M_1 or M_2 is the magnetization in the absence of spin waves and is calculable.

Let's calculate this energy due to spin wave motion for $T > 0$ K. We had before that

$$E = -2J \sum_{i \neq j} \vec{S}_i \cdot \vec{S}_j, \tag{5.12}$$

where $J = J_{nn}$ (nearest-neighbor value).

We may write the dot product explicitly:

$$\vec{S}_i \cdot \vec{S}_j = S_{ix} \cdot S_{jx} + S_{iy} \cdot S_{jy} + S_{iz} \cdot S_{jz}.$$

We assume $\vec{S}_i$ and $\vec{S}_j$ to be classical vectors and not quantum mechanical operators. Expanding S_{jx} in a Taylor series expansion in terms of S_{ix}, we obtain

$$S_{jx} = S_{ix} + \left(\frac{\partial}{\partial x} S_{ix} \right) \Delta x + \left(\frac{\partial}{\partial y} S_{ix} \right) \Delta y + \left(\frac{\partial}{\partial z} S_{ix} \right) \Delta z$$

$$+ \frac{1}{2} \left[\left(\frac{\partial^2}{\partial x^2} S_{ix} \right) \Delta x^2 + \left(\frac{\partial^2}{\partial x \partial y} S_{ix} \right) \Delta x \Delta y + \left(\frac{\partial^2}{\partial x \partial z} S_{ix} \right) \Delta x \Delta z \right.$$

$$+ \left(\frac{\partial^2}{\partial y \partial x} S_{ix} \right) \Delta y \Delta x + \left(\frac{\partial^2}{\partial y^2} S_{ix} \right) \Delta y^2 + \left(\frac{\partial^2}{\partial y \partial z} S_{ix} \right) \Delta y \Delta z$$

$$\left. + \ldots + \left(\frac{\partial^2}{\partial z^2} S_{ix} \right) \Delta z^2 \right] + \ldots \tag{5.13}$$

For cubic symmetry, the linear terms cancel out and we have

$$\sum_j S_{j\alpha} \approx z S_{i\alpha} + 2 \cdot \frac{1}{2} \left[\frac{\partial^2 S_{i\alpha}}{\partial x^2} \Delta x^2 + \frac{\partial^2 S_{i\alpha}}{\partial y^2} \Delta y^2 + \frac{\partial^2 S_{i\alpha}}{\partial z^2} \Delta z^2 \right],$$

where $\Delta x = \Delta y = \Delta z = a$, $\alpha = x, y, z$, and $\Delta x^2 = \Delta y^2 = \Delta z^2 = a^2$, with a being the lattice constant. Thus,

$$\sum_{i,j} S_{i\alpha} S_{j\alpha} \cong \frac{1}{2} \sum_{i=1}^{N} \left(z S_{i\alpha}^2 + a^2 S_{i\alpha} \nabla^2 S_{i\alpha} \right), \tag{5.14}$$

where z is the number of nearest-neighbor spins. We can approximate the energy expression of Equation 5.12 as follows:

$$E = -2J \sum_{i \neq j} \vec{S}_i \cdot \vec{S}_j \approx -J \sum_{i}^{N} \left[z S_{i\alpha}^2 + a^2 S_{i\alpha} \nabla^2 S_{i\alpha} \right]. \tag{5.15}$$

We rewrite E as

$$E = -JzNS_{i\alpha}^2 - J \sum_{i}^{N} [\cdots].$$

Simple substitutions show us that E may be written as

$$E = -\frac{Jz}{Ng^2\beta^2} (Ng\beta S_{i\alpha})(Ng\beta S_{i\alpha}) - J \sum_{i}^{N} [\cdots].$$

Finally, E may be put into a recognizable form

$$E = -\lambda M_\alpha M_\alpha - Na^2 J S_{i\alpha} \nabla^2 S_{i\alpha}, \tag{5.16}$$

where
$\lambda = Jz/Ng^2\beta^2$
$M_\alpha = Ng\beta S_{i\alpha}$

Writing the first term as a dot product and summing over the three values of α, we have that

$$E = -\lambda \vec{M} \cdot \vec{M} - \left(\frac{J}{a} S^2 \right) \sum_{\alpha}^{3} \frac{Ng\beta S_{i\alpha}}{N^2 g^2 \beta^2 S^2} \nabla^2 Ng\beta S_{i\alpha}. \tag{5.17}$$

Recognizing that $M_\alpha = Ng\beta S_{i\alpha}$, we may write Equation 5.17 as

$$E = -\lambda \vec{M} \cdot \vec{M} - \frac{A}{M^2} \vec{M} \cdot \nabla^2 \vec{M}, \tag{5.18}$$

where

$A = (J/a)S^2$ for simple cubic

$A = (2J/a)S^2$ for BCC

$A = (4J/a)S^2$ for FCC

A (exchange stiffness constant) may also be expressed as Na^2S^2J

Let's estimate the exchange stiffness constants A for iron and YIG. For iron, $J = 2.16 \times 10^{-14}$ erg, $a = 2.86 \times 10^{-8}$ cm, $S \approx 1.1$, and it is a BCC material. We calculate

$$A \approx 1.83 \times 10^{-6} \, \text{erg/cm}.$$

For YIG, which is a ferrimagnetic material, the expression for A is a bit complicated because there is more than one J to contend with. The reader is referred to Harris (1963): here for this expression

$$A = \frac{(2.5J_{ad} - 4J_{aa} - 1.5J_{dd})S^2}{a}.$$

where

J_{aa} is a constant between two magnetic ions in the tetrahedral sites

J_{dd} in octahedral sites

J_{ad} between different magnetic sites

Quoting Harris for the values of J's, $a = 12.65 \times 10^{-8}$ cm and $S = 5/2$ at $T = 0$ K, we obtain

$$A = 0.47 \times 10^{-6} \, \text{erg/cm}.$$

The first term in Equation 5.18 is associated with the potential energy of the spontaneous exchange magnetizing energy or self-energy. The second term is associated with any fluctuations of M, which varies with distance, such as spin waves. A convenient representation of Equation 5.18 is given by

$$\vec{E} = -\vec{M} \cdot \left(\vec{H}_{MF} + \vec{h}_{spinwave} \right),$$

where $\vec{H}_{MF} = \lambda \vec{M}$ and

$$\vec{h}_{spinwave} = \frac{A}{M^2} \nabla^2 \vec{M}. \tag{5.19}$$

The implication here is that a magnetic fluctuation may be represented by a dynamic field in the magnetic potential energy.

EFFECTS OF THERMAL SPIN WAVE EXCITATIONS

With the excitation of thermal spin waves, one can approximate the transverse component of the magnetization, $m_k(T)$, to spin wave excitation as

$$m_k(T) = \sqrt{M^2(T) - M_s^2(T)} B\left[\frac{g\beta S}{kT}\left(\frac{A}{M^2}m_k(T)k^2\right)\right], \quad (5.20)$$

where

$m_k(T) \equiv$ transverse component of the magnetization at temperature T

$M(T) \equiv$ magnetization at temperature T in the absence of thermal fluctuations

$M_S(T) \equiv$ static magnetization in the presence of thermal spin wave excitations (see Figure 5.7). In the previous section it was denoted as $M_0(T)$.

B is the Brillouin function. It is noted that the spin wave magnetic field is not collinear with the static magnetization. The only field acting normal to the static magnetization is indeed the spin wave field, as implied in Equation 5.20. The coefficient in front of the Brillouin function represents the highest value of the transverse magnetization at temperature T. This type of approach to the calculation of the spin wave amplitude for a given wave number, k, as a function of temperature is somewhat novel, but worthy of an attempt. It is no different in the approach used as in Figure 5.4 Here, we outline the calculation methodology. (1) M_S is a quantity that can be measured at temperature T, but M is calculable as in Figure 5.4 (2) The variable x as in Figure 5.4 may again be defined inside the Brillouin function. However, one has to be careful in introducing T_C together with the y variable.

At $T=0$ K, thermal spin wave excitations have no effect on the static magnetization in contrast to finite temperatures. Improved calculation model (involving Holstein–Primakoff second quantization schemes) shows a reduction in magnetization scaling like $T^{3/2}$.

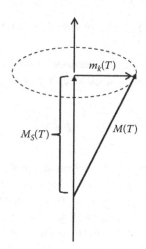

FIGURE 5.7 Excitation of thermal spin waves.

Let's now introduce magnetic fluctuation fields in a ferrimagnetic material. The total energy of the ordered system in a field, H, is then

$$E = -\vec{M}_1 \cdot \vec{H}_{MF}^{(1)} - \vec{M}_2 \cdot \vec{H}_{MF}^{(2)} - \left(\vec{M}_1 + \vec{M}_2 \right) \cdot \vec{H}. \tag{5.21}$$

We substitute Equations 5.8 and 5.9 into Equation 5.21, and obtain

$$E = -\lambda_{11} M_1^2 - \lambda_{22} M_2^2 + 2\lambda_{12} \vec{M}_1 \cdot \vec{M}_2 - \vec{H} \cdot \left(\vec{M}_1 + \vec{M}_2 \right). \tag{5.22}$$

FREE MAGNETIC ENERGY

In general, there are two thermodynamic free energies: Helmholtz free energy and Gibbs free energy. The Helmholtz free energy is defined as follows:

$$F = U - skT, \tag{5.23}$$

where
  U = internal energy
  s = entropy
  k = Boltzmann constant
  T = temperature

The second term is a constant at a given temperature. The internal energy U is related to all interaction energy terms within a magnet. Specifically, U is defined as

U = Magnetization energy + anisotropy energy + magnetostatic energy

  + exchange energy + magnetoelastic energy + etc.

In the free energy expression, it is understood that $\vec{M}$ is the average magnetization and it is averaged over temperature, space, and spin states. The Gibbs free energy is defined as

$$G = U - skT + \vec{M} \cdot \vec{H}. \tag{5.24}$$

We will henceforth use exclusively the Helmholtz form of the free magnetic potential energy minus the entropy term or the second term in Equation 5.23.

The conduit from microscopic to macroscopic magnetism is the so-called partition function. The magnetic free energy is a thermodynamic quantity representing or describing the properties of macroscopic entities. We review where we are in the order of things. We started with macroscopic bodies (the size of earth) and scaled down to atomic scale magnetism. Now, we are reversing the order to build macroscopic bodies from atomic size excitations. The free energy, which is a macroscopic quantity, may be related to atomic excitations via the partition function, Z:

$$F = -kT \ln Z,$$

where

$$Z = \left[\sum_{m=-1/2}^{1/2} e^{-(E_m/kT)} \right]^N$$

N = number of microscopic or ionic sites.

The reader is to be reminded that m usually refers to specific atomic energy levels. For simplicity, we have considered two energy levels. As a specific example, let's choose the Zeeman energy splitting for $m = \pm 1/2 = \pm S$ ($S = 1/2$).

$$E_m = -g\beta mH.$$

If only the ground state is occupied, then F is

$$F = -Ng\beta SH = -\vec{M} \cdot \vec{H},$$

where $\vec{M}$ is the magnetization. One can always hide various excitations under the guise of H (including exchange excitations). However, the inclusion of magnetic anisotropy energy is not trivial, but it is do-able (see Wolf, 1957). Let's proceed with the inclusion of magnetic anisotropy very slowly. These are two models to consider: single ion and pair models.

SINGLE ION MODEL FOR MAGNETIC ANISOTROPY

The reader is referred to Wolf (1957) and Hutchings (1964) for more details and precise definitions of terms. Let's consider a magnetic ion in the environment of six oxygen ions arranged in octahedral coordination (Figure 5.8).

The potential $V(x, y, z)$ due to the six oxygen ions with charge q at point $P(x, y, z)$ may be calculated from

$$V(x,y,z) = q\left[\frac{1}{\sqrt{(x+d)^2 + y^2 + z^2}} + \frac{1}{\sqrt{(x-d)^2 + y^2 + z^2}} + \frac{1}{\sqrt{x^2 + (y+d)^2 + z^2}} \right.$$

$$\left. + \frac{1}{\sqrt{x^2 + (y-d)^2 + z^2}} + \frac{1}{\sqrt{x^2 + y^2 + (z+d)^2}} + \frac{1}{\sqrt{x^2 + y^2 + (z-d)^2}} \right].$$

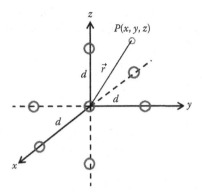

FIGURE 5.8 Octahedral configuration of ions.

The distance d is about 2.10–2.20 Å in most ferrites and is in the order of 1/4 of the crystal unit cell distance. Clearly, we want $V(x, y, z)$ to be applicable in the vicinity of the magnetic ion (origin). This means that we make the approximation that $d > r$. This means that we are dealing with a very localized crystal field. After much algebra (see Hutchings, 1964), we obtain

$$V(x,y,z) = \frac{6q}{a} + \left(\frac{35q}{4d^5}\right)\left[x^4 + y^4 + z^4 - \frac{3}{5}r^4\right] + \text{higher order terms (sixth order).}$$

The first term in V is isotropic and it corresponds to the Madelung term for ionic bonding. Thus, in essence, we are assuming ionic bonding in the single ion model. It is interesting to note that, for tetrahedral coordination of the oxygen ions, $V(x, y, z)$ is modified in the second term to $(-35q/4d^5)$. The potential energy of the magnetic ion in the crystalline field is

$$E_{ion} = q_{ion}V(x,y,z).$$

Clearly, E_{ion} exhibits cubic symmetry with respect to the coordinate system and the physical origin of E_{ion} is purely electrostatic. Based on the arguments of Heisenberg who represented electric Coulomb interaction by spin variables, Bethe and Wigner argued that indeed E_{ion} can also be expressed in terms of spin variables as long as the symmetry is maintained. That is, one may write E_{ion} (see Wolf, 1957) for an S-state ion as

$$E_{ion} = \frac{a}{6}\left(S_x^4 + S_y^4 + S_z^4\right) + \text{higher order terms,} \qquad (5.25)$$

where a is a crystal field parameter (not to be confused with the lattice constant). For example, in tetrahedral coordination, the (a) *parameter* changes sign for an S-state ion. There is much research reported for the case of $L \neq 0$ in terms of the so-called operator equivalent in the literature (see Stevens, 1997). We will not delve into that.

For more details, the reader is referred to Steven's work, see References. For uniaxial symmetry, a term of the following form would be added to E_{ion}:

$$D\left(S_\alpha^2\right)+...,$$

where $\alpha = x$, y, and z.

The single ion model basically adds all of the local crystal field energies to give rise to the total anisotropy energy. As such, one can always argue that since symmetry is conserved, one may assume that the symmetries of the local and total systems to be the same. It would be inconceivable to have a situation where the crystal field energy exhibits cubic symmetry locally, but the total anisotropy energy showed uniaxial symmetry unless there are unaccounted strains in the crystal. As one can see, this topic can become quite complex very rapidly. We are not interested in special situations with mixed symmetry terms but only interested in simple situations where symmetry is conserved locally as well as macroscopically. For most cases of interest in this book, cubic and uniaxial symmetries are assumed. The free energy of a macroscopic body exhibiting cubic symmetry may be written in the following form:

$$F_A = \frac{K_1}{M^4}\left(M_x^4 + M_y^4 + M_z^4\right)+.... \tag{5.26}$$

Note the progression: $\left(X^4 + Y^4 + Z^4\right) \rightarrow \left(S_x^4 + S_y^4 + S_z^4\right) \rightarrow \left(M_x^4 + M_y^4 + M_z^4\right)$!

It is understood that Equation 5.26 applies to a macroscopic body, whereas Equation 5.25 to one atomic site. Although the conclusion or jump from Equation 5.25 to Equation 5.26 is correct, a quantitative explanation for K_1 is still lacking at this time. Wolf provided a classical presentation on the connection between K_1 and crystal field parameters via the partition function approach. The reader is referred to his paper in the References section.

PAIR MODEL

In this model, the interaction between two spins at a time is considered and all the pair interactions within a solid are added. Pairs at long distance contribute much less to pairs which are nearest neighbors. For a single pair, we may write the interaction (according to Neel) pair along chain of spins as (see Chikazumi, 1964 for derivation)

$$E_{ij} = C_0 + C_1\left(\cos^2\alpha - \frac{1}{3}\right) + C_2\left(\cos^4\phi - \frac{6}{7}\cos^2\phi + \frac{3}{35}\right)+...,$$

where α is the angle between the two spin directions relative to the chain axis or bonding axis. E_{ij} takes the form of a Legendre polynomial expansion. Summing up all pair interactions, one obtains the total energy as

$$E = \sum_{i \neq j} E_{ij} = NC_2\left[\alpha_1^4 + \alpha_2^4 + \alpha_3^4 - \frac{3}{5}\right]+....$$

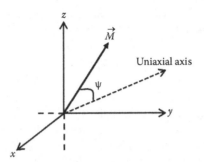

FIGURE 5.9   Definition of uniaxial axis.

The α's are directional cosines between the bonding axes and the local magnetization or spin direction. In solids, there are three orthogonal directions or axes to consider. Hence, the indexing over the angle α must correspond to three possible directions of the chains. Terms of the form α^2 add to a constant term in cubic crystals, since $\alpha_1^2 + \alpha_2^2 + \alpha_3^2 = 1$. Clearly, E assumes cubic symmetry as for the case of the single ion model.

Let's summarize what we have up to now in terms of writing the free energy consisting of Zeeman, exchange, and anisotropy energies:

$$F\left(\text{erg/cm}^3\right) = -\vec{M} \cdot \vec{H} + \lambda M^2 - \frac{A}{M^2}\, \vec{M} \cdot \nabla^2 \vec{M}$$

$$+ \frac{K_1}{M^4}\left(M_x^4 + M_y^4 + M_z^4\right) + \frac{K_u}{M^2}\left(M_1^2\right) + \ldots,$$

where $i = x, y$, and z. Usually, the constant term in the free energy is omitted (including the isotropic exchange term) and written in this form:

$$F\left(\text{erg/cm}^3\right) = -\vec{M} \cdot \vec{H} - \frac{A}{M^2}\, \vec{M} \cdot \nabla^2 \vec{M} + K_1\left(\alpha_1^2 \alpha_2^2 + \alpha_2^2 \alpha_3^2 + \alpha_3^2 \alpha_1^2\right) + K_u \cos^2 \psi + \ldots,$$

where
   α_i's are the directional cosines of $\vec{M}$ with respect to the cubic axes
   ψ is the angle between $\vec{M}$ and the uniaxial anisotropy axis (Figure 5.9).

Of course, there are higher order anisotropy terms of sixth order or more. For now, let's ignore them.

DEMAGNETIZING FIELD CONTRIBUTION TO FREE ENERGY

For simplicity, we assume a magnetized medium with uniform magnetization. Starting with Maxwell equation, we may write

$$\vec{\nabla}\cdot\vec{B}=0,$$

where
$\vec{B}=\vec{H}_D+4\pi\,\vec{M}$ in CGS units
$\vec{H}_D=$ demagnetizing field
$\vec{M}=$ magnetization vector

Since $\vec{\nabla}\cdot\vec{H}_D=0, \vec{H}_D=-\vec{\nabla}\phi$, where $\phi=$ magnetic potential. This means that $\nabla^2\phi=-4\pi\vec{\nabla}\cdot\vec{M}$. This is a classical differential equation whose solution is well known:

$$\phi=-\int\frac{\vec{\nabla}\cdot\vec{M}\left(\vec{r}'\right)}{\left|\vec{r}-\vec{r}'\right|}dv'+\int\frac{d\vec{s}\cdot\vec{M}}{\left|\vec{r}-\vec{r}'\right|}.$$

However, we are assuming a uniform magnetization so that contributions to the scalar potential, ϕ, are derived only at the surface of the medium. Hence, it is shown that ϕ simplifies to

$$\phi=\int\frac{\vec{M}\cdot d\vec{s}}{\left|\vec{r}-\vec{r}'\right|} \text{ and } \vec{H}_D=-\vec{\nabla}\phi=-N_\alpha\vec{M}_\alpha,$$

where $N_\alpha=$ demagnetizing factor. In general (assuming uniform $\vec{M}$), $\vec{H}_D$ is of the form

$$\vec{H}_D=-\left(N_x M_x\vec{a}_x+N_y M_y\vec{a}_y+N_z M_z\vec{a}_z\right),$$

where $N_x+N_y+N_z=4\pi$ in CGS units. The free energy contribution may be derived from

$$F_D=-\int\vec{H}_D\cdot d\vec{M},$$

or $F_D=-\int-\left(N_x M_x\vec{a}_x+N_y M_y\vec{a}_y+N_z M_z\vec{a}_z\right)\left(dM_x\vec{a}_x+dM_y\vec{a}_y+dM_z\vec{a}_z\right)$ which yields

$$F_D=\frac{1}{2}\left(N_x M_x^2+N_y M_y^2+N_z M_z^2\right).$$

It is noted that F_D scales as M^2, which implies that the result could have been derived from the dipole–dipole interaction between a pair of magnetic moments. However, this procedure is rather tedious. There are now in the literature (see References) many citations that calculate the demagnetizing factors N_α ($\alpha=x$, y, and z). Here, we provide an empirical way to calculate them. We assume that M is uniform throughout a non-ellipsoidal-shaped sample (see figure below). Clearly, this may not be the case near the corners of the sample. Our approach is based on the assumption that $\vec{H}_D$ is inversely proportional to distance from the source (Ampere's law). Thus, we write that (see figure below)

$$\frac{N_x}{N_y} = \frac{b}{a} \text{ and } \frac{N_x}{N_z} = \frac{c}{a}.$$

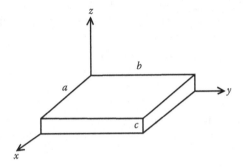

It can be shown that $N_x + (a/b)N_x + (a/c)N_x = 4\pi$ since $N_x + N_y + N_z = 4\pi$. Solving for N_x, we obtain

$$N_x = 4\pi\left[\frac{c/a}{1+(c/l)}\right], \ N_y = 4\pi\left[\frac{c/b}{1+(c/l)}\right], \ N_z = 4\pi\left[\frac{1}{1+(c/l)}\right], \text{ and } l = \frac{ab}{a+b}.$$

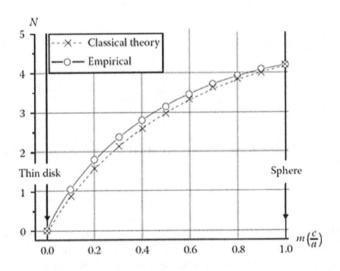

In Table 5.1, we compare demagnetizing factors calculated by the empirical way in this book with other numerical calculations of the integral equation for ϕ.

TABLE 5.1
Demagnetization Factors for Ellipsoids of Revolution

m	N*	D_a	N	m	N*	D_a	N
0.00[a]	0[a]	12.57[a]	0[a]	1.40		3.13	
0.01	—	12.38	—	1.50		2.92	
0.02	0.24	12.17	0.2	1.60		2.75	
0.03	—	11.98	—	1.80		2.44	
0.04	0.046	11.81	0.38	2.00		2.18	
0.05	—	11.62	—	2.50		1.70	
0.06	0.67	11.46	0.56	3.0		1.37	
0.07	—	11.30	—	3.5		1.13	
0.08	0.87	11.13	0.72	4.0		0.95	
0.09	—	10.97	—	4.5		0.81	
0.1	1.04	10.82	0.88	5.0		0.70	
0.125	—	10.42	—	6.0		0.54	
0.167	—	9.84	—	7.0		0.44	
0.200	1.8	9.41	1.58	8.0		0.36	
0.25	—	8.84	—	9.0		0.30	
0.30	2.4	8.30	2.14	10.0[c]	5.98[c]	0.25[c]	6.16[c]
0.40	2.8	7.39	2.59	15.0		0.13	
0.50	3.14	6.61	2.98	20.0		0.085	
0.60	3.4	5.98	3.3	30.0		0.043	
0.70	3.66	5.42	3.6	40.0		0.026	
0.80	3.86	4.95	3.8	50.0		0.018	
0.90	4.09	4.54	4.01	70		0.0098	
1.00[b]	4.19[b]	4.18[b]	4.19[b]	100		0.0053	
1.10		3.96		200		0.0015	
1.20		3.59		500[d]	6.2769[d]	0.00029[d]	6.2830[d]
1.30		3.34					

Note: See Morrish (2001); $N = (1/2)(4\pi - D_a)$. $m = c/a$ and $N^* = 4\pi m/(1 + 2m)$ (empirical result).

[a] Thin disk.
[b] Sphere.
[c] Cylinder.
[d] Needle.

Finally, the total free energy, F, may be written, including magnetizing, exchange, anisotropy, and demagnetizing energies, as follows:

$$F\left(\text{erg/cm}^3\right) = -\vec{M}\cdot\vec{H} - \frac{A}{M^2}\,\vec{M}\cdot\nabla^2\vec{M} + K_1\left(\alpha_1^2\alpha_2^2 + \alpha_2^2\alpha_3^2 + \alpha_3^2\alpha_1^2\right)$$

$$+ K_u\cos^2\psi + \frac{1}{2}\left(N_x M_x^2 + N_y M_y^2 + N_z M_z^2\right) + \ldots.$$

We have omitted higher order cubic anisotropy terms. The term containing K_u represents uniaxial magnetic anisotropy energy, where ψ is the angle between $\vec{M}$ and the uniaxial anisotropy axis. For crystals with hexagonal symmetry, the uniaxial axis is usually the c-axis or hexagonal axis. In addition, hexagonal ferrites may contain anisotropy energy terms representing sixfold symmetry of the basal plane, for example, of the form

$$K_\varphi\cos 6\varphi,$$

where φ is the in-plane azimuth angle.

NUMERICAL EXAMPLES

Let's calculate the magnetization at $T = 0$ of various magnetic materials. The magnetization may be calculated from the expression $M(0) = Ng\beta S$.

Let's calculate $M(0)$ for YIG ($Y_3Fe_5O_{12}$). YIG has a garnet cubic structure and contains three magnetic sublattices, but only two are occupied, see Figure 5.10.

Lattice constant of YIG is $a = 12.365$ Å. Here is a list of magnetic sublattices.

Sublattice (1): 16 a-sites (tetrahedral coordination)
Sublattice (2): 24 d-sites (octahedral coordination)
Sublattice (3): (dodecahedral coordination); no occupancy

The electronic configuration of metallic Fe is $1s^2 2s^2 2p^6 3s^2 3p^6 3d^6 4s^2$.
The electronic configuration for Fe^{3+} is $1s^2 2s^2 2p^6 3s^2 3p^6 3d^5$.
According to Hund's rule, $S = 5/2$ and $L = 0$. Thus,

$$g = 2,\ S = 5/2,\ \beta = 9.27\times 10^{-21}\ \text{emu}.$$

There are eight "net" spins aligned in one direction per unit cell of YIG. Thus, $N = 8/a^3$ and the magnetization is then

$$M = \frac{8\times 10^{24}}{(12.65)^3}\,g\beta S,$$

$$M = 183\,\text{emu/cm}^3,$$

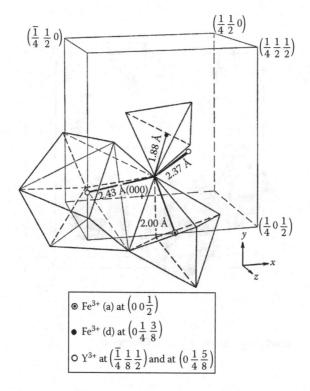

$$Fe^{3+} (a) \text{ at } \left(0\,0\,\tfrac{1}{2}\right)$$

$$Fe^{3+} (d) \text{ at } \left(0\,\tfrac{1}{4}\,\tfrac{3}{8}\right)$$

$$Y^{3+} \text{ at } \left(\tfrac{\bar{1}}{4}\,\tfrac{1}{8}\,\tfrac{1}{2}\right) \text{ and at } \left(0\,\tfrac{1}{4}\,\tfrac{5}{8}\right)$$

FIGURE 5.10 $(Y^{3+})_3(Fe^{3+})_5(O^{2-})_{12}$—YIG (yttrium iron garnet).

$$4\pi M = 2,300\,G; 2,400\,G\,(\text{experiment}).$$

Let's consider the spinel cubic structure as related to ferrite materials such as $MnFe_2O_4$. This magnetic structure consists of two magnetic sublattices, A and B magnetic sites. The A-sites are tetrahedrally coordinated and the B-sites octahedrally coordinated. The valence state of the A-sites is 2+ and that of the B-sites 3+. There are 8 A-sites and 16 B-sites within a unit cell of the spinel cubic structure (see Figure 5.11). Typically, the lattice constant is about $a = 8.5\,\text{Å}$. In the "normal" spinel structure, there are 8 Mn^{2+} ions in the A-sites and 16 Fe^{3+} in the B-sites giving rise to 8 "net" spins accounting for the magnetization. Both Mn^{2+} and Fe^{3+} are S-state ions with $S = 5/2$ and $L = 0$. As such, the magnetization at $T = 0$ may be estimated as follows:

$$M(0) = Ng\beta S,$$

where
$N = 8/a^3$
$g = 2$

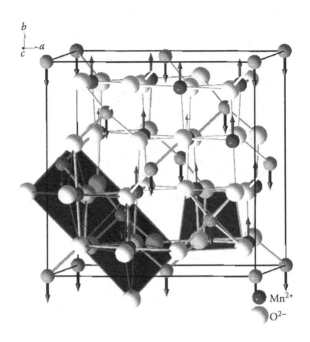

FIGURE 5.11   Spinel magnetic structure (see Chapter 4).

We calculate $M(0)$ to be $M(0) \approx 650$ emu/cm^3 or $4\pi M(0) \approx 8,000$ G.

Experimentally, $4\pi M(0)$ is measured to be $\approx 7,500$ G. Clearly, the comparison with our estimate is reasonable, but it still needs some improvement. One factor to consider is that $MnFe_2O_4$ is an "inverse" spinel. This means that, for example, 20% of the Mn^{2+} ions thermally transfer from A-sites to B-sites "forcing" some Fe ions to take on a valence state of Fe^{2+}, which is not an S-state ion. In order to be more specific, it means that on the average that 6.4 Mn^{2+} and 1.6 Fe^{2+} are on the A-sites and 14.4 Fe^{3+} and 1.6 Mn^{3+} ions are on the B-sites. The analysis of this magnetic state is rather complex, the result of which is a reduction in net magnetic moment. We will not delve much into these types of analyses.

Finally, we consider hexagonal ferrite materials. In particular, we consider the so-called magnetoplumbite hexagonal structure of barium ferrite, $BaFe_{12}O_{19}$ (see Figure 5.12). This structure contains four spinel blocks oriented with the (111) plane parallel to the c-axis or hexagonal axis. The reader is encouraged into understanding the intricacies or subtleties of this structure. It suffices to say that the valence state of Fe ions is 3+, although it is quite common for Fe to take on a 2+ valence state. This may be a result of oxygen defects or vacancies. The presence of Fe^{2+} induces electron hopping. However, the "net" number of spins giving rise to magnetism is still 8. The volume of the unit cell is now

$$\Delta V = \frac{\sqrt{3}}{2} a^2 c,$$

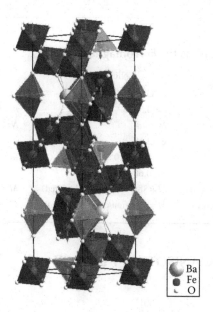

Ba
Fe
O

FIGURE 5.12 Hexagonal ferrite crystal structure (see Chapter 4).

where

$a = 5.89\,\text{Å}$

$c \sim 23.2\,\text{Å}$

Thus, the magnetization at $T=0$ is then

$$M(0) = \frac{8}{\Delta V} g\beta S, \; S = \frac{5}{2}.$$

The above equation yields

$$M(0) = 540\,\text{emu/cm}^3.$$

The reader is reminded that the introduction of Fe^{2+} in this structure introduces complications in the calculation of the magnetization at $T=0$ as well as any other temperature. Again, it is not the purpose of this presentation to delve in these types of details. The reader should consult specialized books on this subject matter. It is remarkable that in all of the magnetic structures considered that there is a commonality: the "net" number of spins partaking in the magnetization process is 8 for the garnet, spinel, and hexagonal magnetic structures.

Finally, we summarize the numerical examples for ferromagnetic metals in Table 5.2.

TABLE 5.2

Summary of Calculations for Ferromagnetic Metals

Iron (Fe)	Nickel (Ni)	Cobalt (Co)
BCC	FCC	FCC/HPC
$a = 2.86\,\text{Å}$	$a = 3.52\,\text{Å}$	$a = 3.52\,\text{Å}$
$N = 2/a^3$	$N = 4/a^3$	$N = 4/a^3$
$g = 2.09$	$g = 2.18$	$g = 2.16$
$\beta = 9.27 \times 10^{-21}$ emu	$\beta = 9.27 \times 10^{-21}$ emu	$\beta = 9.27 \times 10^{-21}$ emu
$S = 1.1$	$S = 0.604$	$S = 1$
$M = 1{,}700$ emu/cm³ (calculated)	$M = 880$ emu/cm³ (calculated)	$M = 1{,}600$ emu/cm³ (calculated)
$M = 1{,}750$ emu/cm³ (measured)	$M = 525$ emu/cm³ (measured)	$M = 1{,}450$ emu/cm³ (measured)

CUBIC MAGNETIC ANISOTROPY ENERGY

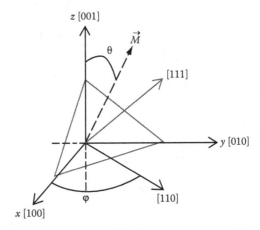

The free energy may be expressed as follows:

$$F_A = K_1\left(\alpha_1^2\alpha_2^2 + \alpha_2^2\alpha_3^2 + \alpha_2^2\alpha_3^2\right)$$

and

$$\alpha_1^2 = \sin^2\theta\cos^2\varphi,\ \alpha_2^2 = \sin^2\theta\sin^2\varphi,\ \text{and}\ \alpha_3^2 = \cos^2\theta.$$

Thus,

$$F_A = K_1\left(\sin^2\theta\sin^2\theta\sin^2\varphi\cos^2\varphi + \sin^2\theta\cos^2\theta\sin^2\varphi + \sin^2\theta\cos^2\theta\cos^2\varphi\right),$$

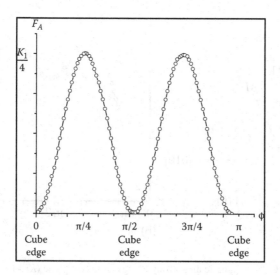

FIGURE 5.13 $K_1 > 0$, cube edge is easy axis of magnetization.

which simplifies to

$$F_A = \frac{K_1}{8}\left[\frac{1}{8}(3 - 4\cos 2\theta + \cos 4\theta)(1 - \cos 4\varphi) + 1 - \cos 4\theta\right].$$

For simplicity, assume [001] plane: $\theta = \pi/2$ (x–y plane).
F_A reduces to

$$F_A = \frac{K_1}{8}(1 - \cos 4\varphi) \quad \text{fourfold symmetry.}$$

For $K_1 > 0$, the easy axis of magnetization is along the cube edge (Figure 5.13).
For $\vec{M}$ in the $(\overline{1}10)$ plane (Figure 5.14), F_A becomes ($\varphi = 45°$)

$$F_A = \frac{K_1}{8}\left[\frac{7}{4} - \cos 2\theta - \frac{3}{4}\cos 4\theta\right].$$

UNIAXIAL MAGNETIC ANISOTROPY ENERGY

The uniaxial magnetic anisotropy energy may be expressed as follows:

$$F_u = K_u \sin^2 \theta \sin^2 \varphi.$$

For simplicity, assume $\varphi = 90$ ($\vec{M}$ in the y–z plane) (Figure 5.15).

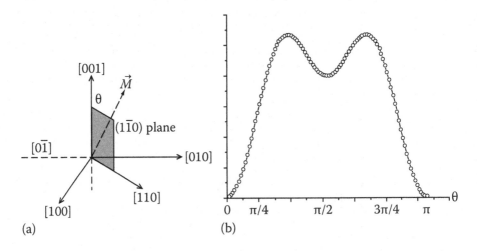

FIGURE 5.14 Cubic magnetic anisotropy energy in ($\overline{11}0$) plane ($K_1 > 0$).

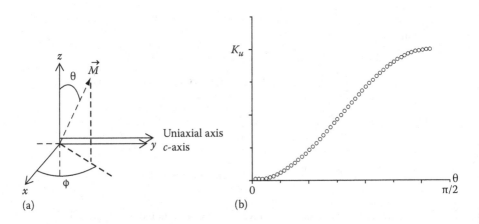

FIGURE 5.15 $F_u = K_u \sin^2 \theta$. $K_u > 0$, F_u is a minimum for $\vec{M}$ perpendicular to the c-axis. $K_u < 0$, F_u is a minimum for $\vec{M}$ parallel to the c-axis, F_u is on the vertical axis.

Example of $K_u > 0$, Y-type hexaferrites:

$$BaZn_2Fe_{12}O_{22}.$$

Example of $K_u < 0$, M-type hexaferrite:

$$BaFe_{12}O_{19}.$$

PROBLEMS

5.1 Show that the expression for M reduces to $M = (Ng\mu_B/2) \tanh y$ for $J = 1/2$.

5.2 Plot $M/Ng\mu_B$ in units of Bohr magneton assuming $S = 7/2$, $L = 0$, $B = 1{,}000$ G, and $g = 2$.

5.3 Do the same for a diamagnetic SC solid.

5.4 Calculate T_c, $M(0)$, λ_{ex}, and H_{MF} for iron. You may assume $J = 2.16 \times 10^{-21}$ J, $g = 2.09$, and $S = 1.1$.

5.5 Assume $M_1 = -fM_2$, $\lambda_{11} = \lambda_{22} = \lambda \ll \lambda_{12}$, $f \le 1$, obtain T_N.

5.6 Obtain A in terms of M_1, M_2, A_1, and A_2.

5.7 Apply dipole–dipole interaction to a chain of dipole moments and calculate the anisotropy energy.

5.8 Include the coupling of spin wave fields in the free energy.

5.9 Plot F_A as a function of θ for $\vec{M}$ in the (110) plane. Assume $K_1 < 0$ and $K_2 = 0$.

5.10 Calculate N_x, N_y, and N_z for a sphere, thin film, and cylinder, where H is perpendicular to the film plane and along the length of the cylinder.

5.11 Assuming that a system of atomic moments do not interact, calculate the potential energy approximately for the moments. Also calculate the approximate energy of thermal motion for a system of atoms. What does this imply for paramagnetic materials?

5.12 What is the physical significance of a positive or negative value for the exchange integral J?

5.13 Calculate the magnetic energy stored in a $1\,\text{cm} \times 1\,\text{cm}$ film that is $5\,\mu\text{m}$ thick with a permeability tensor defined below. The applied field, H, is along the z-axis perpendicular to the film plane and is constant (10^6 A/m).

$$[\mu_R] = \begin{bmatrix} 2 & 0 & 0 \\ 0 & 1 & 0 \\ 0 & 0 & 2 \end{bmatrix}.$$

5.14 Calculate the free energy of a long thin rod with the magnetic field applied along its axis. Assume no anisotropy field and H along axis.

5.15 Calculate $4\pi M$ of $MgFe_2O_4$ at $T = 0$ K.

5.16 Formulate the free energy of a magnetic nanowire of extremely small dimensionality.

BIBLIOGRAPHY

A. Abraham, *The Principles of Nuclear Magnetism*, Clarendon Press, Oxford, UK, 1961.

E.E. Anderson, *Phys. Rev.*, **134**, A1581, 1964.

B.I. Bleaney and B. Bleaney, *Electricity and Magnetism*, Clarendon Press, Oxford, UK, 1965.

S. Chikazumi, *Physics of Magnetism*, John Wiley & Sons, Inc., New York, 1964.

B.D. Cullity and C.D. Graham, *Introduction to Magnetic Materials*, John Wiley & Sons, Inc., New York, 2008.

G.F. Dionne, *Magnetic Oxides*, Springer Verlag Gmbh, Berlin, Germany, 2009.

J.S. Griffiths, *The Theory of Transition Metal-Ions*, Cambridge University Press, Cambridge, UK, 1961.

A.B. Harris, *Phys. Rev.*, **132**, 2398, 1963.

M.T. Hutchings, *Sol. Stat. Phys.*, **16**, 227, 1964.

D.C. Jiles, *Introduction to Magnetism and Magnetic Materials*, Academic Press, New York, 1998 (paperback).

F. Keffer, *Spin Waves, Handbook der Physik*, Vol. XVIII2, Springer, Berlin, Germany, 1966.

C. Kittel, *Introduction to Solid State Physics*, John Wiley & Sons, Inc., New York, 1956.

C. Kittel, *Elementary Statistical Physics*, John Wiley & Sons, Inc., London, UK, 1958.

H. Kronmuller and S. Parkin, *Handbook of Magnetism and Advanced Magnetic Materials*, John Wiley & Sons, Inc., New York, 2007.

B. Lax and K. Button, *Microwave Ferrites and Ferrimagnetics*, McGraw-Hill Book Co., Inc., New York, 1962.

A.H. Morrish, *The Physical Principles of Magnetism*, John Wiley & Sons, Inc., New York, 2001.

L. Neel, *Ann. Phys. Paris*, **5**, 256, 1936.

R.C. O'Handley, *Modern Magnetic Materials: Principles and Applications*, John Wiley & Sons, Inc., New York, 1999.

R. Skomski and J.M.D. Coey, *Permanent Magnetism*, Taylor & Francis, Boca Raton, FL, 1999.

E. Smart, *Effective Field Theories of Magnetism*, W.B. Saunders Co., Philadelphia, 1966.

J. Smit and H.P.J. Wijn, *Ferrites*, John Wiley & Sons, New York, 1959.

K.W.H. Stevens, *Magnetic Ions in Crystals*, Princeton University Press, Princeton, NJ, 1997.

J.H. Van Vleck, *The Theory of Electric and Magnetic Susceptibilities*, Clarendon Press, Oxford, UK, 1932.

W.P. Wolf, *Phys. Rev.*, **108**, 1152, 1957.

SOLUTIONS

5.1 $M = N \langle m \rangle$

$$m = \frac{\sum_{-J}^{+J} g\beta J e^{-E/kT}}{\sum_{-J}^{+J} e^{-E/kT}}.$$

For $J = 1/2$,

$$m = \frac{1}{2} g\beta \left[\frac{e^{g\beta B/2kT} - e^{g\beta B/2kT}}{e^{g\beta B/2kT} + e^{g\beta B/2kT}} \right] = \frac{1}{2} g\beta \tanh\left(\frac{g\beta B}{2kT} \right),$$

$$\boxed{M = \frac{1}{2} Ng\beta \tanh(y) = Nm_0 \tanh(y) = M_0 \tanh(y), \text{ where } y = \frac{g\beta B}{2kT}.}$$

5.2

$$M(T) = Ng\beta \left\{ \left(\frac{2J+1}{2} \right) \times \coth\left[(2J+1)\left(\frac{g\beta H}{2k} \right)\left(\frac{1}{T} \right) \right] - \left(\frac{1}{2} \right) \times \coth\left[\left(\frac{g\beta H}{2k} \right)\left(\frac{1}{T} \right) \right] \right\},$$

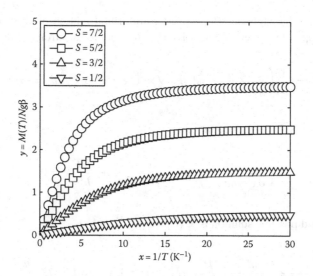

FIGURE S5.2 Normalized magnetization versus $1/T$.

$M(T) = Ng\beta B(X)$, where $B(X) =$ Brillouin function and $X = \dfrac{Jg\beta H}{kT}$.

Plot $M(T)/Ng\beta$ versus $1/T$, where $g = 2$. Also, $L = 0$ ($J = S$) and $B = 1,000$ G (Figure S5.2).

5.3 Diamagnetism contains negative susceptibility (Figure S5.3a)

$$\text{Partition function,} \, P_E = \sum_i e^{-E_i/kT},$$

$$\text{Magnetic moment as } M = NkT\frac{\partial \ln(P_E)}{\partial H} = \frac{N\sum_i \beta e^{-E_i/kT}}{\sum_i e^{-E_i/kT}}.$$

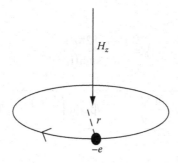

FIGURE S5.3a Particle in a field.

The diamagnetic effects arise from orbital motion of electrons in the applied magnetic field. Solve Hamiltonian for electron gas model:

$$\hat{H} = \frac{1}{2m}\left(\vec{p} - \frac{e}{c}\vec{A}\right)^2 \text{ and } \vec{B} = \vec{\nabla} \times \vec{A}, \text{ where } \vec{A} = A_x\vec{a}_x = -Hy\vec{a}_x.$$

The Schrödinger equation is

$$\left(\frac{\partial}{\partial x} - \frac{je}{\hbar c}Hy\right)^2\psi + \frac{\partial^2\psi}{\partial y^2} + \frac{\partial^2\psi}{\partial z^2} + \frac{2mE}{\hbar^2}\psi = 0,$$

and possible solution is $\psi(\vec{r}) = e^{-j(k_x x + k_z z)}\varphi(y)$

$$\frac{\partial^2\varphi}{\partial y^2} + \left\{\frac{2mE'}{\hbar^2} + \left(\frac{eH}{\hbar c}y - k_x\right)^2\right\}\varphi = 0, \text{ where } E' = E - \frac{\hbar^2 k_z^2}{2m},$$

which is similar to one-dimensional harmonic oscillator (equilibrium at y_0).

$$E' = \left(n + \frac{1}{2}\right)\hbar\omega = (2n+1)\beta H, \quad n = 0,1,2,\ldots.$$

The number of states in the momentum range dp at p is

$$N(p)\,dp = \frac{8\pi m\beta HV dp_z}{h^3} = \frac{-2VeH}{ch}dk_z.$$

The partition function is

$$P_E = \sum_i e^{-E_i/kT} = \sum_{n=0}^{\infty}\frac{2VeH}{ch}\int_{-\infty}^{+\infty}\frac{\exp\left[-\left(\hbar^2 k_z^2/2m\right) - \left((2n+1)\beta H/kT\right)\right]}{kT}dk_z,$$

$$P_E = \frac{VeH}{ch^2}\frac{(2\pi mkT)^{1/2}}{\sinh(\beta H/kT)},$$

$$\boxed{M = NkT\frac{\partial\ln(P_E)}{\partial H} = -N\beta\coth\left(\frac{\beta H}{kT}\right) = -N\beta L(X).}$$

and $N = 1/a^3$ for simple cubic.
  At $H = 1{,}000$ G (Figure S5.3b):

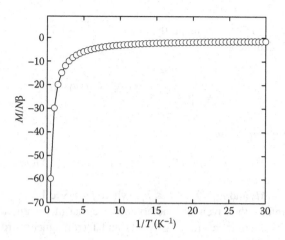

FIGURE S5.3b Normalized magnetization versus $1/T$.

5.4 Fe has a BCC structure (Figure S5.4) with $a = 3.21$ Å
1. Calculate T_c

$$\chi(T) = \frac{M(T)}{H} \cong \frac{Ng^2\beta^2 S(S+1)/3k}{T - JzS(S+1)/3k}.$$

Curie–Weiss's law:

$$\chi(T) = \frac{C}{T - T_c}, \text{ where } \boxed{T_c = \frac{JzS(S+1)}{3k}}.$$

For BCC with $J = 2.16 \times 10^{-21}$, $S = 1.1$, $z = 8$, $T_c = 963.5$ K.

2. Calculate M_0

$\boxed{M_0 = Ng\beta S}$, where β is a Bohr magneton, $N = 2/a^3 \cong 6.05 \times 10^{22}$ atoms/cm³

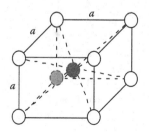

FIGURE S5.4 BCC structure.

$M_0 = 1{,}290$ G, experimentally, $M_0 \cong \dfrac{22{,}000}{4\pi} \cong 1{,}750$ G.

3. Calculate λ_{ex}

$$\lambda_{ex} = \frac{Jz}{Ng^2\beta^2}, \; g = 2.09, \; \lambda_{ex} = 7.60 \times 10^3.$$

4. Calculate H_{MF}

$$H_{MF} = \lambda_{ex} \cdot M_0, \; H_{MF} = 9.80 \times 10^6 \; Oe \sim 10 \text{ MOe}.$$

5.5 $M_1 = -fM_2$ and $\lambda_{11} = \lambda_{22} = \lambda \ll \lambda_{12}$, where $f \le 1$ Site 1
Therefore, the given parameters indicate that the magnetic ordering is antiferromagnetic ordering ($f = 1$) and ferromagnetic ordering ($f < 1$) (Figure S5.5).
Site 1.

$$M_1(T) = M_0^{(1)} B\left(X^{(1)}\right) \text{ where } B\left(X^{(1)}\right) = \text{Brillouin function,}$$

$$M_1(T) = M_0^{(1)} B\left(\frac{g\beta S_1\left[H_{MF}^{(1)} + H\right]}{kT}\right) \cong \frac{C_1}{T}\left(\lambda_{11}M_1 - \lambda_{12}M_2 + H\right).$$

Site 2.

$$M_2(T) \cong \frac{C_2}{T}\left(\lambda_{22}M_2 - \lambda_{12}M_1 + H\right)$$

Site 1   Site 2
M_1 M_2

FIGURE S5.5   Chain of spins aligned antiparallel.

1. Antiferromagnetic ordering $(f=1)$, $M_1 = -M_2$

For simplicity, $C_1 = C_2 = C = \dfrac{M_0(S+1)g\beta}{3kT}$

$$\chi(T) = \frac{M_1(T)}{H} = \frac{C}{T - \lambda_{12}C},$$

$$\boxed{T_N = C\lambda_{12}}.$$

2. Ferrimagnetic ordering $(f<1)$ $M_{1=} -fM_2$, but $\lambda_{11} = \lambda_{22} = \lambda \ll \lambda_{12}$

$$C_1 = \frac{M_0^{(1)}\left(S^{(1)}+1\right)g\beta}{3kT} \text{ and } C_2 = \frac{M_0^{(2)}\left(S^{(2)}+1\right)g\beta}{3kT},$$

$$\chi(T) = \frac{M_1(T)}{H} = \frac{C_1}{T - \left(\lambda_{12}C_1/f\right)} \text{ and } \chi(T) = \frac{M_2(T)}{H} = \frac{C_2}{T - \lambda_{12}C_2 f},$$

$$\boxed{T_N = \frac{C_1\lambda_{12}}{f} = C_2\lambda_{12}f}.$$

5.6   Total energy in two sites

$$E = -\vec{M}_1 \cdot \vec{H}_{MF}^{(1)} - \vec{M}_2 \cdot \vec{H}_{MF}^{(2)} - \left(\vec{M}_1 + \vec{M}_2\right) \cdot \vec{H}_{ext},$$

$$E = -\lambda_{11}M_1^2 - \lambda_{22}M_2^2 + 2\lambda_{12}\vec{M}_1 \cdot \vec{M}_2 - \left(\vec{M}_1 + \vec{M}_2\right) \cdot \vec{H}_{ext},$$

$$-\frac{A_1}{M_1^2}\vec{M}_1 \cdot \nabla^2 \vec{M}_1 - \frac{A_2}{M_2^2}\vec{M}_2 \cdot \nabla^2 \vec{M}_2.$$

Let $\vec{M} = \left(\vec{M}_1 + \vec{M}_2\right)$

$$E = -\vec{M} \cdot \vec{H} - \frac{A}{M^2}\vec{M} \cdot \nabla^2 \vec{M} + \text{extra term}.$$

Assume $A_{11} = A_1 M^2/M_1^2$ and $A_{22} = A_2 M^2/M_2^2$

$$\frac{A_1}{M_1^2}\vec{M}_1 \cdot \nabla^2 \vec{M}_1 + \frac{A_2}{M_2^2}\vec{M}_2 \cdot \nabla^2 \vec{M}_2 = \frac{A_{11}}{M^2}\vec{M} \cdot \nabla^2 \vec{M} + \frac{A_{22}}{M^2}\vec{M} \cdot \nabla^2 \vec{M}$$

$$-\left(\frac{A_{11}}{M^2}\vec{M}_2 \cdot \nabla^2 \vec{M}_1 + \frac{A_{22}}{M^2}\vec{M}_1 \cdot \nabla^2 \vec{M}_2\right)$$

$$-\left(\frac{A_{11}}{M^2}\vec{M}_1\cdot\nabla^2\vec{M}_2 + \frac{A_{22}}{M^2}\vec{M}_2\cdot\nabla^2\vec{M}_1\right)$$

$$A = A_{11} + A_{22}, \quad \frac{A}{M^2}\vec{M}\cdot\nabla^2\vec{M} - (\text{extra term})$$

$$\boxed{A = M^2\left(\frac{A_1}{M_1^2} + \frac{A_2}{M_2^2}\right).}$$

Extra term =

$$\frac{A_{11}}{M^2}\vec{M}_2\cdot\nabla^2\vec{M}_1 + \frac{A_{22}}{M^2}\vec{M}_1\cdot\nabla^2\vec{M}_2 + \frac{A_{11}}{M^2}\vec{M}_1\cdot\nabla^2\vec{M}_2 + \frac{A_{22}}{M^2}\vec{M}_2\cdot\nabla^2\vec{M}_1.$$

5.7 Total energy due to dipole–dipole interaction (Figure S5.7):

$$E = \frac{1}{4\pi r^3}\left\{\vec{m}_1\cdot\vec{m}_2 - \frac{3}{r^2}(\vec{m}_1\cdot r)(\vec{m}_2\cdot r)\right\},$$

$$\vec{m}_1\cdot\vec{m}_2 \cong m_1 m_2 \cos\varphi = m_1 m_2 \ for \ \varphi \sim 0,$$

$$\vec{m}_1\cdot r = m_1 r \cos\theta, \text{ and } \vec{m}_2\cdot r = m_2 r \cos\theta,$$

a is the distance between $\vec{m}_1$ and $\vec{m}_2$.

For $m_1, m_2, m_3, \ldots, m_N$ (same magnitude) $\boxed{E = \frac{3m^2}{4\pi a^3}N\left(\frac{1}{3} - \cos^2\theta\right).}$

5.8 The total energy including molecular field at two different sites is given by

$$E = -\vec{M}_1\cdot\vec{H}_{MF}^{(1)} - \vec{M}_2\cdot\vec{H}_{MF}^{(2)} - \left(\vec{M}_1 + \vec{M}_2\right)\cdot\vec{H}.$$

After we include coupling of spin wave,

$$E = -\vec{M}_1\cdot\left(\vec{H}_{MF}^{(1)} + \vec{h}_{sw}^{(1)}\right) - \vec{M}_2\cdot\left(\vec{H}_{MF}^{(2)} + \vec{h}_{sw}^{(2)}\right) - \left(\vec{M}_1 + \vec{M}_2\right)\cdot\vec{H},$$

where

$$\vec{H}_{MF}^{(1)} = \lambda_{11}\vec{M}_1 - \lambda_{12}\vec{M}_2 \text{ for } \lambda_{11}, \lambda_{12} > 0 \text{ and } \vec{h}_{sw}^{(1)} = \frac{A_1}{M_1^2}\nabla^2\vec{M}_1$$

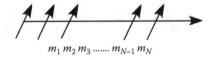

$$m_1\ m_2\ m_3 \ldots\ldots m_{N-1}\ m_N$$

FIGURE S5.7 Spin arrangement in a chain.

$$\vec{H}_{MF}^{(2)} = \lambda_{22}\vec{M}_2 - \lambda_{12}\vec{M}_1 \text{ for } \lambda_{22}, \lambda_{12} > 0 \text{ and } \vec{h}_{sw}^{(2)} = \frac{A_2}{M_2^2}\nabla^2\vec{M}_2.$$

$$E = -\vec{M} \cdot H + \lambda_{12}\vec{M}_1 \cdot \vec{M}_2 - \left(\lambda_{11}M_1^2 + \lambda_{22}M_2^2\right)$$

+spinwave terms, where the spinwave terms

$$= \frac{A_1}{M_1^2}\vec{M}_1 \cdot \nabla^2\vec{M}_1 + \frac{A_2}{M_2^2}\vec{M}_2 \cdot \nabla^2\vec{M}_2$$

5.9 $E_A = K_1\left(\alpha_1^2\alpha_2^2 + \alpha_1^2\alpha_3^2 + \alpha_2^2\alpha_3^2\right) + K_2\left(\alpha_1^2\alpha_2^2\alpha_3^2\right)$ (Figure S5.9a)

$$\alpha_1 = \sin\theta\cos\varphi = \frac{1}{\sqrt{2}}\sin\theta,$$

$$\alpha_2 = \sin\theta\cos\varphi = \frac{1}{\sqrt{2}}\sin\theta,$$

$$\alpha_3 = \cos\theta.$$

1. $K_1 < 0$, 2. $K_2 = 0$ for M

in $\langle 110 \rangle$ plane $\boxed{E_A = -|K_1|\left(\sin^2\theta - \frac{3}{4}\sin^4\theta\right) + \frac{K_2}{4}\left(\sin^4\theta - \sin^6\theta\right)}$

usually $K_1 \gg K_2$ (Figure S9.9b).

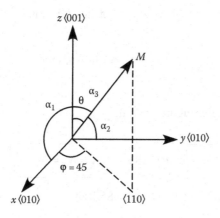

FIGURE S5.9A Geometry of the problem.

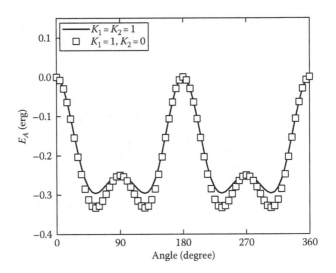

FIGURE S5.9B Angular variation of the free energy.

5.10

$$H_{Dx} = N_x M_x,$$

$$H_{Dy} = N_y M_y,$$

$$H_{Dz} = N_z M_z,$$

$$N_x + N_y + N_z = 4\pi \text{ in CGS unit,}$$

$$N_x + N_y + N_z = 1 \text{ in MKS unit.}$$

1. For sphere (Figure S5.10a)

$$\boxed{N_x = N_y = N_z = \frac{4\pi}{3}} \text{ in CGS and } \boxed{N_x = N_y = N_z = \frac{1}{3}} \text{ in MKS unit.}$$

2. For thin film (slab) (Figure S5.10b)

$$W, L \gg t,$$

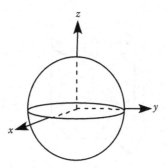

FIGURE S5.10A Example of a spherical sample.

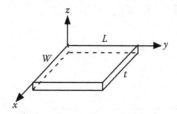

FIGURE S5.10B Example of platelet shaped sample.

$$\frac{N_x}{N_z} = \frac{t}{W}, \frac{N_y}{N_z} = \frac{t}{L},$$

$$\frac{t}{W}N_z + \frac{t}{L}N_z + N_z = 4\pi,$$

$$N_x = \left(\frac{t}{W}\right)\frac{4\pi}{t\left((1/W)+(1/L)\right)+1}, N_y = \left(\frac{t}{L}\right)\frac{4\pi}{t\left((1/W)+(1/L)\right)+1}, \text{ and}$$

$$N_z = \frac{4\pi}{t\left((1/W)+(1/L)\right)+1}$$

In the limit $(t \ll 1)$, $\boxed{N_x \sim N_y \approx 0 \text{ and } N_z \cong 4\pi}$ in CGS unit.
In MKS unit,

$$N_x = \left(\frac{t}{W}\right)\frac{4\pi}{t\left((1/W)+(1/L)\right)+1}, N_y = \left(\frac{t}{L}\right)\frac{4\pi}{t\left((1/W)+(1/L)\right)+1}, \text{ and}$$

$$N_z = \frac{4\pi}{t\left((1/W)+(1/L)\right)+1}.$$

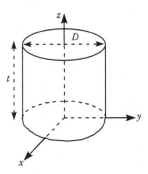

FIGURE S5.10C Geometry of the platelet with dimensions.

in $(t \ll 1)$, $\boxed{N_x \sim N_y \approx 0 \text{ and } N_z \cong 1}$.

3. For cylinder (Figure S5.10c)

Case 1. $\dfrac{D}{t} \gg 1$

$\boxed{N_x = N_y = 0 \text{ and } N_z = 4\pi}$ in CGS unit,

$\boxed{N_x = N_y = 0 \text{ and } N_z = 1}$ in MKS unit.

Case 2. $\dfrac{D}{t} \ll 1$

$\boxed{N_x = N_y = 2\pi \text{ and } N_z = 0}$ in CGS unit,

$\boxed{N_x = N_y = \dfrac{1}{2} \text{ and } N_z = 0}$ in MKS unit.

5.11 $V = +q((1/d_1) + (1/d_2))$ voltage between $+q$ and $-e$ (Figure S5.11)
Far zone approximation

$$V = +\frac{q}{a}\left(\frac{1}{\sqrt{1 + \left((r^2 - 2az)/a^2\right)}} + \frac{1}{\sqrt{1 + \left((r^2 + 2az)/a^2\right)}} \right).$$

After binomial expansion, then

$$V \cong +\frac{2q}{a} + \frac{q}{a^3}\left(3z^3 - r^2\right) + \frac{2q}{a^5}(\cdots) + \ldots$$

$W(\text{potential energy}) = -eV \cong C[3Jz^2 - (J(J+1)) + \text{constant terms}]$

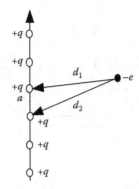

FIGURE S5.11   Ions distributed in a chain.

where $C = f(r, a)$.
Since $L = 0$, $J_z = S_z$ and $J = S$

$$W \cong D \sum_i S_{iz}^2 - \left(S\left(S + \frac{1}{3} \right) \right) + \text{constant terms}.$$

For dipole–dipole interaction energy with moment, we calculate Problem 5.7.

$$E = \frac{3\left(Ng\beta S_z \right)^2}{4\pi a^3} N\left(\frac{1}{3} - \cos^2 \theta \right) = kT.$$

Assuming spin polarization due to the dipole–dipole interaction, the transition temperature is $T \sim 10^{-4}$ K ($E \sim 10^{-20}$ erg).

5.12 If J (exchange integral) is positive, then the spins are parallel (ferromagnetic ordering). Otherwise, spins are antiparallel (antiferromagnetic ordering).

5.13 Total magnetic energy stored in the system (Figure S5.13):

$$W_m = \frac{1}{2} \int_{\text{Volume}} \vec{B} \cdot \vec{H} dx^3,$$

$$\vec{B} = \vec{H} + 4\pi \vec{M} = \mu_0 \ddot{\mu}_r \cdot \vec{H}.$$

Using CGS unit,

$$\vec{B} = \begin{bmatrix} 2 & 0 & 0 \\ 0 & 1 & 0 \\ 0 & 0 & 2 \end{bmatrix} \begin{bmatrix} 0 \\ 0 \\ 4\pi \times 10^3 \end{bmatrix},$$

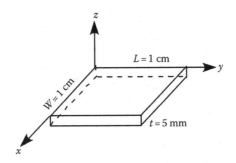

FIGURE S5.13   Platelet geometry.

where $\vec{H} = 10^6 \,(\text{A/m}) = 4\pi \times 10^6 \text{ Oe } \vec{a}_z$

$$\boxed{W_m = \frac{1}{2}\left\{32\pi^2 V\right\} = 8\pi^2 \times 10^3 \text{ (ergs)},}$$

where $V = 5 \times 10^{-4} \text{cm}^3$.

5.14 Total free energy is (see Figure S5.14 as a reference)

$$F\left(\text{erg/cm}^3\right) = -\vec{M}\cdot\vec{H} + \frac{1}{2}\left(N_x M_x^2 + N_y M_y^2 + N_z M_z^2\right) + K_u + E_{me} + E_c,$$

E_{me} (magneto-elastic) and E_c (crystal energy) are neglected. $K_u \sim 0$, then

$$F\left(\text{erg/cm}^3\right) = -\vec{M}\cdot\vec{H} + \frac{1}{2}\left(N_x M_x^2 + N_y M_y^2 + N_z M_z^2\right),$$

$$F\left(\text{erg/cm}^3\right) = -M_z H_z + \pi\left(M_x^2 + M_y^2\right).$$

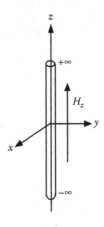

FIGURE S5.14   Needle geometry.

5.15 $MgFe_2O_4$ magnesium spinel ferrite at $T = 0$ K:

$Mg^{2+}Fe_2^{3+}O_4^{2-}$ in spinel structure $8 = A$ site(tetrahedral site),

$16 = B$ site(octahedral site).

Electron configuration for metallic Fe and Fe^{3+} is

$1s^2 2s^2 2p^6 3s^2 3p^6 3d^6 4s^2$ and $1s^2 2s^2 2p^6 3s^2 3p^6 3d^5$.

Electron configuration for Mg and Mg^{2+} is

$1s^2 2s^2 2p^6 3s^2$ and $1s^2 2s^2 2p^6$ (no net moment).

Since $MgFe_2O_4$ is an inverse spinel, the ferrite can be represented as

$$Mg_x{}^{2+}Fe_{1-x}{}^{3+}\left[Mg_{1-x}{}^{2+}Fe_{1+x}{}^{3+}\right]O_4.$$

For $g = 2$, $\beta = 9.271 \times 10^{-21}$, $S = 5/2$, $a = 8.36$ Å, and also $N_{net} = 2x/a^3$

x	$N_{net} = 2x/a^3$ ($\times 10^{20}$)	$4\pi M_{net} = 4\pi N_{net} g\beta\langle S\rangle$ in G
0	0	0
0.1	3.42	~199
0.25	8.56	~499
0.5	17.12	~997
0.75	25.67	~1,495
1	34.23	~1,994

$x = 1$ normal and $x = 0$ inverse spinel.
From the book Smit and Wijn (1959),

$$4\pi M_{net} = 1,800 \text{ G at } T = 0 \text{ K}.$$

Therefore, ~90% Fe^{3+} in B site and 10% in A site.
Also, from the book Smit and Wijn (1959),

$$4\pi M_{net} = 1,500 \text{ G at } T = 20°C.$$

6 Phenomenological Theory

SMIT AND BELJERS FORMULATION

From the previous arguments, the free energy, F, may be expressed in terms of directional cosines of $\vec{M}$ with respect to x, y, and z-axes, which can represent the bonding axes or simply an arbitrary coordinate system. In general, we may write the free energy as

$$F = F(\alpha_1, \alpha_2, \alpha_3), \tag{6.1}$$

where

$$\alpha_1 = \sin\theta\cos\phi,$$
$$\alpha_2 = \sin\theta\sin\phi,$$

and

$$\alpha_2 = \cos\theta.$$

Thus,

$$F = F(\theta, \phi).$$

If the magnetization direction is changing with time, then F must also change with time. Assuming for simplicity that the fluctuations or changes in $\vec{M}$ are small, we can write at any time that

$$\theta = \theta_0 + \Delta\theta$$

and

$$\phi = \phi_0 + \Delta\phi,$$

where θ_0 and ϕ_0 are not time dependent, but $\Delta\theta$ and $\Delta\phi$ may be. Thus, θ_0 and ϕ_0 designate the equilibrium angular positions of the magnetization relative to a chosen coordinate system.

$$M_x = M_0 \sin\theta_0 \cos\phi_0,$$
$$M_y = M_0 \sin\theta_0 \sin\phi_0,$$
$$M_z = M_0 \cos\theta_0,$$

DOI: 10.1201/9781003431244-6

and

$$M_0^2 = M_x^2 + M_y^2 + M_z^2.$$

In this book, uppercase $\vec{M}$ implies the static magnetization and lowercase $\vec{m}$ the dynamic component of the magnetization, which may be time dependent. We can determine approximately the time-dependent free energy terms from the knowledge of time-independent free energy using the Taylor series expansion

$$F = F_0 + \frac{\partial F}{\partial \theta} \Delta\theta + \frac{\partial F}{\partial \phi} \Delta\phi + \frac{1}{2} \frac{\partial^2 F}{\partial \theta^2} (\Delta\theta)^2 + \frac{1}{2} \frac{\partial^2 F}{\partial \phi^2} (\Delta\phi)^2 + \frac{\partial^2 F}{\partial \theta \partial \phi} \Delta\theta\Delta\phi + \cdots.$$

All of the derivatives are evaluated at equilibrium θ_0 and ϕ_0. We assume that for $\theta = \theta_0$ and $\phi = \phi_0$, at equilibrium, the free energy is in steady state or has reached a minimum or maximum value or

$$\partial F / \partial \theta = \partial F / \partial \phi = 0. \tag{6.2}$$

Thus, we may mathematically express the free energy as

$$F = F_0 + \frac{1}{2} \frac{\partial^2 F}{\partial \theta^2} (\Delta\theta)^2 + \frac{1}{2} \frac{\partial^2 F}{\partial \phi^2} (\Delta\phi)^2 + \frac{\partial^2 F}{\partial \theta \partial \phi} \Delta\theta\Delta\phi + \cdots.$$

By keeping terms up to second order, we have linearized the free energy expression. Of course, this approximation is appropriate for small-field excitations relative to internal static fields. Let us now translate this into real-time dependence of $\vec{M}$. In spherical coordinates, the equation of motion becomes

$$\frac{1}{\gamma} \frac{dm_\phi}{dt} = M_0 h_\theta,$$

$$\frac{1}{\gamma} \frac{dm_\phi}{dt} = -M_0 h_\phi, \tag{6.3}$$

where

$$h_\theta = -\frac{1}{M_0} \frac{\partial F}{\partial(\Delta\theta)} = -\frac{1}{M_0} \left(F_{\phi\theta} \Delta\phi + F_{\theta\theta} \Delta\theta \right)$$

and

$$h_\phi = -\frac{1}{M_0 \sin\theta_0} \frac{\partial F}{\partial(\Delta\phi)} = -\frac{1}{M_0} \left(F_{\phi\phi} \Delta\phi + F_{\phi\theta} \Delta\theta \right),$$

where m_θ and m_ϕ are the two components of the magnetization transverse to M_0. This means that

$$m_\theta = M_0 \Delta\theta$$

and

$$m_\phi = M_0 \sin\theta_0 \Delta\phi.$$

We assume $e^{j\omega t}$ solution, and combining Equations 6.2 and 6.3, we obtain

$$0 = F_{\theta\theta}\Delta\theta + \left(F_{\phi\theta} + \frac{M_0 \sin\theta_0\, j\omega}{\gamma} \right)\Delta\phi$$

and

$$0 = F_{\phi\phi}\Delta\phi + \left(F_{\phi\theta} + \frac{M_0 \sin\theta_0\, j\omega}{\gamma} \right)\Delta\theta.$$

Nontrivial solutions for $\Delta\theta$ and $\Delta\phi$ exist, if

$$\left(\frac{\omega}{\gamma} \right)^2 = \left[F_{\theta\theta}F_{\phi\phi} - \left(F_{\theta\phi}\right)^2 \right] \frac{1}{M_0^2 \sin^2\theta_0}. \tag{6.4}$$

This is the classical condition for ferromagnetic resonance (FMR) and it was derived by Smit and Beljers in 1954.

EXAMPLES OF FERROMAGNETIC RESONANCE

Example 6.1: Semi-Infinite Magnetic Medium

For this case, the free energy is simply (see Figure 6.1)

$$F = -\vec{M} \cdot \vec{H} \sin\theta \cos\phi.$$

In this example, the external applied field, H, is arbitrarily chosen to be along the x-axis. The magnetization direction is uniquely defined by specifying both θ and ϕ, as shown in Figure 6.1.

Implicit in the derivation of Equation 6.4 is that the free energy emulates the energy of a harmonic oscillator for small magnetic perturbations of the system. This means that one needs to establish that there is indeed a minimum of the free energy before applying Equation 6.4. Hence, the procedure in applying Equation 6.4 is as follows:

a. Write appropriate free energy in spherical coordinates (in terms of θ and ϕ).
b. Establish equilibrium conditions so that the position of the static magnetization in real space may be known. This is accomplished by taking the

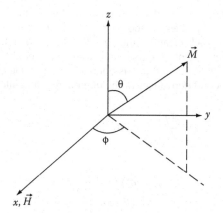

FIGURE 6.1   Semi-infinite medium with H along the x-axis.

derivatives of the free energy with respect to θ and ϕ and setting them equal to zero. A set of trigonometric transcendental equations result for which equilibrium angles θ_0 and ϕ_0 may be solved.

c. Finally, apply Equation 6.4.

For part (a), the free energy consists only of the magnetizing energy. For part (b), we need to determine θ_0 and ϕ_0. Thus, applying the following equilibrium conditions, we obtain

$$\frac{\partial F}{\partial \theta} = -MH \cos\theta \cos\phi = 0,$$

which gives $\theta_0 = \pi/2$. From

$$\frac{\partial F}{\partial \phi} = MH \sin\theta \sin\phi = 0,$$

we obtain $\phi_0 = 0$ (Figure 6.2).

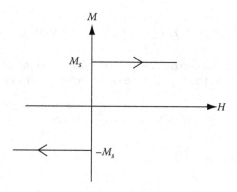

FIGURE 6.2 M versus H for semi-infinite medium.

A point of clarification is needed at this juncture. We use the symbols M, M_S, and M_0 to represent the static magnetization, which is also the saturation magnetization in other books. It is equally clear that the static magnetization is temperature dependent but not time dependent; see Chapter 5 where we used the same symbols.

Finally, for part (c), FMR condition is obtained by applying Equation 6.4. Thus, let's evaluate the second derivatives of the free energy at equilibrium. Thus,

$$F_{\theta\theta} = \left.\frac{\partial^2 F}{\partial \theta^2}\right|_{\theta_0\phi_0} = MH \sin\theta_0 \cos\phi_0 = MH,$$

$$F_{\phi\phi} = \left.\frac{\partial^2 F}{\partial \theta^2}\right|_{\theta_0\phi_0} = MH,$$

and

$$F_{\theta\phi} = \left.\frac{\partial^2 F}{\partial\theta\,\partial\phi}\right|_{\theta_0\phi_0} = 0.$$

Henceforth, we omit the θ_0 and ϕ_0 designation in the derivations because it is understood. From Equation 6.4, we obtain

$$\frac{\omega^2}{\gamma^2} = H^2$$

or

$$\frac{\omega}{\gamma} = H.$$

Example 6.2: Spherical Magnetic Sample

Since the sample is of finite size, demagnetizing energy terms must be introduced in the free energy. Again, we apply the field, H, in the x-direction and write F as

$$F = -\vec{M}\cdot\vec{H} + \frac{1}{2}\left(N_X M_X^2 + N_Y M_Y^2 + N_Z M_Z^2\right).$$

The first term is the magnetizing or Zeeman energy and the second term is the demagnetizing energy. For a sphere, the demagnetizing factors are

$$N_X = N_Y = N_Z = \frac{4\pi}{3}, \quad \text{CGS}$$

Expressing F in terms of θ and ϕ

$$F = -MH \sin\theta\cos\phi + \frac{1}{2}\left(\frac{4\pi}{3}\right)M^2\left(\sin^2\theta\cos^2\phi + \sin^2\theta\sin^2\phi + \cos^2\theta\right)$$

or

$$F = -MH \sin\theta\cos\phi + \frac{2\pi}{3}M^2.$$

Thus, the free energy is of the same form as that of the semi-infinite medium (see Example 6.1) with an additional constant term of $(2\pi/3)M^2$. The constant term does not affect the resonance, since angular derivatives are involved in its derivation. As such, the resonance condition may be written as

$$\frac{\omega}{\gamma} = H, \quad H \geq H_D.$$

However, magnetic resonance may be excited only if the external magnetic field, H, is greater than the demagnetizing field $H_D = 4\pi M/3$. Demagnetizing field diagram is shown in Figure 6.3.

For $H < H_D$, the spherical sample decomposes into many magnetic domains, depending on the size of the sample. As such, there is no uniform precessional magnetic resonance.

Example 6.3: Needle-Shaped Magnetic Sample

For a needle-shaped sample, the demagnetizing field along the axis of the needle may be assumed to be small so that

$$N_X \cong 0,$$

and

$$N_Y = N_Z.$$

Since $N_x + N_y + N_z = 4\pi$, $N_y = N_z = 2\pi$. Hence, the free energy expression becomes

$$F = -MH \sin\theta\cos\phi + \pi M^2 \left(\sin^2\theta\sin^2\phi + \cos^2\theta\right).$$

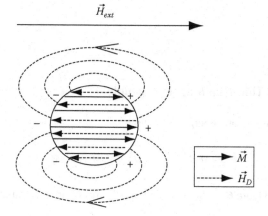

FIGURE 6.3 Demagnetizing field in a sphere.

The equilibrium conditions now become

$$\frac{\partial R}{\partial \theta} = 0 = -MH\cos\theta\cos\phi - \pi M^2\left(\cos^2\phi\sin 2\theta\right)$$

and

$$\frac{\partial F}{\partial \phi} = 0 = MH\sin\theta\sin\phi + 2\pi M^2\sin^2\theta\sin 2\phi,$$

yielding

$$\phi_0 = 0,$$

$$\theta_0 = \frac{\pi}{2}.$$

This means that in a single-domain excitation, the magnetization is constant with H. Also, M is along the axis of the needle.

For now, we have omitted any hysteresis effects. The magnetic resonance condition may now be evaluated assuming $\phi_0 = 0$ and $\theta_0 = \pi/2$ yielding

$$F_{\theta\theta} = F_{\phi\phi} = M(H + 2\pi M)$$

and

$$F_{\theta\phi} = 0.$$

Thus,

$$\frac{\omega^2}{\gamma^2} = \frac{1}{M^2}\left[M^2(H + 2\pi M)^2\right]$$

or

$$\frac{\omega}{\gamma} = H + 2\pi M.$$

Example 6.4: Thin-Film Magnetic Sample

Assuming no anisotropy energies, we write

$$F = -\vec{M}\cdot\vec{H} + \frac{1}{2}\left(N_X M_X^2 + N_Y M_Y^2 + N_Z M_Z^2\right).$$

For thin films (Figure 6.4), the following approximation is reasonable:

$$N_Y \cong 4\pi \quad \text{and} \quad N_X \cong N_Z \cong 0.$$

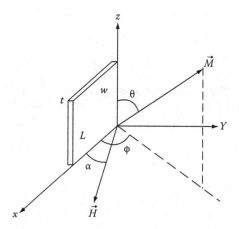

FIGURE 6.4   Thin-film case.

Estimates of N_x, N_y, and N_z may be found in Chapter 5. Under the above approximations,

$$F = -MH \sin\theta \cos(\phi - \alpha) + 2\pi M_Y^2.$$

Writing F in terms of θ and ϕ, we obtain

$$F = -MH \sin\theta \cos(\phi - \alpha) + 2\pi M^2 \sin^2\theta \sin^2\phi.$$

θ_0 may be solved from

$$\frac{\partial F}{\partial \theta} = -MH \cos\phi \cos(\phi - \alpha) + 4\pi M^2 \sin\theta \cos\theta \sin^2\phi = 0.$$

This equilibrium condition gives $\theta_0 = \pi/2$, which implies that $\vec{M}$ lies in the x–y plane. ϕ_0 may be solved from

$$\frac{\partial F}{\partial \phi} = MH \sin\theta \sin(\phi_0 - \alpha) + 4\pi M^2 \sin^2\theta_0 \sin\phi_0 \cos\phi_0 = 0.$$

which gives

$$\sin(\phi_0 - \alpha) = \frac{-4\pi M}{H} \sin\phi_0 \cos\phi_0.$$

This is a transcendental equation, which can be solved numerically by the computer once α is specified. For simplicity, let's choose α to be either 0 (H in the film plane) or $\pi/2$ (H normal to the film plane).

For $\alpha = 0$, we have

$$\sin\phi_0 \left(1 + \frac{4\pi M}{H} \cos\phi_0 \right) = 0.$$

There are two possible solutions for ϕ_0 and they are

$$\cos\phi_0 = -\frac{H}{4\pi M}.$$

This is an unphysical solution, since M is not aligned along H. The other solution is $\phi_0 = 0$, which is indeed a proper physical solution, since M and H are parallel to each other at equilibrium.

For $\alpha = \pi/2$, we have

$$\left(1 - \frac{4\pi M}{H}\sin\phi_0\right)\cos\phi_0 = 0,$$

which gives

$$\sin\phi_0 = \frac{H}{4\pi M}$$

or

$$\phi_0 = \sin^{-1}\left(\frac{H}{4\pi M}\right).$$

Typical solutions of the above equation are illustrated in Figure 6.5.

Let the component of M along H equal to M_H. Then

$$M_H = M\sin\phi_0 = \frac{H}{4\pi}.$$

Thus, the DC susceptibility M_H/H for $H \leq 4\pi M$ is equal to $1/4\pi$ (see Figure 6.6).

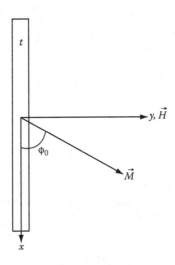

FIGURE 6.5 H is less than the saturation magnetization.

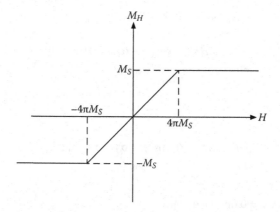

FIGURE 6.6   Magnetization versus H applied normal to the film plane.

For $H \geq 4\pi M$ (Figure 6.7), we have the other equilibrium condition that

$$\cos\phi_0 = 0,$$

which gives

$$\phi_0 = \frac{\pi}{2}.$$

Again, this is a physical solution for saturating magnetic fields since M and H align perpendicular to the film plane. Application of the FMR condition requires evaluating

$$\frac{\partial^2 F}{\partial\theta\,\partial\phi} = 0,$$

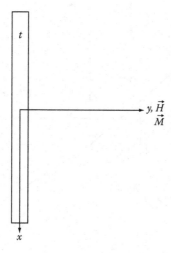

FIGURE 6.7 $H \geq 4\pi M$ saturating fields.

as well as the following second derivatives:

$$\frac{\partial^2 F}{\partial \theta^2} = \frac{MH \sin\theta_0 \cos(\phi_0 - \alpha) + 4\pi M^2 \cos 2\theta_0}{\times \sin^2 \phi_0}$$

and

$$\frac{\partial^2 F}{\partial \phi^2} = \frac{MH \sin\theta_0 \cos(\phi_0 - \alpha) + 4\pi M^2 \sin^2 \theta_0}{\times \cos 2\phi_0}.$$

We see that for $\alpha = 0$ (in-plane FMR),

$$\frac{\omega^2}{\gamma^2} = H(H + 4\pi M)$$

and that for $\alpha = \pi/2$ (perpendicular FMR) and $H \geq 4\pi M$, we have

$$\frac{\omega}{\gamma} = H - 4\pi M.$$

Example 6.5: Inclusion of Uniaxial Magnetic Anisotropy Energy in Films

As shown in Figure 6.8, the uniaxial axis is along the x-axis and H is applied at an angle α from the uniaxial axis. Both H and the uniaxial axis are in the film plane. The free energy in this case is then

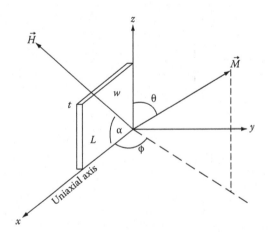

FIGURE 6.8 Thin-film case with uniaxial magnetic anisotropy.

$$F = -MH\left(\sin\theta\cos\phi\cos\alpha + \cos\theta + \sin\alpha\right) + 2\pi M^2 \sin^2\theta\sin^2\phi$$
$$+K_U\left(\sin\theta\cos\phi\right)^2.$$

(6.5)

As in the previous examples, the two equilibrium conditions are obtained as follows:

$$\frac{\partial F}{\partial \phi} = 0,$$

yielding

$$\phi_0 = 0.$$

Also, from the other equilibrium condition, we have

$$\frac{\partial F}{\partial \theta} = -MH\cos(\theta + \alpha) + 2K_U\sin\theta_0\cos\phi_0 = 0.$$

For simplicity, let's assume $\alpha = \pi/2$ and $K_u < 0$. This assumes that the uniaxial axis is the easy axis of magnetization. There are two solutions for θ_0. One solution of θ_0 gives

$$\cos\theta_0 = \frac{H}{2|K_U|/M} = \frac{H}{H_A},$$

where $H_A = 2\,|K_u|/M$. Meaningful solution of θ_0 may be obtained for $H \le H_A$ from the above equation. The other solution is

$$\sin\theta_0 = 0, \quad \theta_0 = 0.$$

The latter solution is applicable for $H \ge H_A$.

We have plotted $M_H = M\cos\theta_0$ as a function of H, where M_H is the component of magnetization along H. The applied field H is perpendicular to the uniaxial axis. Clearly, the slope M_H/H is M_S/H_A, which is the DC susceptibility. The DC permeability is simply

$$\mu_{DC} = 1 + \frac{4\pi M_S}{H_A}, \quad H \le H_A.$$

For soft magnetic materials in which $H_A \ll 4\pi M_S$, $\mu_{DC} \to 4\pi M_S/H_A$.

In real magnetic materials, hysteresis loops are traced out as H is swept for positive and negative values of field. However, the slope from the origin to the onset of magnetic saturation is proportional to M_H/H_A, as shown in Figure 6.9.

By evaluating the second derivatives at θ_0 and ϕ_0 as defined in Equation 6.4, we obtain the FMR condition

$$\left(\frac{\omega}{\gamma}\right)^2 = \left(H\cos\theta_0 - H_A\cos2\theta_0\right)\left(4\pi M_S + H_A\right).$$

(6.6)

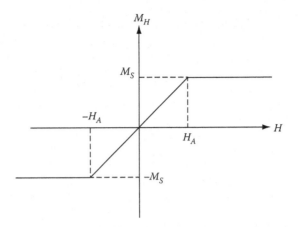

FIGURE 6.9   Magnetization versus H applied normal to the uniaxial axis.

Substituting $\cos \theta_0$ and $\cos 2\theta_0$ into the above equation, we obtain

$$\left(\frac{\omega}{\gamma}\right)^2 = \frac{\left(H_A^2 - H^2\right)\left(4\pi M_S + H_A\right)}{H_A}, \quad H \le H_A. \tag{6.7}$$

For $\theta_0 = 0$ and $H \ge H_A$, we have a singularity, since

$$\left(\frac{\omega}{\gamma}\right)^2 = \frac{\left(H - H_A\right)\left(4\pi M_S + H_A\right)}{\sin^2 \theta_0}, \quad \theta_0 \to 0. \tag{6.8}$$

This is a mathematical singularity, not a physical one. One way of avoiding these types of singularities is to "place" the film in the x–y plane rather than in the x–z plane and apply H along the y-axis, for example. The uniaxial axis is still along the x-axis. The nature or the physics of the problem has not changed, but the mathematical singularity has been removed. The free energy is then simply (see Figure 6.10)

$$F = -MH \sin \theta \cos(\phi - \alpha) + 2\pi M^2 \cos^2 \theta + K_U \left(\sin^2 \theta \cos^2 \phi\right), \quad K_U < 0. \tag{6.9}$$

For $\alpha = \pi/2$ and $H > H_A = 2 \left|K_u\right|/M$, we have from the equilibrium conditions that

$$\theta_0 = \phi_0 = \frac{\pi}{2}.$$

The FMR condition may be obtained as before and the result is

$$\left(\frac{\omega}{\gamma}\right)^2 = \left(H - H_A\right)\left(4\pi M_S + H_A\right), \quad H \ge H_A. \tag{6.8'}$$

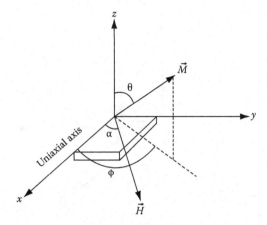

FIGURE 6.10   Thin-film case avoiding mathematical singularity.

For $H \leq H_A$, the equilibrium conditions become

$$\frac{\partial F}{\partial \theta} = 0 = -MH\cos\theta\cos(\phi - \alpha) + 4\pi M^2 \sin\theta\cos\theta + 2K_U \sin\theta\cos\theta\cos^2\phi.$$

This yields the condition that

$$\cos\theta_0 = 0$$

or

$$\theta_0 = \pi/2.$$

This means that $\vec{M}$ lies in the film plane only. Let's now determine where in the plane $\vec{M}$ is oriented. The in-plane equilibrium condition may be obtained from

$$\frac{\partial F}{\partial \phi} = 0 = MH\sin(\phi - \alpha) - 2K_U \sin\phi\cos\phi. \qquad (6.10)$$

For $\alpha = \pi/2$ ($\vec{H}$ applied perpendicular to the uniaxial axis) (Figure 6.11)

$$H\cos\phi_0 = H_A \sin\phi_0 \cos\phi_0.$$

There are two solutions and they are

$$\phi_0 = \frac{\pi}{2}, \quad H \geq H_A$$

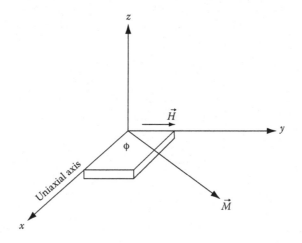

FIGURE 6.11   Static field geometry.

and

$$\sin\phi_0 = \frac{H}{H_A}, \quad H \le H_A,$$

where

$$H_A = \frac{2|K_U|}{M}, \quad K_U < 0.$$

Returning to the case of $H \le H_A$ (non-saturation case), let's define the magnetization measured along H as

$$M_H$$

or

$$M_H = M_S \sin\phi_0 = M_S \frac{H}{H_A}.$$

Thus, M_H is linearly proportional to H and the slope being M_S/H_A, as shown in Figure 6.9.

SIMPLE MODEL FOR HYSTERESIS

As it is well known, there is curvature around saturation fields, as shown in Figure 6.12 for a typical experimental measurement of the magnetization versus magnetic field, H.

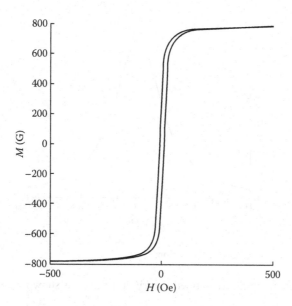

FIGURE 6.12 Typical hysteresis curve.

Let's model the hysteresis loop by calculating the average of $\sin \phi_0$ in a polycrystalline sample, where H_A can vary from point to point in a film. The field geometry is illustrated in Figure 6.11. Again, we are assuming the case of H in the plane and perpendicular to the uniaxial axis in the non-saturation limit (see discussions above). Applying simple mathematical averaging arguments, we have that

$$\sin \phi_0 = H \frac{1}{H_A} = H \int P\left(\frac{1}{H_A}\right) d\left(\frac{1}{H_A}\right), \tag{6.11}$$

where $P(1/H_A) \equiv$ probability distribution function. Redefining variables of integration, we have that

$$\sin \phi_0 = \int_0^{H/H_A} P\left(\frac{H}{H_A}\right) d\left(\frac{H}{H_A}\right).$$

There are many forms for P, but we simply assume a form that lends itself to simple integration. Thus, let $P(H/H_A) = N/(1+(H/H_A)^2) \equiv$ Cauchy probability distribution, where $N \equiv$ normalizing factor. In Figure 6.13, we plot a typical Cauchy probability distribution. The reader should not invoke anything unique about this distribution as related to thin films.

Let's determine N:

$$\int_{-\infty}^{\infty} P\left(\frac{H}{H_A}\right) d\left(\frac{H}{H_A}\right) = 1 = N \int_{-\infty}^{\infty} \frac{d(H/H_A)}{1+(H/H_A)^2} = N \frac{\pi}{2}.$$

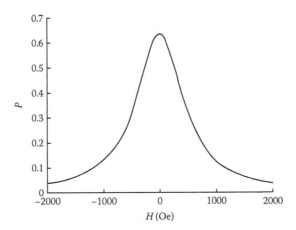

FIGURE 6.13 Cauchy probability function.

In evaluating the integral, we applied theory of residues as shown below:

$$N \oint \frac{dx}{(x+j)(x-j)} = 2\pi j \sum \text{residues} \rightarrow 2\pi j N \left(\frac{1}{2j} \right) = \pi N,$$

$$N \int_{-\infty}^{\infty} \frac{dx}{1+x^2} = N\pi,$$

or

$$N \int_{0}^{\infty} \frac{dx}{1+x^2} = N \frac{\pi}{2} = 1.$$

Thus,

$$N = \frac{2}{\pi}.$$

Finally, we obtain

$$\sin \phi_0 = \frac{2}{\pi} \int_{0}^{H/H_A} \frac{d(H/H_A)}{1+(H/H_A)^2} = \frac{2}{\pi} \tan^{-1} \left(\frac{H}{H_A} \right). \tag{6.12}$$

Define

$$M_H = M_S \sin \phi_0 = \frac{2M_S}{\pi} \tan^{-1} \left(\frac{H}{H_A} \right).$$

In Figure 6.14, M_H is plotted as a function of H for parameters assumed in Figure 6.13, for example.

The flux density is related to the average measured magnetization as follows:

$$B_H = H + 4\pi M_H$$

or

$$B_H = H + 4\pi M_S \sin \phi_0. \tag{6.13}$$

At this point, there is no hysteresis in M or B unless we introduce the concept of the coercive field, H_C. If one expects H_C to play the same role as H, then one may include H_C in the expression of the measured M as follows:

$$M_H = \frac{2M_S}{\pi} \tan^{-1}\left(\frac{H \pm H_C}{H_A}\right).$$

In Figure 6.15, we plot M_H as a function of H assuming $H_C = 500$ Oe, $M_S = 800$ G, and $H_A = 500$ Oe.

H_C is usually due to voids, defects, and internal imperfections. For single crystals, it is very small $\ll 0.1$ Oe. We have arbitrarily chosen the values of the coercive field and saturation magnetization in this example. The FMR condition may now be calculated for $H \le H_A$.

$$\frac{\partial^2 F}{\partial \theta^2} = MH \sin \theta_0 \sin \phi_0 - 4\pi M^2 \cos 2\theta_0 - 2K_u \cos 2\theta_0 \cos^2 \phi_0,$$

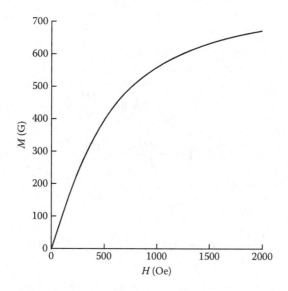

FIGURE 6.14 Effect of averaging the saturation magnetization field.

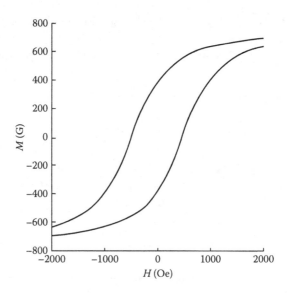

FIGURE 6.15   Calculated hysteresis loop based on Cauchy probability function.

where $\theta_0 = \pi/2$ and $\sin \phi_0 = H/H_A$. Thus,

$$F_{\theta\theta} = MH\left(\frac{H}{H_A}\right) + 4\pi M^2 + 2K_u\left(1 - \left(\frac{H}{H_A}\right)^2\right),$$

$$F_{\theta\theta} = M[4\pi M + H_A].$$

Also,

$$\frac{\partial^2 F}{\partial \phi^2} = MH\sin\phi_0 + 2|K_u|\cos 2\phi_0$$

$$= MH\frac{H}{H_A} + 2|K_u|\left(1 - \frac{2H^2}{H_A^2}\right),$$

$$F_{\phi\phi} = M\left(H_A - \frac{H^2}{H_A}\right) = M\frac{H_A^2 - H^2}{H_A}.$$

Finally,

$$\left(\frac{\omega}{\gamma}\right)^2 = (H_A + 4\pi M)\left(\frac{H_A^2 - H^2}{H_A}\right), \quad H \le H_A.$$

However, in the resonance expression shown above, one has to be careful in taking account of hysteretic behavior of M, since M is varying with H in a nonlinear way.

For $H \geq H_A$, $\theta_0 = \pi/2$ and $\phi_0 = \pi/2$.

$$\frac{\partial^2 F}{\partial \theta^2} = MH + 4\pi M^2 = M(H + 4\pi M)$$

and

$$\frac{\partial^2 F}{\partial \phi^2} = MH - 2|K_u| = M(H - H_A).$$

Thus,

$$\left(\frac{\omega}{\gamma}\right)^2 = (H - H_A)(H + 4\pi M), \qquad H \geq H_A.$$

On may assume M to be the saturation magnetization. Consider now the case where $\alpha = 0$ (parallel to the uniaxial axis). For $\alpha = 0$, $\theta_0 = \pi/2$ and $\phi_0 = 0$. This means that $\vec{M}$ and $\vec{H}$ are parallel to each other for all values of H. Nevertheless, one needs to apply a field H to be much greater than the coercive field in order to assure M to be parallel to H.

The FMR condition for $\alpha = 0$ is then

$$\frac{\partial^2 F}{\partial \theta^2} = MH \sin\theta_0 \cos\phi_0 - 4\pi M^2 \cos 2\theta - 2|K_u| \cos 2\theta_0 \cos^2 \phi_0,$$

$$F_{\theta\theta} = M(H + 4\pi M + H_A).$$

$$\frac{\partial^2 F}{\partial \phi^2} = MH + 2|K_u| \sin^2 \theta_0 \cos 2\phi_0,$$

$$F_{\phi\phi} = M(H + H_A).$$

Thus, the FMR condition is then

$$\frac{\omega}{\gamma} = \sqrt{(H + H_A)(H + H_A + 4\pi M)}.$$

Again, the reader should pay attention to the hysteretic behavior of M.

In Figure 6.16, we plot a typical curve of ω/γ versus H for $\alpha = 0$ and $\pi/2$.

GENERAL FORMULATION

If time-dependent perturbations to magnetic materials are not small, then the harmonic oscillator approximation to the free energy is not a good one. This is due to the fact that magnetization is finite in magnitude and not infinitely large. Equivalently, the depth of the energy "well" of the free energy is shallow in real magnetic materials.

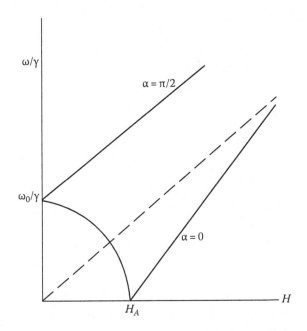

FIGURE 6.16 ω/γ versus H.

Hence, an alternative formulation is derived here in order to consider nonlinear perturbations of magnetic materials. However, even in the linear regime (small perturbations), the Smit and Beljers method is not a convenient one to calculate the permeability of magnetic materials, for example.

Let's now develop the formalism that allows for nonlinear effects in ferrite materials and allows routinely for the calculations of the microwave permeability of anisotropic materials. This method was introduced in the sixties by this author. In order to illustrate the method, let's consider the same examples as in the previous sections, calculate the same quantities as before, and then extend the calculations to permeability calculations. First, we express F in terms of M_x, M_y, and M_z instead of θ and ϕ. As such, we write for the cases studied in the previous section the following free energy expression:

$$F = -M_x H_x - M_z H_z + 2\pi M_y^2 + K_u \left(\frac{M_x}{M}\right)^2, \quad K_u < 0. \tag{6.14}$$

Clearly, the film occupies the x–z plane and uniaxial axis is along the x-direction, as shown in Figure 6.8. Second, the internal fields are separated into time-dependent and time-independent variables in the equation of motion:

$$\frac{1}{\gamma}\frac{d\vec{M}}{dt} = \vec{M} \times \vec{H} = \left(\vec{M}_0 + \vec{m}\right) \times \left(\vec{H}_0 + \vec{h}\right). \tag{6.15}$$

For simplicity, the magnetization and fields denoted by subscript "0" denote time-independent quantities. The other quantities are time dependent. The dynamic magnetic field $\vec{h}$ is further redefined as

$$\vec{h} = \vec{h}_a + \vec{h}_F,$$

where

$\vec{h}_a$ = external microwave field (time dependent)
$\vec{h}_F$ = internal intrinsic dynamic field (time dependent)
$\vec{m}$ = microwave magnetic field (time dependent)
$\vec{M}_0$ = static magnetization (time independent) or saturation magnetization
$\vec{H}_0$ = static internal magnetic field (time independent)

The equation of motion is a nonlinear differential equation since it involves products of m and h. The free energy and the equation of motion are sufficient to calculate the magnetic response of a magnetic material to an external perturbation of arbitrary strength.

CONNECTION BETWEEN FREE ENERGY AND INTERNAL FIELDS

Van Vleck postulated in 1937 that it may be possible to calculate an internal field in a paramagnet provided that the spin Hamiltonian of the system is known by simply taking the gradient of the spin Hamiltonian. The gradient was defined in terms of spin operators and not the usual spatial coordinates. No one paid attention to this until 1967. The idea was adopted by us to an ordered magnetic material in which the gradient was defined in terms of magnetization components rather than spin operators. Specifically, the connection in 1967 was made as follows:

Given a free energy, F, the total internal field H_i is simply

$$\vec{H}_i = -\vec{\nabla}_M F, \tag{6.16}$$

where

$$\vec{\nabla}_M = \frac{\partial}{\partial M_x}\vec{a}_x \frac{\partial}{\partial M_y}\vec{a}_y + \frac{\partial}{\partial M_z}\vec{a}_z.$$

The internal field may be separated into two terms which are time dependent, $\vec{h}_F$, and time independent, $\vec{H}_0$:

$$\vec{H}_i = \vec{H}_0 + \vec{h}_F.$$

Substituting F as defined in Equation 6.14 into Equation 6.16 yields the following:

$$\vec{H}_i = \vec{a}_x\left[H_x - 2\left(\frac{K_u}{M}\right)\left(\frac{M_x}{M}\right)\right] - \vec{a}_y 4\pi M_y + \vec{a}_z H_z, \tag{6.17}$$

where

$$M_\alpha = M_{0\alpha} + m_\alpha \qquad \alpha = x, y, z$$

and

$$M_0^2 = M_{0x}^2 + M_{0y}^2 + M_{0z}^2.$$

The subscript "0" is to designate magnetization components which are not time dependent. For most cases of interest, M_0 is simply the saturation magnetization. As in the previous section, let's assume that the external applied field, H, is along the z-direction (that coincides with $\alpha = \pi/2$). As such, the internal field becomes

$$\vec{H}_i = \vec{a}_x \left[\left(\frac{H_A}{M_0} \right) M_{0x} + \left(\frac{H_A}{M_0} \right) m_x \right] + \vec{a}_y \left[-4\pi M_{0y} - 4\pi m_y \right] + \vec{a}_z H, \qquad (6.18)$$

where
$$H_A = 2|K_u|/M$$
$$K_u < 0$$

Clearly, $\vec{H}_i$ contains terms which are time dependent as well as terms not time dependent. Let's identify terms which are not time dependent under one common term $\vec{H}_0$ and the time-dependent ones as $\vec{h}_F$. Thus,

$$\vec{H}_0 = \vec{a}_x \left(\frac{H_A}{M_0} \right) M_{0x} - \vec{a}_y 4\pi M_{0y} + \vec{a}_z H \qquad (6.19)$$

and

$$\vec{h}_F = \vec{a}_x \left(\frac{H_A}{M_0} \right) m_x - \vec{a}_y 4\pi m_y. \qquad (6.20)$$

It is unconventional to define H_A as $2|K_u|/M_0$. The subscript "0" is understood to denote the saturation magnetization, M_S. Henceforth, we drop either subscript "0" or "S" in the definition of H_A. The conventional definition is simply $2|K_u|/M$, no subscript on M.

STATIC FIELD EQUATIONS

$\vec{H}_0$ is the internal static field and it is assumed to be parallel to the internal static magnetization, $\vec{M}_0$. The definition of $\vec{H}_0$ allows us to obtain the equilibrium condition of magnetized bodies. In this example, parallelism means that

$$\frac{M_{0y}}{M_{0x}} = \frac{H_{0y}}{H_{0x}} \qquad (6.21)$$

and

$$\frac{M_{0x}}{M_{0z}} = \frac{H_{0x}}{H_{0z}}.$$ (6.22)

Let's consider the ramification of these two static equations. By definition, we have that

$$M_{0x} = M_0 \sin\theta_0 \cos\phi_0,$$

$$M_{0y} = M_0 \sin\theta_0 \sin\phi_0,$$

and

$$M_{0z} = M_0 \cos\theta_0.$$

From Equation 6.21, we obtain

$$\tan\phi_0 \left(1 + \frac{4\pi M_0}{H_A}\right) = 0,$$

which yields

$$\phi_0 = 0.$$

From Equation 6.22, we obtain

$$\cos\theta_0 = \frac{H}{H_A}.$$ (6.22a)

An alternative point of view is to say that the internal field along the magnetization direction is H_A and the projection of this field along the external field direction is given in the above expression.

Thus, the solutions for θ_0 and ϕ_0 are identical to the solutions obtained from $\partial F/\partial\phi$ and $\partial F/\partial\theta = 0$, as outlined before. We are now in a position to consider the dynamic equations of motion in which both time dependent and static fields are included in the equations of motion.

DYNAMIC EQUATIONS OF MOTION

For simplicity, let's assume that $H \geq H_A$ (saturation fields) and $\alpha = \pi/2$ so that

$$\theta_0 = 0.$$

Furthermore, this implies that

$$M_{0z} = M_0$$

and

$$M_{0x} = M_{0y} = 0.$$

The equation of motion yields the following (see Figure 6.17 and assuming $\alpha = \pi/2$):

$$\frac{1}{\gamma}\frac{dm_x}{dt} = \left(m_y H - M h_y\right) + \left[m_y h_z - m_z h_y\right], \tag{6.23}$$

$$\frac{1}{\gamma}\frac{dm_y}{dt} = \left(-m_x H + M h_x\right) + \left[-m_x h_z + m_z h_x\right], \tag{6.24}$$

and

$$\frac{1}{\gamma}\frac{dm_z}{dt} = \left[m_x h_y - m_y h_x\right]. \tag{6.25}$$

Terms inside the square bracket are nonlinear terms, since it involves the product of two variables which are time dependent. Thus, this formulation contains the basis for nonlinear analysis. However, we will not do any nonlinear analysis here. Let's now define further h_x and h_y and substitute into the above linearized equations. The above equations may be linearized by simply dropping terms contained in the square bracket. From Equation 6.20, we have

$$h_x = \left(\frac{H_A}{M}\right) m_x \tag{6.26}$$

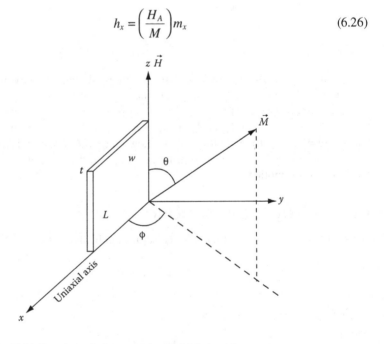

FIGURE 6.17   Field direction relative to magnetic field geometry.

and

$$h_x = -4\pi m_y. \tag{6.27}$$

We have dropped the subscript F on h in Equation 6.20, since the applied microwave field is omitted in this derivation. Thus, the linearized equations of motion reduce to the following forms:

$$\frac{1}{\gamma}\frac{dm_x}{dt} = (H + 4\pi M)m_y = H_1 m_y \tag{6.23}$$

and

$$\frac{1}{\gamma}\frac{dm_y}{dt} = -(H - H_A)m_x = -H_2 m_x. \tag{6.24}$$

Assuming solutions of the $e_j^{\omega t}$ form, Equations 6.23 and 6.24 become

$$j\frac{\omega}{\gamma}m_x = (H + 4\pi M)m_y = H_1 m_y$$

and

$$j\frac{\omega}{\gamma}m_y = -(H - H_A)m_x = -H_2 m_x.$$

There are nontrivial solutions of m_x and m_y, if the following dispersion law is obeyed:

$$\left(\frac{\omega}{\gamma}\right)^2 = (H + 4\pi M)(H - H_A) = H_1 H_2, \tag{6.28}$$

$$\frac{\omega^2}{\gamma^2} = H_1 H_2,$$

or

$$\frac{\omega}{\gamma} = \sqrt{H_1 H_2}.$$

This is a convenient way to express the FMR condition for it can be applied to most or all cases of interest (see, for example, Table 6.1). We recognize that Equation 6.28 is exactly the same as the one derived by the Smit and Beljers method. The advantage of this method is that it allows one to choose the degree of nonlinearity to be excluded or included in the equation of motion.

TABLE 6.1

Permeability Parameters H_1 and H_2 for Various Magnetic Field Excitation Configurations

Field Configuration	H_1	H_2
 M-type hexaferrites (saturated case)	$H + 4\pi M.$	$H - H_A,$ $H \geq H_A,$ $H_A = 2\lvert K_u \rvert/M,$ $K_u < 0.$
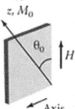 M-type hexaferrites (Unsaturated case)	$H_A + 4\pi M,$ $H_A = 2\lvert K_u \rvert/M,$ $K_u < 0.$	$\dfrac{H_A^2 - H^2}{H_A},$ $H \leq H_A,$ $\theta_0 = \cos^{-1}(H/H_A).$
 M-type hexaferrites	$H + H_A + 4\pi M,$ $H_A = 2\lvert K_u \rvert/M,$ $K_u < 0.$	$H + H_A$
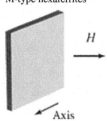 M-type hexaferrites	$H - H_A - 4\pi M,$ $H \geq H_A + 4\pi M,$ $H_A = 2\lvert K_u \rvert/M,$ $K_u < 0.$	$H - 4\pi M$
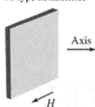 M-type hexaferrites	$H - H_A + 4\pi M,$ $H_A = 2\lvert K_u \rvert/M,$ $K_u < 0,$ $H - H_A - 4\pi M.$	H

(Continued)

TABLE 6.1 (*Continued*)

Permeability Parameters H_1 and H_2 for Various Magnetic Field Excitation Configurations

Field Configuration	H_1	H_2
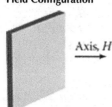 Axis, H M-type hexaferrites	$H + H_A - 4\pi M,$ $H + H_A - 4\pi M,$ $H_A = 2\lvert K_u \rvert / M,$ $K_u < 0.$	$H + H_A - 4\pi M$
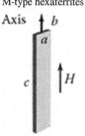 Axis, b, a, c, H Nanowires	$H + H_A + N_1 M,$ $N_1 \cong \dfrac{4\pi}{1 + b/a}$ $c - a > b.$	$H + H_A + N_2 M,$ $N_2 \cong \dfrac{4\pi}{1 + a/b},$ $H_A = 2\lvert K_u \rvert / M,$ $K_u < 0.$
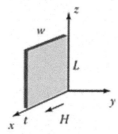 w, L, t, H Nanoparticles	$H + (N_y - N_x)M,$ $N_y \cong \dfrac{4\pi}{1 + t/L + t/w},$ $N_x \cong \dfrac{4\pi(t/w)}{1 + t/L + t/w}.$	$H + (N_z - N_x)M,$ $N_z \cong \dfrac{4\pi(t/L)}{1 + t/L + t/w}.$
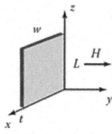 w, L, H, t Nanoparticles	$H + (N_x - N_y)M,$ $N_y \cong \dfrac{4\pi}{1 + t/L + t/w},$ $N_x \cong \dfrac{4\pi(t/w)}{1 + t/L + t/w}.$	$H + (N_z - N_y)M,$ $N_z \cong \dfrac{4\pi(t/L)}{1 + t/L + t/w}.$

(Continued)

TABLE 6.1 (*Continued*)

Permeability Parameters H_1 and H_2 for Various Magnetic Field Excitation Configurations

Field Configuration	H_1	H_2
 M-type hexaferrites (oriented compacts)	$\dfrac{(H_A - 4\pi M)^2 - H^2}{H_A - 4\pi M}$, $H \leq (H_A - 4\pi M)$.	$H_A - 4\pi M$, $H_A = 2\lvert K_u \rvert / M$, $K_u < 0$.
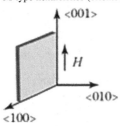 Cubic crystals	$H + H_C + 4\pi M$, $H_C = 2\lvert K_1 \rvert / M$, $K_1 < 0$.	$H + H_C$
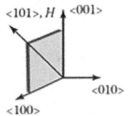 Cubic crystals	$H + H_C/2 + 4\pi M$, $H_C = 2\lvert K_1 \rvert / M$, $K_1 < 0$.	$H - H_C$
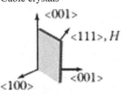 Cubic crystals	$H + 2/3\,H_C + 4\pi M$, $H_C = 2\lvert K_1 \rvert / M$, $K_1 < 0$.	$H + 2/3\,H_C$
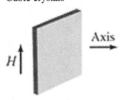 Y-type hexaferrites	$H + H_A + 4\pi M$, $H_A = 2\lvert K_u \rvert / M$, $K_u > 0$.	H

(Continued)

TABLE 6.1 (*Continued*)

Permeability Parameters H_1 and H_2 for Various Magnetic Field Excitation Configurations

Field Configuration	H_1	H_2
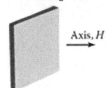 Axis, H	$H - H_A - 4\pi M,$ $H > (H_A + 4\pi M),$ $K_u > 0.$	$H - H_A - 4\pi M$

Y-type hexaferrites

MICROWAVE PERMEABILITY

In this derivation, we assume $H \geq H_A$ and $\alpha = \pi/2$ as an example. In order to calculate the microwave permeability, we need to modify Equations 6.26 and 6.27 in order to include the external microwave field $\vec{h}_a$.

$$h_x = \left(\frac{H_A}{M}\right)m_x + h_{ax} \tag{6.26'}$$

and

$$h_x = -4\pi m_y + h_{ay}. \tag{6.27'}$$

The subscript "*a*" is to denote an external microwave drive, and let's drop the subscript "*a*." In this case, the equations of motion become

$$j\frac{\omega}{\gamma}m_x = H_1 m_y - M h_y$$

and

$$j\frac{\omega}{\gamma}m_y = -H_2 m_x - M h_x.$$

Let's solve for m_x and m_y in terms of h_x and h_y. We have

$$\Omega^2 m_x = H_1 M h_x - j\frac{\omega}{\gamma}M h_y \tag{6.29}$$

and

$$\Omega^2 m_y = H_2 M h_y - j\frac{\omega}{\gamma}M h_x, \tag{6.30}$$

where

$$\Omega^2 = H_1 H_2 - \left(\frac{\omega}{\gamma}\right)^2.$$

We combine Equations 6.29 and 6.30 into matrix form so that

$$\begin{pmatrix} m_x \\ m_y \end{pmatrix} = \begin{pmatrix} \chi_{xx} & \chi_{xy} \\ \chi_{yx} & \chi_{yy} \end{pmatrix} \begin{pmatrix} h_x \\ h_y \end{pmatrix}, \tag{6.31}$$

where

$$\chi_{xx} = H_1 M / \Omega^2,$$

$$\chi_{yy} = H_2 M / \Omega^2,$$

$$\chi_{xy} = -j(\omega/\gamma) M / \Omega^2,$$

and

$$\chi_{yx} = -\chi_{xy}.$$

In general, we define a susceptibility tensor of the following form for most cases of interest (see Table 6.1):

$$[\chi] = \begin{bmatrix} \chi_{11} & \chi_{12} \\ \chi_{21} & \chi_{22} \end{bmatrix},$$

where we purposely choose χ_{11} to be a function of microwave fields in the plane of a magnetic film only. As such, the microwave in-plane demagnetizing fields are negligible so that $\chi_{11} > \chi_{22}$. Thus, our convention is to choose χ_{11} to be proportional to H_1 which is greater than H_2. For example, the in-plane microwave fields are m_x and h_x. Clearly, m_x "sees" negligible demagnetizing field. As a result, H_1 is greater than H_2, since it contains the microwave demagnetizing factor of $4\pi M$. In general, in cases where the demagnetizing factors are different, because of the shape of the sample, χ_{11} will always be chosen to represent microwave fields in a direction such that the demagnetizing fields are minimized. Commercial codes or software consider only susceptibility tensors where the diagonal elements are assumed to be equal to each other (the perpendicular demagnetizing factor is assumed to be zero). By applying Maxwell boundary conditions via the codes, the perpendicular demagnetizing factor is "automatically" included in whatever analysis. Here, we have included it (Table 6.1) explicitly, since we are not developing any code. If one were to develop a code from our expressions in Table 6.1, the perpendicular demagnetizing factor should be omitted. Nevertheless, in some cases, we have $\chi_{11} \neq \chi_{22}$ even with the demagnetizing factor omitted. In summary, the susceptibility tensor may be written as follows:

$$[\chi] = \frac{M}{\left(H_1 H_2 - \left(\omega^2/\gamma^2\right)\right)} \begin{bmatrix} H_1 & -j\omega/\gamma \\ j\omega/\gamma & H_2 \end{bmatrix} = \begin{bmatrix} \chi_{11} & \chi_{12} \\ \chi_{21} & \chi_{22} \end{bmatrix}.$$

The 2×2 matrix is recognized as the susceptibility tensor and it is Hermitian. The permeability tensor may be expressed in terms of the susceptibility tensor as follows (CGS):

$$[\mu] = [I] + 4\pi[\chi]. \tag{6.32}$$

Explicitly, we have

$$[\mu] = \begin{pmatrix} 1 + 4\pi\chi_{11} & 4\pi\chi_{12} & 0 \\ 4\pi\chi_{21} & 1 + 4\pi\chi_{22} & 0 \\ 0 & 0 & 1 \end{pmatrix}.$$

Let's introduce the concept of anti-resonance FMR (AFMR) frequency. The AFMR frequency is defined for frequency where

$$\mu = 0.$$

For simplicity, let's assume μ_{11} of this form

$$\mu_{11} = 1 + 4\pi\chi_{11} = 1 + \frac{4\pi M H_1}{\Omega^2}$$

or

$$\mu_{11} = 1 + \frac{4\pi M H_1}{H_1 H_2 - \left(\omega^2/\gamma^2\right)}.$$

We have ignored magnetic damping for now. Setting $\mu_{11} = 0$ (condition for AFMR) yields the following:

$$H_1 H_2 + 4\pi M H_1 = \frac{\omega^2}{\gamma^2}.$$

Solving for ω/γ, we obtain

$$\frac{\omega}{\gamma} = \sqrt{H_1 H_2}\sqrt{1 + \frac{4\pi M}{H_2}}.$$

Thus,

$$f_{AFMR} = f_{FMR}\sqrt{1 + \frac{4\pi M}{H_2}},$$

where

$$f_{FMR} = (\gamma/2\pi)\sqrt{H_1 H_2}.$$

Although we have specialized our results to a specific case ($H \geq H_A$ and $\alpha = \pi/2$), the mathematical procedure is applicable in general to any case of interest. Indeed, the permeability tensor is always of the same form irrespective of the shape of the sample and magnetic composition. In Table 6.1, we provide results for various geometries and magnetic structures.

NORMAL MODES

Clearly, the normal modes of precession are not linearly polarized, since the tensor [μ] is non-diagonal in the h_x and h_y representation. Let's determine the normal modes of the previous example in order to drive the precessional excitation in a normal mode. In the previous example, the following equation with no external drive was derived:

$$j\frac{\omega}{\gamma}m_x = (H + 4\pi M)m_y = H_1 m_y. \tag{6.33}$$

We also determined that the precessional resonant frequency was given as

$$\frac{\omega}{\gamma} = \sqrt{(H + 4\pi M)(H - H_A)} = \sqrt{H_1 H_2}. \tag{6.34}$$

Substituting the above resonance condition into Equation 6.33, we obtain

$$\frac{m_x}{m_y} = -j\sqrt{\frac{H_1}{H_2}} \tag{6.35}$$

or

$$m_y = j\sqrt{\frac{H_2}{H_1}}m_x.$$

Let the solution of m_x in real time be simply

$$m_x(t) = A\cos\omega t \tag{6.36}$$

so that

$$m_y(t) = \mathrm{Re}\left[A\sqrt{\frac{H_2}{H_1}}e^{j\pi/2}e^{j\omega t}\right] = -A\sqrt{\frac{H_2}{H_1}}\sin\omega t. \tag{6.37}$$

By combining Equations 6.36 and 6.37, we obtain

$$m_x^2(t) + \frac{m_y^2(t)}{a^2} = A^2, \tag{6.38}$$

where

$$a^2 = \frac{H_2}{H_1}.$$

Thus, m_x and m_y trace out an ellipsoidal motion with the largest magnitude along the x-direction (see Figure 6.18).

The sense of rotation is counterclockwise. The other normal mode can be described similarly by Equation 6.38. However, the sense of rotation is clockwise.

Let's now drive the film with an external microwave magnetic field. From the susceptibility expression, we may write

$$m_x = \frac{M}{\Omega^2}\left(H_1 h_x - j\frac{\omega}{\gamma}h_y\right) \tag{6.39}$$

and

$$\pm j\frac{m_y}{a} = \frac{M}{\Omega^2}\left(\mp\frac{\omega}{\gamma}\frac{h_x}{a} \pm jH_2\frac{h_y}{a}\right). \tag{6.40}$$

Adding the two equations yields

$$m_x \pm j\frac{m_y}{a} = \frac{M}{\Omega^2}H_1\left(h_x \pm j\frac{H_2}{aH_1}h_y\right) \mp \frac{M}{\Omega^2}\frac{\omega}{\gamma}\frac{1}{a}\left(h_x \pm jah_y\right)$$

or

$$m_\pm = m_x \pm j\frac{m_y}{a} = \frac{M}{\Omega^2}\left(H_1 \mp \frac{\omega}{a\gamma}\right)\left(h_x \pm jah_y\right). \tag{6.41}$$

Substituting

$$\Omega^2 = H_1 H_2 - \frac{\omega^2}{\gamma^2} = \left(\sqrt{H_1 H_2} - \frac{\omega}{\gamma}\right)\left(\sqrt{H_1 H_2} + \frac{\omega}{\gamma}\right),$$

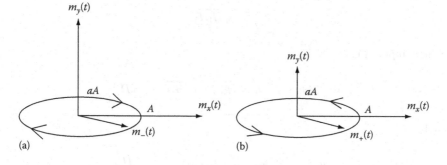

(a)   (b)

FIGURE 6.18   Resonant and non-resonant normal mode precessional motions.

we obtain

$$m_\pm = \frac{M}{a} \frac{\left(\sqrt{H_1 H_2} \mp (\omega/\gamma)\right) h_\pm}{a\left(\sqrt{H_1 H_2} - (\omega/\gamma)\right)\left(\sqrt{H_1 H_2} + (\omega/\gamma)\right)},$$

where

$$h_\pm = h_x \pm j a h_y$$

and

$$m_\pm = m_x \pm j \frac{m_y}{a}.$$

Thus,

$$m_+ = \frac{(M/a) h_+}{\sqrt{H_1 H_2} + (\omega/\gamma)}, \quad \text{non-resonant mode,}$$

$$m_- = \frac{(M/a) h_-}{\sqrt{H_1 H_2} - (\omega/\gamma)}, \quad \text{resonant mode.}$$

Introduce a susceptibility in normal excitations, $h_\pm$, as

$$\chi_\pm = \frac{M/a}{\sqrt{H_1 H_2} \pm (\omega/\gamma)} = \frac{h_\pm}{m_\pm}$$

or

$$[\chi] = \begin{bmatrix} \chi_+ & 0 \\ 0 & \chi_- \end{bmatrix}.$$

Let's see what the sense of precessional motion for each excitation is. For non-resonant mode excitation, h_+, we have the resonant condition as

$$\frac{\omega}{\gamma} = -\sqrt{H_1 H_2},$$

which implies that

$$m_x = H_1 h_x - j \frac{\omega}{\gamma} h_y = H_1 h_x + j \sqrt{H_1 H_2} h_y = H_1 h_+$$

and

$$m_y = H_2 h_y + j \frac{\omega}{\gamma} h_x = H_2 h_y - j \sqrt{H_1 H_2} h_x = -j \sqrt{H_1 H_2} h_+.$$

Thus,

$$\frac{m_x}{m_y} = j\sqrt{\frac{H_1}{H_2}} = \frac{j}{a}$$

or

$$m_y = -jam_x.$$

The sense of rotation is then clockwise, since

$$m_x(t) = A\cos\omega t,$$

$$m_y(t) = \mathrm{Re}\left[aAe^{-j\pi/2}e^{j\omega t}\right] = aA\sin\omega t.$$

For the resonant mode, m_-, we have that

$$\frac{m_x}{m_y} = -ja,$$

which is a counterclockwise motion.

   In Figure 6.19, we show the normal mode response and excitation for the resonant mode, m_- and h_-, respectively.

   The reader is reminded that the actual physical motion is opposite of what we have just determined, since the sign of γ depends on the electron charge. For convenience, we have assumed γ to be positive.

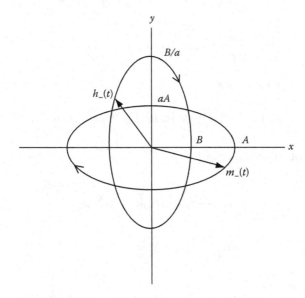

FIGURE 6.19   Normal mode motions for m_- and h_-, the resonant mode.

MAGNETIC RELAXATION

Let's consider the effect of magnetic damping in steady-state motion of a magnetic system. Specifically, let's consider a magnetic system excited by a microwave circularly polarized magnetic field, h, defined as

$$h_\pm = h_x \pm jh_y.$$

We consider the equation of motion assuming first Gilbert damping, and second, Landau–Lifshitz damping. For the Gilbert form of damping, we have that

$$\frac{1}{\gamma}\frac{d\vec{M}}{dt} = \vec{M} \times \vec{H} - \frac{\alpha}{\gamma M}\vec{M} \times \frac{d\vec{M}}{dt}.$$

The Gilbert's equation of motion at steady state is then

$$j\frac{\omega}{\gamma}\vec{m} = \vec{M}_0 \times \vec{h} + \vec{m} \times \vec{H}_0 - \frac{\alpha}{M}\vec{M}_0 \times j\frac{\omega}{\gamma}\vec{m},$$

where

$\vec{M}_0 \equiv$ static internal magnetization
$\vec{H}_0 \equiv$ static internal magnetization
α is the dimensionless Gilbert damping parameter

Clearly, we have linearized the equation of motion. Putting the above equation in components form, we have

$$j\frac{\omega}{\gamma}m_x = -Mh_y + m_yH_0 + \alpha j\frac{\omega}{\gamma}m_y,$$

$$j\frac{\omega}{\gamma}m_y = Mh_x - m_xH_0 + \alpha j\frac{\omega}{\gamma}m_x.$$

Adding the two equations yields

$$j\frac{\omega}{\gamma}\left(m_x \pm jm_y\right) = \pm jM\left(h_x \pm jh_y\right) \mp jH_0\left(m_x \pm jm_y\right) \mp j\alpha j\frac{\omega}{\gamma}\left(m_x \pm jm_y\right).$$

The above equation simplifies to

$$\frac{\omega}{\gamma}m_\pm = \pm Mh_\pm \mp H_0 m_\pm \mp j\alpha\frac{\omega}{\gamma}m_\pm,$$

$$m_\pm\left(\frac{\omega}{\gamma} \pm H_0 \pm j\alpha\frac{\omega}{\gamma}\right) = \pm Mh_\pm,$$

$$m_- = \frac{-Mh_-}{(\omega/\gamma) - H_0 - j\alpha(\omega/\gamma)} = \frac{-Mh_-}{H_0\left((\omega/\gamma) - j\alpha(\omega/\gamma)\right)}.$$

And finally, we have the resonant mode expressed as follows:

$$m_- = \frac{Mh_-}{H_0 - (\omega/\gamma)(1 - j\alpha)}. \tag{6.41a}$$

Assuming the Landau–Lifshitz form for damping, we have

$$\frac{1}{\gamma}\frac{d\vec{M}}{dt} = \vec{M} \times \vec{H} - \frac{\lambda}{\gamma M^2}\vec{M} \times (\vec{M} \times \vec{H}),$$

where λ is the Landau–Lifshitz damping parameter in units of radians per second. The equation of motion becomes

$$j\frac{\omega}{\gamma}m_x = -Mh_y + m_yH_0 + CM_0^2h_x - CM_0H_0m_x,$$

$$j\frac{\omega}{\gamma}m_y = Mh_x + m_xH_0 + CM_0^2h_y - CM_0H_0m_y,$$

$$j\frac{\omega}{\gamma}(m_x \pm jm_y) = \pm jM(h_x \pm jh_y) \mp jH_0(m_x \pm jm_y) + CM_0^2(h_x \pm jh_y)$$

$$-CM_0H_0(m_x \pm jm_y),$$

where

$$C = \frac{\lambda}{\gamma M_0^2}, \quad CM_0^2 = \frac{\lambda}{\gamma}, \quad CM_0H_0 = \left(\frac{\lambda}{\gamma}\right)\frac{H_0}{M_0}.$$

For the resonant mode, we have

$$\frac{\omega}{\gamma}m_- = -Mh_- + H_0m_- - j\frac{\lambda}{\gamma}h_- + j\frac{\lambda}{\gamma}\frac{H_0}{M_0}m_-,$$

$$\left(\frac{\omega}{\gamma} - H_0 - j\frac{\lambda}{\gamma}\frac{H_0}{M_0}\right)m_- = \left(-M - j\frac{\lambda}{\gamma}\right)h_-,$$

$$\left(H_0 - \frac{\omega}{\gamma} + j\frac{\lambda}{\gamma}\frac{H_0}{M_0}\right)m_- = M\left(1 + j\frac{\lambda}{\gamma M}\right)h_-,$$

$$\left(H_0 - \frac{\omega}{\gamma} + j\frac{\lambda}{\gamma M}H_0\right)m_- = M\left(1 + j\frac{\lambda}{\gamma M}\right)h_-,$$

$$\left[H_0\left(1 + j\frac{\lambda}{\gamma M}\right) - \frac{\omega}{\gamma}\right]m_- = M\left(1 + j\frac{\lambda}{\gamma M}\right)h_-.$$

As in analogous presentation of the Gilbert formulation, we finally have

$$\left[H_0 - \frac{\omega/\gamma}{1 + j(\lambda/\gamma M)} \right] m_- = M_0 h_-,$$

$$\boxed{m_- = \frac{M_0}{H_0 - \dfrac{\omega/\gamma}{1 + j(\lambda/\gamma M)}} h_-.}$$

(6.41b)

Although Equation 6.41a uses M and 6.41b M_0 (being the same), the reader should notice the similarity and the difference between the two equations.

In summary, we may include magnetic damping in our previous expressions for permeability by making the following substitutions:

$$\frac{\omega}{\gamma} \to \frac{\omega}{\gamma}(1 - j\alpha),$$

if one assumes Gilbert damping form or

$$\frac{\omega}{\gamma} \to \frac{\omega/\gamma}{1 + j(\lambda/\gamma M)},$$

if one assumes Landau–Lifshitz form for damping, see appendix for the calculation of magnetic damping from first principles.

Note that for small damping or losses, the two damping parameters may be related to each other:

$$\alpha = \frac{\lambda}{\gamma M}.$$

In this limit, the two equations of motion may be rewritten as follows:

$$\frac{1}{\gamma}\frac{d\vec{M}}{dt} = \vec{M} \times \vec{H} - \frac{\alpha}{M}\left((\vec{M} \cdot \vec{H})\vec{M} - M^2 \vec{H} \right).$$

(6.42)

Let's look at the physical implications of the above equation. For simplicity, assume a semi-infinite medium and H along the z-axis. Thus, write

$$\frac{1}{\gamma}\frac{dM_x}{dt} = M_y H - \frac{\alpha}{M}(M_z H)M_x,$$

(6.43)

$$\frac{1}{\gamma}\frac{dM_y}{dt} = -M_x H - \frac{\alpha}{M}(M_z H)M_y,$$

(6.44)

and

$$\frac{1}{\gamma}\frac{dM_z}{dt} = \frac{\alpha}{M}\left(M^2 - M_z^2\right)H. \tag{6.45}$$

Using the relationship

$$\frac{\vec{M}\cdot d\vec{M}}{dt} = M_x\frac{dM_x}{dt} + M_y\frac{dM_y}{dt} + M_z\frac{dM_z}{dt}$$

and the equations of motion for each component of the magnetization, we show that

$$\frac{d\vec{M}}{dt} = 0.$$

Thus, the magnetic system relaxes or damps out in a manner that keeps the magnitude of the magnetization a constant of the motion. A spiral motion is induced, as shown in Figure 6.20. For most ordered magnetic materials, the two forms of damping are appropriate. For disordered and/or paramagnetic materials, there are other forms of magnetic relaxation that are more appropriate. For example, the Bloch–Bloembergen model assumes that the magnetic relaxation damping parameter is not isotropic as in Gilbert and Landau–Lifshitz form of damping. We do not discuss the latter models, since we deal only with ordered magnetic materials.

Let's consider Equation 6.45 and make the approximation for small deviations from equilibrium as follows:

$$M^2 - M_z^2 \cong 2M\left(M - M_z\right).$$

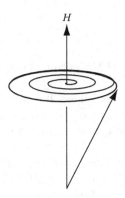

FIGURE 6.20   Relaxation mechanism.

Then, Equation 6.45 becomes

$$\frac{1}{\gamma}\frac{dM_z}{dt} \cong 2\alpha(M - M_z)H \tag{6.45'}$$

whose solution becomes

$$M_z(t) \cong M - (M - M_0)e^{-t/\tau},$$

where

$$\frac{1}{\tau} = 2\alpha H\gamma.$$

At $t = 0$, $M_z(t) = M_0$, but as $t \to \infty$, $M_z(t) \to M$, which is the equilibrium value of the magnetization.

In the early 1950s, τ was measured by measuring transients of the magnetic system upon the application of pulsating magnetic fields. In the late 1950s and afterward, steady-state techniques were developed to measure τ, such as magnetic resonance techniques. The FMR linewidth (ΔH) was measured and it was related to τ as follows:

$$\Delta H = \frac{1}{\gamma}\left(\frac{1}{\tau}\right) = 2\alpha H = 2\frac{\lambda}{\gamma M}H. \tag{6.46}$$

The FMR linewidth may be related to the FMR frequency bandwidth ($\Delta\omega$) as follows:

$$\Delta\omega = \gamma\frac{\partial(\omega/\gamma)}{\partial H}\Delta H,$$

where

$$\frac{\omega}{\gamma} = H$$

for this example. In general, the FMR condition may not be as simple as the above expression, see Table 6.1.

The basic definition of ΔH is as follows. It is the difference in DC magnetic field values for which the imaginary part of the susceptibility is reduced by a factor of two. Also, at these values, the microwave power absorbed by the magnetic sample is one-half power points.

We have plotted μ_{11} and μ_{22} as a function of frequency for H applied in the plane of a permalloy film (Figure 6.21). H was fixed at 1,100 Oe. The Gilbert damping parameter was assumed to be equal to 0.00043. Note $\mu_{11} = \mu_{xx}$, $\mu_{22} = \mu_{yy}$, and $\mu_{xy} = j\chi$; $4\pi M_S = 10,000$ G; and $g = 2.09$.

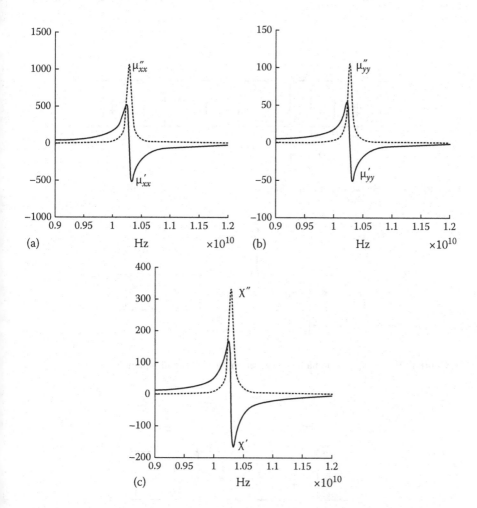

FIGURE 6.21 (a) Real and imaginary parts of one-diagonal permeability tensor element. (b) Real and imaginary parts of the other diagonal permeability tensor element. (c) Real and imaginary parts of the off-diagonal permeability tensor element.

FREE ENERGY OF MULTI-DOMAINS

In non-saturation field conditions, the sample is usually decomposed into magnetic multi-domains. The results of Table 6.1 are based on single-domain assumption. Hence, one must consider a new analysis for multi-domain excitations in non-saturation fields. Let's consider two cases of interest: (1) H in the film plane (//) and (2) H normal to the film plane ($\perp$). For both cases, we assume that the ferromagnetic film is characterized by uniaxial magnetic anisotropy energy with an easy axis of magnetization normal to the film plane. Further, we assume that the domain periodic structures in the film consist of either parallel stripes or bubble domains, as shown in Figures 6.22–6.24.

For H in the film plane, we assume that the magnetic moment in each domain rotates toward H with no net moment in the z-direction. In writing the free energy, it is assumed that the magnetization amplitude, M, is constant from point to point and equal to M_S. Thus,

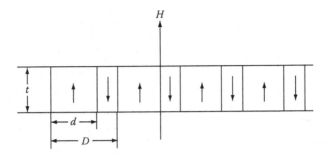

FIGURE 6.22 Magnetic domain configuration for H normal to the film plane.

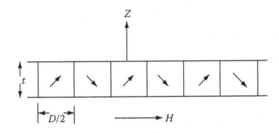

FIGURE 6.23 Magnetic domain configuration for H in the film plane.

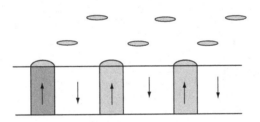

FIGURE 6.24 Magnetic bubble domains configuration.

$$F\left(\text{erg/cm}^3\right) = -\frac{H}{2}\left(M_{1y} + M_{2y}\right) = \frac{K}{2M_S^1}\left(M_{1x}^2 + M_{1y}^2 + M_{2x}^2 + M_{2y}^2\right)$$

$$+\frac{1}{8}\left[N_x\left(M_{1x} + M_{2x}\right)^2 + N_y\left(M_{1y} + M_{2y}\right)^2 + N_z\left(M_{1z} + M_{2z}\right)^2 + 4\pi\left(M_{1z} + M_{2z}\right)^2\right] + \sigma^2.$$

$$(6.47)$$

Terms of order $K^2 \sin^4\theta$ are omitted for simplicity. They are found to be relatively unimportant for the purpose on hand. The film is assumed to have a relatively small ratio of thickness to the lateral dimension of the film.

The first term in the free energy expression is the Zeeman energy for the two domain regions. The two domain regions are designated by the subscripts 1 and 2. H is in the y-direction. The platelet plane forms the x–y plane. The factor of 2 is due to dividing the sample into two equal volumes with different magnetization orientations. The second term is the uniaxial magnetic anisotropy energy in the two domain regions. Since the easy axis of magnetization is taken to be along the z-axis, K is positive. The terms in the square bracket include all of the demagnetizing energy terms. σ'_ω indicates an appropriate surface wall energy density (erg/cm^2) as distinguished from the wall volume energy density σ_ω (erg/cm^3). The demagnetization factors N_x, N_y, and N_z can be calculated for both stripe and bubble domains (see references). However, N_x, N_y, and N_z are a function of size and shape of the magnetic domain. It is in a sense a "local" demagnetizing factor. This is to be contrasted with the last demagnetizing energy term

$$4\pi\left(M_{1z} + M_{2z}\right)^2,$$

in which the demagnetizing factor of 4π applies for the whole shape of the sample. For example, if we had an ellipsoidal sample instead of a film, we would need terms of the following form besides the first three terms in F:

$$N_x^s\left(M_{1x} + M_{2x}\right)^2 + N_y^s\left(M_{1y} + M_{2y}\right)^2 + N_z^s\left(M_{1z} + M_{2z}\right)^2.$$

Those three terms would replace the above term. The superscript "s" is to indicate that the demagnetizing factor applies over the whole shape or size of the sample. For stripe domains, we assume that the domain wall is either aligned with the static field, H, or perpendicular to it, either $N_y = 0$ or $N_x = 0$, respectively. We also assume that bubble domains do not immediately vanish upon the application of an in-plane H in accordance with observations of ferrite single crystal films (Figures 6.24 and 6.25).

The application of H normal to the platelet plane induces a net magnetization along H. We define the ratio of the volume containing magnetic moments aligned along H to the total sample volume as $v = d/D$ (see Figure 6.24).

For bubble domains, $v = 2\pi R^2/pD^2$. We are assuming a simple bubble domain lattice in which the periodicity in the x–y plane is two-dimensional. Along the x-direction, the period is D and along the y-direction pD. The radius of the bubble domain is R. The film thickness is t. The free energy for this field configuration is written as

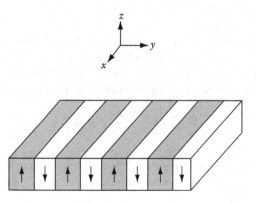

FIGURE 6.25   Magnetic stripe domains.

$$F\left(\text{erg/cm}^3\right) = -H\left[vM_{1z} + (1-v)M_{2z}\right]$$

$$+\frac{K}{M_S^2}\left[\left(M_{1x}^2 + M_{1y}^2\right)v + \left(M_{2x}^2 + M_{2y}^2\right)(1-v)\right]$$

$$+8\pi\left[vM_{1z} + (1-v)M_{2z}\right]^2$$

$$+\frac{v(1-v)}{2}\left[N_x\left(M_{1x} - M_{2x}\right)^2 + N_y\left(M_{1y} - M_{2y}\right)^2 + N_z\left(M_{1z} - M_{2z}\right)^2\right] - \sigma_\omega.$$

$$(6.48)$$

The first two terms are recognized as being the Zeeman and uniaxial magnetic anisotropy energies (see Equation 6.47 for comparison). The remaining terms represent the total demagnetizing energy. We note, for example, that for $v = 1/2$, we obtain Equation 6.47. The factor $v(1 - v)/2$ appearing in the next term is somewhat artificial but it serves two purposes: (1) it puts this term in direct correspondence with other terms in the free energy and (2) it expresses this term as a function of the measurable quantity v or $(1 - v)$.

The effective internal fields in each domain region are obtained by taking the negative gradient of the free energy:

$$\vec{H}_1 = -\vec{\nabla}_1 F$$

and

$$\vec{H}_2 = -\vec{\nabla}_2 F,$$

where

$$\vec{\nabla}_n = \frac{2}{M_n}\left(\vec{a}_x\frac{\partial}{\partial a_x} + \vec{a}_y\frac{\partial}{\partial a_y} + \vec{a}_z\frac{\partial}{\partial a_z}\right)_n,$$

$$M_1 = vM_s, \text{ and}$$

$$M_2 = (1-v)M_s.$$

where α_x, α_y, and α_z are the directional cosines of $\vec{M}_n$ with respect to the x-, y-, and z-axes and $n = 1, 2$. The subscripts 1, 2 outside of the parentheses indicate the regions of definition for the directional cosines. The set of nonlinear coupled equations is linearized by keeping terms in the equation of motion only to orders of h_n and m_n or

$$j\frac{\omega}{\gamma}\vec{m}_n = \left(\vec{m}_n \times \vec{H}_n + \vec{M}_n \times \vec{h}_n\right).$$

$\vec{H}_n$ is the static internal field for region n. The analysis follows exactly as outlined in the single-domain analysis of the previous section. The reader is referred to the references listed in the appendix.

PROBLEMS

6.1 In 1947, the first FMR experiment on nickel films or foils was performed by Griffiths. He applied H in the plane of the film and measured resonance at 1,500 Oe. Calculate g value, if the operating frequency was $\approx 9\,\text{GHz}$. Assume $4\pi M_S = 6{,}000$ G.

6.2
   a. Calculate FMR condition for a sphere.
   b. Calculate FMR condition on a needle for H along the needle axis.

6.3 Assume the earth's molten lava is paramagnetic and the temperature of the earth's core is $1{,}400°\text{C}$.
   a. Determine the magnetic moment m and the magnetization M.
   b. Use $M = Ng\beta SB(T)$, where $B(T)$ is the Brillouin function, find S.
   c. Identify possible ion corresponding to S.

6.4 Derive χ_+ and χ_- for H applied along the uniaxial axis in the film plane. You may assume $H_A \neq 0$.

6.5 Derive $\chi+$ and $\chi-$ for H applied perpendicular to the film plane. You may assume $H_A = 0$ for this problem.

6.6 Calculate $\Delta\omega$ assuming ΔH is known for the case H is applied in the film plane and perpendicular to the uniaxial axis.

6.7 Calculate χ_{xx}, χ_{yy}, and χ_{xy} for the film case in which H is applied in the film plane and perpendicular to the uniaxial axis. Include magnetic damping in the calculation.

APPENDIX 6.A: MAGNETIC DAMPING PARAMETER CALCULATED FROM FIRST PRINCIPLES

A relaxation mechanism for ferrites and ferromagnetic metals was postulated whereby the coupling between the magnetic motion and lattice was based purely on continuum arguments concerning magnetostriction (see ref. by Vittoria and Widom). This theoretical approach contrasts with the previous mechanisms based on microscopic formulations of spin–phonon interactions employing a discrete lattice. The model explains for the first time the scaling of the intrinsic FMR linewidth with frequency, M^{-1} temperature dependence, the anisotropic nature of magnetic relaxation in ordered magnetic materials, where M is the magnetization, and the square of the magnetostriction constants. Without introducing adjustable parameters, the model is in qualitative and quantitative agreement with experimental measurements of the intrinsic magnetic resonance linewidths of important class of ordered magnetic materials, insulators, or metals.

The experimental value of the Gilbert damping parameter α_{exp} may be deduced from the FMR linewidth ΔH at frequency f as

$$\alpha_{\text{exp}} = \frac{\sqrt{3}}{2}\left(\frac{\gamma\Delta H}{2\pi f}\right). \tag{6.A.1}$$

The factor $\sqrt{3}/2$ assumes Lorentzian line shape of the resonance absorption curve. The theoretical Gilbert damping parameter α_{th} value is expressed in terms of known parameters, as shown in Table 6.A.1. The theoretical prediction for the Gilbert damping parameter is

$$\alpha_{th} = \frac{36\rho\gamma}{M\tau}\left[\frac{\lambda_{100}^2}{q_L^2} + \frac{\lambda_{111}^2}{q_T^2}\right], \qquad (6.A.2)$$

wherein ρ is the mass density, $q_T \approx v_T \dfrac{M}{2\gamma A}$ is the transverse acoustic propagation constant, q_L is the longitudinal acoustic propagation constant, v_T is the transverse sound velocity, A is the exchange stiffness constant, and λ_{100} and λ_{111} are magnetostriction constants for a cubic crystal magnetic material. The transverse acoustic propagation constant was approximated on the basis that the relaxation process conserved energy and wave vector. Since the acoustic frequency is fixed in the process, the longitudinal propagation constant may also be calculated to be $q_L = q_T(v_T/v_L)$ for magnetic materials, wherein v_L is the longitudinal sound wave velocity.

The agreement between theory and experiments is remarkable in view of the fact that any of the cited parameters could differ from the ones listed in Table 6.A.1 by as much as 20%–30%.

TABLE 6.A.1

Calculated and Measured Gilbert Damping (α) Parameters

Materials	q_T (10^6 cm^{-1})	λ_{100} (10^{-6})	λ_{111} (10^{-6})	M (G/4π)	A (10^{-6} erg/cm)	ΔH (Oe)	f (GHz)	τ (10^{-13})	α_{th} (10^{-5})	α_{exp} (10^{-5})
$Y_3Fe_5O_{12}$[a]	3.8	1.25	2.8	139	0.4	0.33	9.53	4.4	5.56	9
$Y_3Fe_4GaO_{12}$[a]	1.46	−1	−1	36	0.28	3	9.53	4.4	51	76
$Li_{0.5}Fe_{2.5}O_4$[b]	8.6	−8	0	310	0.4	2	9.5	1.5	26	50
$NiFe_2O_4$[b]	7.49	−63	−26	270	0.4	35	24	1.5	710	350
$MgFe_2O_4$[b]	9.3	−10	−1	90	0.1	2.3	4.9	1.5	120	120
$MnFe_2O_4$[b]	6.6	−30	−5	220	0.4	38	9.2	1.5	930	1,040
$BaFe_{12}O_{19}$[c]	9.6	~	15	350	0.4	6	55	1.5	18	26
Ni[d]	6.3	−46	25	484	0.75	102	9.53	1.8	770	2,600
Fe[d]	8.75	20	−20	1,690	1.9	9	9.53	1.8	30	220
Co[d]	5.1	~	80	1,400	2.78	15	9.53	1.8	530	370

[a] Garnets.
[b] Spinels.
[c] Hexagonal ferrite.
[d] Ferromagnetic metals.

Note: Longitudinal acoustic wave constant is $q_L = (v_T/v_L)q_T$.

BIBLIOGRAPHY

B.I. Bleaney and B. Bleaney, *Electricity and Magnetism*, Clarendon Press, Oxford, U.K., 1965;
    L.H. Thomas, *Nature*, **117**, 514, 1926.
F. Bloch, *Z. Physik*, **74**, 295 (1932).
H. B. Callen, *Fluctuations, Relaxation, and Resonance in Magnetic Systems*, Oliver & Boyd
    Ltd., London, (1962).
T. L. Gilbert, *IEE Trans. Mag.*, **40**, 3443, 2004.
C. Kittel, *J. Phys. Soc. Jpn.*, Suppl., **17**(B1), 396, 1962.
L.D. Landau and E.M. Lifshitz, *Quantum Mechanics*, Pergamon Press, Oxford, U.K., 1958.
L.D. Landau and E.M. Lifshitz, *Mechanics*, Pergamon Press, Oxford, U.K., 1962.
L.D. Landau and E. M. Lifshitz, *Phys. Z. Sowjet.*, **8**, 153, 1935.
R. Orbach, *Proc. Phys. Soc. (London)*, **77**, 821, 1961.
F.J. Rachford, P. Lubitz, and C. Vittoria, *J. Appl. Phys.*, **53**, 8949, 1982.
C.H. Slichter, *Principles of Magnetic Resonance*, Harper & Row, New York, 1963.
J. Smit and H.P.J. Wijn, *Ferrites*, John Wiley & Sons, Inc., New York, 1959.
L.H. Thomas, *Nature*, **117**, 514, 1926.
J.H. Van Vleck, *The Theory of Electric and Magnetic Susceptibility*, Oxford University Press,
    Oxford, U.K., 1932.
C. Vittoria, S. D. Yoon and A. Widom, *Phys. Rev.*, **81**, 014412, 2010.

SOLUTIONS

6.1 For Ni film, the magnetic free energy can be written as (Figure S6.1)

$$(1)\ F = -M_y H_{ext} + 2\pi M_x^2.$$

The internal field $\vec{H}_i$ is given by

$$(2)\ \vec{H}_i = -\nabla_{\vec{M}} F,$$

i.e.,

$$(3)\ H_{ix} = -4\pi M_x,\ H_{iy} = H_{ext},\ H_{iz} = 0.$$

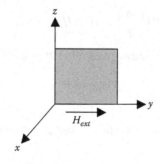

FIGURE S6.1 Field geometry.

If the sample is magnetically saturated, the static magnetization is

$$(4) \ M_x = 0, \ M_y = M_s, \ M_z = 0.$$

The static part of $\vec{H}_i$ can be written as

$$(5) \ H_x = 0, \ H_y = H_{ext}, \ H_z = 0.$$

The dynamic part of $\vec{H}_i$ can be written as

$$(6) \ h_x = -4\pi m_x, \ h_y = 0, \ h_z = 0.$$

Neglecting higher order terms, the equation of motion can be written as

$$(7) \ \frac{1}{\gamma}\frac{d\vec{m}}{dt} = \vec{m} \times \vec{H} + \vec{M} \times \vec{h},$$

where $\vec{m}$ is the dynamic magnetization.

Substituting (4–6) into (7), we obtain

$$(8) \ \begin{cases} i(\omega/\gamma)m_x = -H_{ext}m_z \\ i(\omega/\gamma)m_y = 0 \\ i(\omega/\gamma)m_z = (H_{ext} + 4\pi M_s)m_x. \end{cases}$$

The condition that there are nonzero solutions for (8) gives

$$(9) \ \begin{vmatrix} i\dfrac{\omega}{\gamma} & H_{ext} \\ H_{ext} + 4\pi M_s & -i(\omega/\gamma) \end{vmatrix} = 0,$$

i.e., the FMR condition:

$$(10) \ \left(\frac{\omega}{\gamma}\right)^2 = H_{ext}(H_{ext} + 4\pi M_s).$$

Substituting the numbers given into (10), we obtain

$$(11) \ g = 1.917.$$

6.2

a)   For sphere, the magnetic free energy can be written as (Figure S6.2a)

$$(a-1) \ F = -M_y H_{ext} + \frac{2\pi}{3}M_s^2.$$

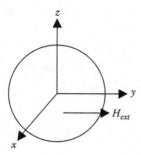

FIGURE S6.2A Magnetic spherical sample.

The internal field $\vec{H}_i$ is given by

$$(a-2)\ \vec{H}_i = -\nabla_{\vec{M}} F,$$

i.e.,

$$(a-3)\ H_{ix} = 0,\ H_{iy} = H_{ext},\ H_{iz} = 0.$$

If the sample is magnetically saturated, the static magnetization is

$$(a-4)\ M_x = 0,\ M_y = M_S,\ M_z = 0.$$

The static part of $\vec{H}_i$ can be written as

$$(a-5)\ H_x = 0,\ H_y = H_{ext},\ H_z = 0.$$

The dynamic part of $\vec{H}_i$ can be written as

$$(a-6)\ h_x = 0,\ h_y = 0,\ h_z = 0.$$

Neglecting higher order terms, the equation of motion can be written as

$$(a-7)\ \frac{1}{\gamma}\frac{d\vec{m}}{dt} = \vec{m}\times\vec{H} + \vec{M}\times\vec{h},$$

where $\vec{m}$ is the dynamic magnetization
   Substituting (4–6) into (7), we obtain

$$(a-8)\ \begin{cases} i(\omega/\gamma)m_x = -H_{ext}m_z \\ i(\omega/\gamma)m_y = 0. \\ i(\omega/\gamma)m_z = H_{ext}m_x \end{cases}$$

The condition that there are nonzero solutions for (8) is

$$(a-9) \quad \begin{vmatrix} i(\omega/\gamma) & H_{ext} \\ H_{ext} & -i(\omega/\gamma) \end{vmatrix} = 0,$$

i.e., the FMR condition:

$$(a-10) \quad \frac{\omega}{\gamma} = H_{ext}.$$

b)   For cylinder, the magnetic free energy can be written as (Figure S6.2b)

$$(b-1) \quad F = -M_x H_{ext} + \pi M_y^2 + \pi M_z^2.$$

The internal field $\vec{H}_i$ is given by

$$(b-2) \quad \vec{H}_i = -\nabla_{\vec{M}} F,$$

i.e.,

$$(b-3) \quad H_{ix} = H_{ext}, \ H_{iy} = -2\pi M_y, \ H_{iz} = -2\pi M_z.$$

The static magnetization is

$$(b-4) \quad M_x = M_S, \ M_y = 0, \ M_z = 0.$$

The static part of $\vec{H}_i$ can be written as

$$(b-5) \quad H_x = H_{ext}, \ H_y = 0, \ H_z = 0.$$

The dynamic part of $\vec{H}_i$ can be written as

$$(b-6) \quad h_x = 0, \ h_y = -2\pi m_y, \ h_z = -2\pi m_y.$$

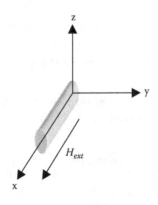

FIGURE S6.2B   Magnetic needle and field direction.

Neglecting higher order terms, the equation of motion can be written as

$$(b-7) \quad \frac{1}{\gamma}\frac{d\vec{m}}{dt} = \vec{m}\times\vec{H} + \vec{M}\times\vec{h},$$

where $\vec{m}$ is the dynamic magnetization

Substituting (4–6) into (7), we obtain

$$(b-8) \quad \left\{ \begin{array}{l} i(\omega/\gamma)m_x = 0 \\ i(\omega/\gamma)m_y = (H_{ext} + 2\pi M_S)m_z. \\ i(\omega/\gamma)m_z = (H_{ext} + 2\pi M_S)m_y \end{array} \right.$$

The condition that there are nonzero solutions for (8) gives

$$(b-9) \quad \begin{vmatrix} i(\omega/\gamma) & H_{ext} + 2\pi M_S \\ H_{ext} + 2\pi M_S & -i(\omega/\gamma) \end{vmatrix} = 0,$$

i.e., the FMR condition:

$$(b-10) \quad \frac{\omega}{\gamma} = H_{ext} + 2\pi M_S.$$

6.3 Assume the lava is a sphere and uniformly magnetized along the z-axis, the magnetic scalar potential outside is (Figure S6.3)

$$(1) \quad \varphi(\vec{R}) = \frac{R_0^3}{3}\frac{\vec{M}_0\cdot\vec{R}}{R^3},$$

where $\vec{M}_0$ is the magnetization of the lava and R_0 is the radius of the lava sphere. The magnetic field outside is given as

$$(2) \quad \vec{H} = -\nabla\varphi = -\frac{R_0^3}{3R^3}\left(\vec{M}_0 - 3\vec{M}_0\cdot\hat{R}\hat{R}\right).$$

On the equator, the second term in (2) is zero, and the magnetic field is

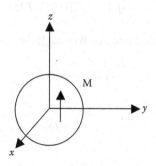

FIGURE S6.3 Model for "magnetic" lava.

$$(3) \ \vec{H} = -\frac{R_0^3}{3R^3} \vec{M}_0 \ \text{or} \ \vec{B} = -\frac{\mu_0 R_0^3}{3R^3} \vec{M}_0.$$

Substituting the values, $B = 5 \times 10^{-5}$ T, $\mu_0 = 4\pi \times 10^{-7}$, $R_0 = 3.5 \times 10^6$ m, and $R = 6.4 \times 10^6$ m, into (3), we obtain

$$(4) \ M_0 = 7.30 \times 10^2 \, \text{A/m} = 0.73 \, \text{emu/cm}^3.$$

Thus,

$$(5) \ m = 1.31 \times 10^{26} \ \text{emu}.$$

Inside the sphere, the magnetic scalar potential is

$$(6) \ \varphi(\vec{R}) = \frac{\vec{M}_0 \cdot \vec{R}}{3}.$$

Thus, the magnetic field is

$$(7) \ \vec{H} = -\nabla\varphi = -\frac{\vec{M}_0}{3}.$$

Substituting for M_0 from (4), we obtain

$$(8) \ H = 243 \, \text{A/m} = 3.05 \ \text{Oe}.$$

From the mass density of iron (Fe) 7.87 g/cm³, the number density of the lava is

$$(9) \ N = 8.4635 \times 10^{25} /\text{cm}^3.$$

Substituting H and N into

$$(10) \ M = Ng\beta SB(S;T),$$

where $g = 2.0$ is assumed and β is the Bohr magneton, we obtain

$$(11) \ S = 1.95.$$

This is between S of Fe^{3+} (5/2) and Ni^{2+} (1).

6.4   The magnetic free energy can be written as (Figure S6.4)

$$(1) \ F = -M_y H_{ext} + 2\pi M_x^2 - K_u \left(\frac{M_y}{M_s}\right)^2, \ K_u > 0.$$

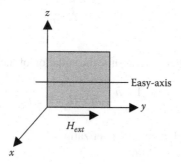

FIGURE S6.4 Direction of uniaxial magnetic axis relative to the external magnetic field.

The internal field $\vec{H}_i$ is given by

$$(2)\ \vec{H}_i = -\nabla_{\vec{M}} F,$$

i.e.,

$$(3)\ H_{ix} = -4\pi M_x,\ H_{iy} = H_{ext} + H_A \frac{2K_u}{M_s}\frac{M_y}{M_s},\ H_{iz} = 0.$$

If we define H_A as

$$(4)\ H_A = \frac{2K_u}{M_s},$$

(3) can be rewritten as

$$(3')\ H_{ix} = -4\pi M_x,\ H_{iy} = H_{ext}\frac{M_y}{M_s},\ H_{iz} = H_0.$$

If the sample is magnetically saturated, the static magnetization is

$$(4)\ M_x = 0,\ M_y = M_S,\ M_z = 0.$$

The static part of $\vec{H}_i$ can be written as

$$(5)\ H_x = 0,\ H_y = H_{ext} + H_A,\ H_i = 0.$$

The dynamic part of $\vec{H}_i$ can be written as

$$(6)\ h_x = -4\pi m_x,\ h_y = \frac{H_A}{M_s} m_y,\ h_z = 0.$$

To calculate the microwave susceptibility and FMR condition, the dynamic applied field is introduced as a perturbation to the dynamic part of $\vec{H}_i$. Thus, (6) can be rewritten as

$$(6') \; h_x = -4\pi m_x + h_{ax}, \; h_y = \frac{H_A}{M_s} m_y + h_{ay}, \; h_z = h_{az}.$$

Neglecting higher order terms, the equation of motion can be written as

$$(7) \; \frac{1}{\gamma} \frac{d\vec{m}}{dt} = \vec{m} \times \vec{H} + \vec{M} \times \vec{h},$$

where $\vec{m}$ is the dynamic magnetization.

Substituting (4–6) into (7), we obtain

$$(8) \begin{cases} i(\omega/\gamma)m_x = -(H_{ext} + H_A)m_z - M_s h_{az} \\ \qquad i(\omega/\gamma)m_y = 0 \\ i(\omega/\gamma)m_z = (H_{ext} + H_A + 4\pi M_s)m_x - M_s h_{ax} \end{cases}$$

$$\Rightarrow \begin{pmatrix} H_{ext} + H_A + 4\pi M_s & -i(\omega/\gamma) \\ i(\omega/\gamma) & H_{ext} + H_A \end{pmatrix} \begin{pmatrix} m_x \\ m_z \end{pmatrix} = M_s \begin{pmatrix} h_{ax} \\ h_{az} \end{pmatrix}.$$

Solving (8), we obtain the magnetic susceptibility

$$(9) \begin{pmatrix} m_x \\ m_z \end{pmatrix} = \frac{M_s}{\Delta} \begin{pmatrix} H_{ext} + H_A & i(\omega/\gamma) \\ -i(\omega/\gamma) & H_{ext} + H_A + 4\pi M_s \end{pmatrix} \begin{pmatrix} h_{ax} \\ h_{az} \end{pmatrix},$$

where

$$(10) \; \Delta = (H_{ext} + H_A)(H_{ext} + H_A + 4\pi M_s) - \left(\frac{\omega}{\gamma}\right)^2.$$

The coefficient matrix in (9) is the susceptibility matrix $\ddot{\chi}$, which can be diagonalized as

$$(11) \; \frac{M_s}{\Delta}$$

$$\begin{pmatrix} H_{ext} + H_A + 2\pi M_s + \sqrt{4\pi^2 M_s^2 + (\omega^2/\gamma^2)} & 0 \\ 0 & H_{ext} + H_A + 2\pi M_s - \sqrt{4\pi^2 M_s^2 + (\omega^2/\gamma^2)} \end{pmatrix}$$

Thus,

$$(12) \; \chi\pm = \frac{M_S}{\Delta}\left(H_{ext} + H_A + 2\pi M_S \pm \sqrt{4\pi^2 M_S^2 + \frac{\omega^2}{\gamma^2}}\right).$$

6.5 The magnetic free energy can be written as (Figure S6.5)

$$(1) \; F = -M_x H_{ext} + 2\pi M_x^2.$$

The internal field $\vec{H}_i$ is given by

$$(2) \; \vec{H}_i = -\nabla_{\vec{M}} F,$$

i.e.,

$$(3) \; H_{ix} = H_{ext} - 4\pi M_x, \; H_{iy} = 0, \; H_{iz} = 0.$$

If the sample is magnetically saturated, the static magnetization is

$$(4) \; M_x = M_S, \; M_y = 0, \; M_z = 0.$$

The static part of $\vec{H}_i$ can be written as

$$(5) \; H_x = H_{ext} - 4\pi M_S, \; H_y = 0, \; H_i = 0.$$

The dynamic part of $\vec{H}_i$ can be written as

$$(6) \; h_x = -4\pi m_x, \; h_y = 0, \; h_z = 0.$$

To calculate the microwave susceptibility, the dynamic applied field is introduced as a perturbation to the dynamic part of $\overline{H}_i$. Thus, (6) can be rewritten as

$$(6') \; h_x = -4\pi m_x + h_{ax}, \; h_y = h_{ay}, \; h_z = h_{az}.$$

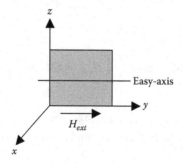

FIGURE S6.5 Field direction relative to the easy axis of magnetization.

Neglecting higher order terms, the equation of motion can be written as

$$(7) \quad \frac{1}{\gamma} \frac{d\vec{m}}{dt} = \vec{m} \times \vec{H} + \vec{M} \times \vec{h},$$

where $\vec{m}$ is the dynamic magnetization.

Substituting (4–6) into (7), we obtain

$$(8) \quad \begin{cases} i(\omega/\gamma)m_x = 0 \\ i(\omega/\gamma)m_y = (H_{ext} - 4\pi M_S)m_z - M_S h_{az} \\ i(\omega/\gamma)m_z = -(H_{ext} - 4\pi M_S)m_y - M_S h_{ay} \end{cases}$$

$$\Rightarrow \begin{pmatrix} H_{ext} - 4\pi M_S & i(\omega/\gamma) \\ -i(\omega/\gamma) & H_{ext} - 4\pi M_S \end{pmatrix} \begin{pmatrix} m_y \\ m_z \end{pmatrix} = M_S \begin{pmatrix} h_{ay} \\ h_{az} \end{pmatrix}.$$

Solving (8), we obtain the magnetic susceptibility

$$(9) \quad \begin{pmatrix} m_x \\ m_z \end{pmatrix} = \frac{M_S}{\Delta} \begin{pmatrix} H_{ext} - 4\pi M_S & -i(\omega/\gamma) \\ i(\omega/\gamma) & H_{ext} - 4\pi M_S \end{pmatrix} \begin{pmatrix} h_{ax} \\ h_{az} \end{pmatrix},$$

where

$$(10) \quad \Delta = (H_{ext} - 4\pi M_S)^2 - \left(\frac{\omega}{\gamma}\right)^2.$$

The coefficient matrix in (9) is the susceptibility matrix $\vec{\chi}$ and can be diagonalized as

$$(11) \quad \frac{M_S}{\Delta} \begin{pmatrix} H_{ext} - 4\pi M_S + (\omega/\gamma) & 0 \\ 0 & H_{ext} - 4\pi M_S - (\omega/\gamma) \end{pmatrix}.$$

Thus,

$$(12) \quad \chi\pm = \frac{M_S}{\Delta}\left(H_{ext} - 4\pi M_S \pm \frac{\omega}{\gamma}\right).$$

6.6

a) The uniaxial axis is in the plane (Figure S6.6a).
   The magnetic free energy can be written as

$$(1) \quad F = -M_z H_{ext} + 2\pi M_x^2 - K_u \left(\frac{M_y}{M_S}\right)^2, \quad K_u > 0.$$

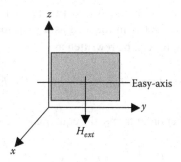

FIGURE S6.6A Field direction relative to the easy axis.

The internal field $\vec{H}_i$ is given by

$$(2)\ \vec{H}_i = -\nabla_{\vec{M}} F,$$

i.e.,

$$(3)\ H_{ix} = -4\pi M_x,\ H_{iy} = \frac{2K_u}{M_s} \frac{M_y}{M_s},\ H_{iz} = H_{ext}.$$

If we define H_A as

$$(4)\ H_A = \frac{2K_u}{M_s},$$

(3) can be rewritten as

$$(3')\ H_{ix} = -4\pi M_x,\ H_{iy} = H_A \frac{M_y}{M_s},\ H_{iz} = H_{ext}.$$

If the sample is magnetically saturated $(H > H_A)$, the static magnetization is

$$(4)\ M_x = 0,\ M_y = 0,\ M_z = M_S.$$

The static part of $\vec{H}_i$ can be written as

$$(5)\ H_x = 0,\ H_y = 0,\ H_i = H_{ext}.$$

The dynamic part of $\vec{H}_i$ can be written as

$$(6)\ h_x = -4\pi m_x,\ h_y = \frac{H_A}{M_s} m_y,\ h_z = 0.$$

To calculate the microwave susceptibility and FMR condition, the dynamic applied field is introduced as a perturbation to the dynamic part of $\vec{H}_i$. Thus, (6) can be rewritten as

$$(6')\ h_x = -4\pi m_x + h_{ax},\ h_y = \frac{H_A}{M_s} m_y + h_{ay},\ h_z = h_{az}.$$

Neglecting higher order terms, the equation of motion can be written as

$$(7)\ \frac{1}{\gamma}\frac{d\vec{m}}{dt} = \vec{m}\times\vec{H} + \vec{M}\times\vec{h},$$

where $\vec{m}$ is the dynamic magnetization.

Substituting (4–6) into (7), we obtain

$$(8)\ \begin{cases} i(\omega/\gamma)m_x = (H_{ext} - H_A)m_y - M_s h_{ay} \\ i(\omega/\gamma)m_y = -(H_{ext} - 4\pi M_S)m_x + M_S h_{ax} \\ i(\omega/\gamma)m_z = 0 \end{cases}$$

$$\Rightarrow \begin{pmatrix} H_{ext} + 4\pi M_S & i(\omega/\gamma) \\ -i(\omega/\gamma) & H_{ext} - H_A \end{pmatrix}\begin{pmatrix} m_x \\ m_y \end{pmatrix} = M_S \begin{pmatrix} h_{ax} \\ h_{ay} \end{pmatrix}.$$

Solving (8), we obtain the magnetic susceptibility

$$(9)\ \begin{pmatrix} m_x \\ m_y \end{pmatrix} = \frac{M_S}{\Delta}\begin{pmatrix} H_{ext} - H_A & -i(\omega/\gamma) \\ i(\omega/\gamma) & H_{ext} + 4\pi M_S \end{pmatrix}\begin{pmatrix} h_{ax} \\ h_{ay} \end{pmatrix},$$

where

$$(10)\ \Delta = (H_{ext} - H_A)(H_{ext} + 4\pi M_S) - \left(\frac{\omega}{\gamma}\right)^2.$$

The FMR condition is given by $\Delta = 0$, i.e.,

$$(11)\ \left(\frac{\omega}{\gamma}\right)^2 = (H_{ext} - H_A)(H_{ext} + 4\pi M_S).$$

Differentiating on both sides of (11), we obtain

$$(12)\ \frac{\Delta\omega}{\gamma} = \frac{H_{ext} - (H_A/2) + 2\pi M_S}{\omega/\gamma}\Delta H.$$

b. The uniaxial axis is normal to the film plane (Figure S.6b). The magnetic free energy can be written as

$$(1) \quad F = -M_z H_{ext} + 2\pi M_x^2 - K_u \left(\frac{M_x}{M_S} \right)^2, \quad K_u > 0.$$

The internal field $\vec{H}_i$ is given by

$$(2) \quad \vec{H}_i = -\nabla_{\vec{M}} F,$$

i.e.,

$$(3) \quad H_{ix} = \frac{2K_u}{M_s} \frac{M_x}{M_s} - 4\pi M_x, \quad H_{iy} = 0, \quad H_{iz} = H_{ext}.$$

If we define H_A as

$$(4) \quad H_A = \frac{2K_u}{M_s},$$

(3) can be rewritten as

$$(3') \quad H_{ix} = H_A \frac{M_x}{M_s} - 4\pi M_x, \quad H_{iy} = 0, \quad H_{iz} = H_{ext}.$$

If the sample is magnetically saturated $(H > H_A - 4\pi M_S)$, the static magnetization is

$$(4) \quad M_x = 0, \quad M_y = 0, \quad M_z = M_S.$$

The static part of $\vec{H}_i$ can be written as

$$(5) \quad H_x = 0, \quad H_y = 0, \quad H_i = H_{ext}.$$

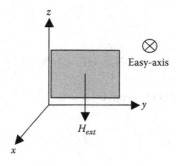

FIGURE S6.6B Field direction relative to the easy axis.

The dynamic part of $\vec{H}_i$ can be written as

$$(6) \ h_x = \frac{H_A}{M_s} m_x - 4\pi m_x, \ h_y = 0, \ h_z = 0.$$

To calculate the microwave susceptibility and FMR condition, the dynamic applied field is introduced as a perturbation to the dynamic part of $\vec{H}_i$. Thus, (6) can be rewritten as

$$(6') \ h_x = \frac{H_A}{M_s} m_y - 4\pi m_x + h_{ax}, \ h_y = h_{ay}, \ h_z = h_{az}.$$

Neglecting higher order terms, the equation of motion can be written as

$$(7) \ \frac{1}{\gamma} \frac{d\vec{m}}{dt} = \vec{m} \times \vec{H} + \vec{M} \times \vec{h},$$

where $\vec{m}$ is the dynamic magnetization.

Substituting (4–6) into (7), we obtain

$$(8) \left\{ \begin{array}{c} i(\omega/\gamma) m_x = H_{ext} - M_s h_{ay} \\ i(\omega/\gamma) m_y = -(H_{ext} - H_A + 4\pi M_s) m_x + M_s h_{ax} \\ i(\omega/\gamma) m_z = 0 \end{array} \right.$$

$$\Rightarrow \left(\begin{array}{cc} H_{ext} - H_A + 4\pi M_s & i(\omega/\gamma) \\ -i(\omega/\gamma) & H_{ext} \end{array} \right) \left(\begin{array}{c} m_x \\ m_y \end{array} \right) = M_s \left(\begin{array}{c} h_{ax} \\ h_{ay} \end{array} \right).$$

Solving (8), we obtain the magnetic susceptibility

$$(9) \left(\begin{array}{c} m_x \\ m_y \end{array} \right) = \frac{M_s}{\Delta} \left(\begin{array}{cc} H_{ext} & -i(\omega/\gamma) \\ i(\omega/\gamma) & H_{ext} - H_A + 4\pi M_s \end{array} \right) \left(\begin{array}{c} h_{ax} \\ h_{ay} \end{array} \right),$$

where

$$(10) \ \Delta = H_{ext}(H_{ext} - H_A + 4\pi M_s) - \left(\frac{\omega}{\gamma} \right)^2.$$

The FMR condition is given by $\Delta = 0$, i.e.,

$$(11) \left(\frac{\omega}{\gamma} \right)^2 = H_{ext}(H_{ext} - H_A + 4\pi M_s).$$

Differentiating on both sides of (11), we obtain

$$(12) \quad \frac{\Delta\omega}{\gamma} = \frac{H_{ext} - (H_A/2) + 2\pi M_S}{\omega/\gamma} \Delta H.$$

6.7

a. The uniaxial axis is in the plane.
The microwave susceptibility is given by (9) of Problem 6.6 part (a), i.e.,

$$(1) \quad \begin{pmatrix} \chi_{xx} & \chi_{xz} \\ \chi_{zx} & \chi_{zz} \end{pmatrix} = \frac{M_S}{\Delta} \begin{pmatrix} H_{ext} - H_A & -i(\omega/\gamma) \\ i(\omega/\gamma) & H_{ext} + 4\pi M_S \end{pmatrix}.$$

b. The uniaxial axis is normal to the plane.
The microwave susceptibility is given by (9) of Problem 6.6 part (b), i.e.,

$$(1) \quad \begin{pmatrix} \chi_{xx} & \chi_{xy} \\ \chi_{yx} & \chi_{yy} \end{pmatrix} = \frac{M_S}{\Delta} \begin{pmatrix} H_{ext} - H_A + 4\pi M_S & -i(\omega/\gamma) \\ i(\omega/\gamma) & H_{ext} \end{pmatrix}.$$

7 Electrical Properties of Magneto-Dielectric Films

BASIC DIFFERENCE BETWEEN ELECTRIC AND MAGNETIC DIPOLE MOMENTS

We saw in Chapter 2 that the magnetic dipole moment is defined as

$$\vec{m} = \frac{1}{2m} \int \vec{r} \times \vec{p} dq.$$

For simplicity, let $\vec{m} = m_z \vec{a}_z$ so that we may write

$$m_z = \frac{1}{2} \int \left(x \frac{\partial y}{\partial t} - y \frac{\partial x}{\partial t} \right) dq.$$

If we replace x and y by $-x$ and $-y$, m_z remains the same. We say that m_z is conserved under an operation of inversion symmetry, or parity is conserved. The basic definition of an electric dipole moment, $\vec{p}$, is as follows:

$$\vec{p} = q\vec{r}; \quad q = \text{charge}.$$

Clearly, if we replace $\vec{r}$ by $-\vec{r}$, the sign of $\vec{p}$ is changed. We say then that parity is not conserved. The implication of this basic difference will be illustrated, as we describe the microwave properties of electrical and magnetic materials. However, the electric dipole moment is conserved under time reversal symmetry operation, but not the dipole moment. As a result, magnetic materials are useful for non-reciprocal applications with respect to wave propagation directions, for example.

ELECTRIC DIPOLE ORIENTATION IN A FIELD

Consider a collection of single electric dipole moments dispersed randomly in a solid. In the absence of an electric field, the net moment is zero for the ensemble. In the presence of an external electric field, the dipole moments will attempt to align with the electric field. It is assumed that the electric field, $\vec{E}$, makes an angle θ with respect to $\vec{p}$. The thermal energy prevents total alignment, since motion imparted by thermal energy is isotropic. At any given time, the potential energy at any point in the ensemble is

$$W = -pE \cos\theta,$$

DOI: 10.1201/9781003431244-7

where

the angle θ can take on any continuous value between 0 and π

p is the magnitude of the electrical dipole moment

The thermal average of $\cos \theta$ may be calculated as follows:

$$\cos \theta = \frac{\displaystyle\int_0^\pi \cos \theta F(W) F(\theta) d\theta}{\displaystyle\int_0^\pi F(W) F(\theta) d\theta},$$

where

$$F(W) = e^{-W/kT}, \text{ where } k = \text{Boltzman's constant,}$$

$$F(\theta) = 2\pi \sin \theta.$$

Let $x = \cos \theta$ and $y = -W$, then

$$x = \frac{\displaystyle\int_{-1}^1 x e^{xy} dx}{\displaystyle\int_{-1}^1 e^{xy} dx}$$

which integrates to

$$x = \coth(y) - \frac{1}{y} = L(y),$$

where $L(y)$ is referred to in the literature as the Langevin function.

If n is the density of dipoles in the field $\vec{E}$, the total average polarization $\vec{P}$ is defined as

$$\vec{P} = n\vec{p}, \text{ where } n = \text{number of charges/cm}^3,$$

$$p \cong p\cos \theta = \frac{p^2 E}{3kT}\left(1 - \frac{p^2 E^2}{15k^2 T^2} + \cdots\right).$$

Thus,

$$\vec{P} \cong \alpha n \vec{E}, \text{ where } \alpha = \frac{p^2}{3kT}\left(1 - \frac{p^2 E^2}{15k^2 T^2} + \cdots\right).$$

In the above expression, we have ignored higher order terms, since they are small compared to unity. α is the polarizability coefficient.

EQUATION OF MOTION OF ELECTRICAL DIPOLE MOMENT IN A SOLID

For simplicity, let us assume that a linear chain of atoms are coupled to each other. They are coupled to each other by electrostatic force such that a periodic lattice along the chain is formed. The separation between atoms is x_0 at equilibrium. This means that any kind of extraneous force will be balanced by a restoring force, Kx, to put the atom back in equilibrium or at a distance x_0 away from neighboring atoms. We will not calculate x_0 from first principles, but simply assume that is how solids, even linear chains, are formed in nature. Thus, the potential energy of a given atom in the chain may simply be written as follows:

$$H = -qxE_x + \frac{1}{2}Kx^2.$$

As implied in the above discussion, we have assumed that the chain is along the x-direction. The equilibrium position of the atom is $x=0$ for $E_x=0$. The next atom is x_0 distance away. Here, x is the displacement relative to the equilibrium value of x_0, as shown in Figure 7.1. The equation of motion may be deduced from application of Hamilton–Jacobi equation:

$$\frac{dp_x}{dt} = -\frac{\partial H}{\partial x} = qE_x - Kx$$

or

$$m\frac{d^2x}{dt^2} = qE_x - Kx. \tag{7.1}$$

Let's drop the subscript on E and write

$$m\frac{d^2x}{dt^2} + m\beta\frac{dx}{dt} + Kx = qE,$$

where

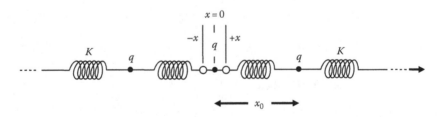

FIGURE 7.1 Lattice motion in a solid.

m = mass of charge or ion
q = electric charge
β = damping parameter (not to be confused with the Bohr magneton for now)
K = restoring elastic constant
x = displacement of charge relative to equilibrium position, x_0

In order to make Equation 7.1 realistic, we have put in a damping term in a phenomenological way (β term). Let's define the total polarization vector $\vec{P}$ as

$$\vec{P} = nqx\vec{a}_x.$$

Rewriting the equation of motion in terms of P, we have

$$\frac{d^2P}{dt^2} + \beta\frac{dP}{dt} + \frac{K}{m}P = \frac{q^2nE}{m}. \tag{7.2}$$

Assuming solutions of the form $e_j^{\omega t}$, we obtain

$$P = \frac{nq^2E/m}{\left(\omega_0^2 - \omega^2 + j\omega\beta\right)},$$

where $\omega_0^2 = K/m$.

The electric susceptibility (χ) may be defined as

$$\chi = \frac{P}{E\varepsilon_0} = \frac{nq^2s/m}{\varepsilon_0\left(\omega_0^2 - \omega^2 + j\omega\beta\right)}. \tag{7.3}$$

The electric displacement, D, is defined in terms of χ as follows:

$$D = \varepsilon_0\left(1+\chi\right)E = \varepsilon E \ (MKS),$$

which implies that

$$\varepsilon = \varepsilon_0 = \frac{nq^2/m}{\left(\omega_0^2 - \omega^2 + j\omega\beta\right)}.$$

FREE ENERGY OF ELECTRICAL MATERIALS

As in ferromagnetism, a free energy expression is derivable from first principles. We will not do so. We will simply write a general free energy appropriate to ferroelectric material whose crystal symmetry is cubic. In real solids, a given charge is coupled in three directions. The coupling obeys symmetry rules as expressed in the free energy. We don't determine the coupling coefficient from first principles, but simply adopt the results found in the literature. Thus,

$$F\left(\text{erg/cm}^3\right) = -\vec{P}\cdot\vec{E} + \frac{1}{2}C_{11}\sum_{i=1}^{3}e_{ii}^2 + C_{12}\sum_{i\neq j}^{3}e_{ii}e_{jj} + \frac{1}{2}C_{44}\sum_{i\neq j}^{3}e_{ij}^2$$

$$+A\sum_{i=1}^{3}P_i^2 + B\sum_{i=1}^{3}P_i^4 + C\sum_{i\neq j}^{3}P_i^2P_j^2, \tag{7.4}$$

where

$$\vec{P} = nq\vec{U},$$

$$e_{ii} = \frac{\partial U_i}{\partial x_i},$$

$$e_{ij} = \frac{\partial U_i}{\partial x_j} + \frac{\partial U_j}{\partial x_i},$$

and

$$x_1 = x;\ x_2 = y;\ x_3 = z.$$

$\vec{U}$ is the lattice displacement relative to its equilibrium position or $\vec{U} = \vec{r} - \vec{r}_0$. The first term in F is very analogous to the magnetizing energy term and is recognized as the electric polarizing energy. The next three terms are the elastic energies of the system and they are identified as the restoring potential energies (as in a harmonic oscillator). It is noted that this energy is represented by three parameters C_{11}, C_{12}, and C_{44}. This is a consequence of cubic symmetry assumed in the free energy. For hexagonal or other crystal structures, one would need a different form for the free energy. The last three terms correspond to cases in which there may be spontaneous electric polarization as in ferroelectric materials. Clearly, the total free energy must include both magnetic and elastic free energies. Therefore, let's include a coupling scheme to the magnetic system.

As an example, let us consider a ferroelectric material whose free energy exhibits uniaxial anisotropy energy of exactly the same form as Equation 6.14 for a magnetic material ordered ferromagnetic. If we identify the following transformation of field variables and anisotropy parameters, the free energy of a ferroelectric would be identical to Equation 6.14.

$$F = -M_xH_x - M_zH_z + 2\pi M_y^2 + K_u\left(\frac{M_x}{M}\right)^2, \qquad \text{except}$$

$$\vec{E}_0 \to \vec{H},$$

$$\vec{P} \to \vec{M},$$

$$K_e \to K_u,$$

where
    E_0 is the external electric field
    H is the external magnetic field
    P is the electric polarization
    M is the magnetization
    Ks are anisotropy parameters in units of ergs/cm^3

The beauty of this analogy is that we can adopt all of the mathematical formalism of Chapter 6. For example, for mathematical convenience, one may introduce the gradient operator defined as

$$\vec{\nabla}_P = \frac{\partial}{\partial P_x}\vec{a}_x + \frac{\partial}{\partial P_y}\vec{a}_y + \frac{\partial}{\partial P_z}\vec{a}_z,$$

which is the analog of the gradient operator $\vec{\nabla}_M$ used in Chapter 6 and introduced by this author in the 1960s. The application of such gradients allows for the internal fields (either electric or magnetic) to be expressed in terms of external fields. In the electric case, the internal field, $\vec{E}$, may be expressed in terms of $\vec{E}_0$ using

$$\vec{E} = -\vec{\nabla}F.$$

Adopting exactly the same analysis as in Chapter 6, the angle θ between $\vec{E}$ and $\vec{E}_0$ is calculated from

$$\cos\theta = \frac{E_0}{E_A},$$

where
    $E_A = 2K_e/P_0$
    P_0 is the spontaneous polarization of the ferroelectric material

The reader is referred to Equation 6.22a for the magnetic case.

MAGNETO-ELASTIC COUPLING

The coupling between elastic and magnetic motions needs to be introduced, if we want to combine free energies. This also applies if we want to include electric polarization effects. We designate the above elastic free energy as F_σ. Magneto-elastic coupling may be introduced by considering a simple argument cited in the literature (Chikazumi, 1997). It is well known experimentally that a magnetic material mechanically distorts upon the application of a strong magnetic field (see Figure 7.2).

For an unmagnetized spherical sample, the sample is not strained (dotted lines). For the sample magnetized along the z-direction, the spherical sample is strained or distorted, as shown in Figure 7.2. We see that the same distortion (or strain) is noted for magnetic polarization in the $\pm z$-directions, for example. The induced strain measured in a plane perpendicular to the z-axis is isotropic in that plane, and of course, possesses the same sign (independent of the azimuth angle, for example). However, the induced strain measured in the $+z$-direction is approximately $+\Delta R/R_0$, but when measured in the $-z$-direction, the strain is $-\Delta R/R_0$. Reversing the polarization direction along the z-axis does not alter this conclusion or result. In effect, this defines a unique axis for the magnetization direction whereby flipping the magnetization gives rise to the same mechanical distortions. We refer to the z-axis, in this example, as the uniaxial axis of the magnetization direction. As such, whatever coupling energy introduced must reflect the uniaxial symmetry. This means that the coupling energy must contain terms proportional to the magnetization squared, since uniaxial energy is expressed in terms only up to the square of the magnetization. However, the symmetry of the induced strain must be unidirectional, since the strain changes sign when measured along the $\pm z$-directions. In summary, we postulate that the magneto-elastic coupling must vary linearly with the strain (e) and the square of the magnetization (M) or

$$F_\lambda \propto eM^2.$$

For cubic crystal symmetry materials, F_λ is of the following form:

$$F_\lambda = B_1 \sum_{i=1}^{3} e_{ii}\alpha_{ii}^2 + B_2 \sum_{i \neq j}^{3} e_{ij}\alpha_i\alpha_j, \tag{7.5}$$

where $\alpha_i = M_i/M$ ($i = x$, y, and z).

The form of F_λ must also conform with the cubic symmetry of the crystal. This means that at equilibrium F_λ must be of cubic symmetry. Let's prove this point. At equilibrium, we have that

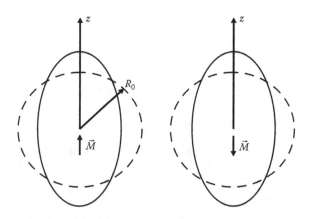

FIGURE 7.2 Strain induced by magnetic polarization along the z-axis.

$$\frac{\partial F_\lambda}{\partial e_{ii}} = 0 = B_1 \alpha_i^2 + C_{11} e_{ii} + C_{12} \sum_{i \neq j}^{3} e_{jj}$$

and

$$\frac{\partial F_\lambda}{\partial e_{ij}} = 0 = B_2 \alpha_i \alpha_j + C_{44} e_{ij}.$$

Solving for the strains at equilibrium, we obtain

$$e_{ii}^0 = \left(\frac{B_1}{C_{12} - C_{11}} \right) \alpha_i^2$$

and

$$e_{ij}^0 = -\left(\frac{B_2}{C_{44}} \right) \alpha_i \alpha_j,$$

where $i, j = x, y$, and z.

F_λ at equilibrium may be calculated as

$$F_\lambda^0 = \left(\frac{B_1^2}{C_{12} - C_{11}} \right) (\alpha_1^4 + \alpha_2^4 + \alpha_3^4) - \left(\frac{B_2^2}{C_{44}} \right) (\alpha_1^2 \alpha_2^2 + \alpha_1^2 \alpha_3^2 + \alpha_2^2 \alpha_3^2).$$

Indeed, the coupling term exhibits cubic symmetry as it should at equilibrium.

B_1 and B_2 may be related to the magnetostriction constants (see Chikazumi, 1997), λ_{100} and λ_{111}, as follows:

$$B_1 = \frac{3}{2}(C_{12} - C_{11})\lambda_{100}$$

and

$$B_2 = -3C_{44}\lambda_{111}.$$

For an isotropic magneto-elastic medium, F_λ may be written as follows (Chikazumi):

$$F_\lambda = -3\lambda_s \mu \left[\sum_{i=1}^{3} e_{ii} \alpha_i^2 + \frac{1}{2} \sum_{i \neq j}^{3} e_{ij} \alpha_i \alpha_j \right],$$

and the elastic energy as

$$F_\sigma = \frac{1}{2}(\lambda + 2\mu) \sum_{i=1}^{3} e_{ii}^2 + \frac{1}{2} \sum_{i \neq j}^{3} \left(\lambda e_{ii} e_{jj} + \frac{1}{2} \mu e_{ij}^2 \right),$$

where

    λ and μ are the Lame constants

    λ_s is the magnetostriction constant for an isotropic medium

Finally, the elastic equations of motion may be derived from the Hamilton–Jacobi principle of motion utilizing the free energy expression, F:

$$m\frac{d^2 U_i}{dt^2} = -\frac{\partial F}{\partial U_i} + \sum_{j=1}^{3} \frac{\partial}{\partial x_j}\frac{\partial F}{\partial e_{ij}},$$

where $x_i = x$, y, and z.

MICROWAVE PROPERTIES OF PERFECT CONDUCTORS

In a constant electric field $\vec{E}$, electrons accelerate according to Lorentz law as

$$m_q \frac{d\vec{v}_s}{dt} = q\vec{E} \tag{7.6}$$

where

    m_q = mass of carrier

    q = charge of carrier

    v_s = velocity of charge carrier

There is no restoring force term, since the charge carrier is free to move. The super-current density $\vec{J}_s$ is defined as

$$\vec{J}_s = n_s q \vec{v}_s, \tag{7.7a}$$

where n_s = carrier density. Solving for $\vec{J}_s$, we get

$$\dot{\vec{J}}_s = \frac{n_s q^2 \vec{E}}{m_q}. \tag{7.7b}$$

If we were to include friction in the flow of current, we would write Equation 7.6 as

$$\dot{\vec{v}}_s + \beta \vec{v}_s = \frac{q\vec{E}}{m_q}. \tag{7.8}$$

Combining Equations 7.7 and 7.8 and assuming sinusoidal electrical field drive of radial frequency, ω, we get

$$\vec{J}_s = \left(\frac{n_s q^2 \tau / m_q}{1 - j\omega\tau}\right)\vec{E}; \quad \tau = \frac{1}{\beta}.$$

The quantity in the bracket is recognized as the conductivity, σ, or

$$\sigma = \frac{n_s q^2 \tau / m_q}{1 - j\omega\tau}.$$

The parameter τ is the relaxation time and is related to the average elapsed time between collisions, for example. There are two limits to consider $\tau \to \infty$ and $\omega \neq 0$ and $\tau \to \infty$ and $\omega = 0$. These limits correspond to special properties of a perfect conductor, not a superconductor! For the first limit, we have

$$\sigma \to \frac{j n_s q^2}{m_q \omega}.$$

This says that the conductivity for a perfect conductor is purely imaginary at finite frequencies in contrast to Ohm's law. For the other limit, we have

$$\sigma \to \frac{n_s q^2 \tau}{m_q}.$$

This represents the DC conductivity, which is infinite as $\tau \to \infty$. A perfect conductor would exhibit infinite conductivity.

PRINCIPLES OF SUPERCONDUCTIVITY: TYPE I

Let's now consider the mechanism that gives rise to superconductivity—the Cooper pair charges. Consider a charge q moving in a closed path shown below. According to the BCS (see "References") theory, the net spin is zero and angular momentum is also zero. One possible way to obtain a net zero spin and orbital angular momentum ($S = L = 0$) is for two electrons to move in opposite velocities in orbit and having their spins oppose each other, see Figure 7.3. The charge of the particles is $2e$.

The energy is totally angular in motion which implies

$$H = \frac{P_\theta^2}{2mR^2},$$

where

P_θ is the angular momentum
R is the radius of the circular motion

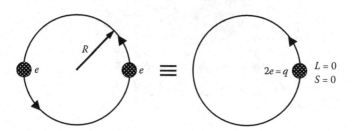

FIGURE 7.3 Hypothetical motion of Cooper pair.

In Figure 7.4, we show more of a detailed conception of the motion. Clearly, the Cooper pair must "see" the lattice providing the $+2|e|$ charge in order to neutralize the total charge. The mechanism of how the lattice motion induces the positive charge is well beyond the scope of this book. In a sense, the lattice and the Cooper pair constitute a three-body interacting system. This situation is very much analogous to the helium atom, whereby the two orbiting electrons form a stable bond with the nucleus. In essence, the induced charge on the lattice stabilizes the motion of the Cooper pair. According to the BCS theory, the spin state of the Cooper pair is $S=0$ and the orbital angular momentum is $L=0$. The excitation of a Cooper pair does not interfere with the motion of other Cooper pairs, since these excitations are boson excitations.

The super carriers involved in the conduction must necessarily be energetic and near the Fermi energy, E_F. However, when they form a stable bond with the lattice, their kinetic energy is lowered relative to the Fermi energy. Otherwise, electrons are "free" to move on singly. Hence, there is condensation of kinetic energy or lowering of the kinetic energy. This is a result of a stable bonding with the lattice. For each stable Cooper excitation, the energy of the Cooper pair is lowered by Δ amount of energy (we will estimate this later). If there are n_s excitations (Cooper pairs/cm³), it means that the kinetic energy of the ensemble of superconducting carriers is lowered by an amount of $(E_F - \Delta)n_s V = E_s$. V is the volume of the superconductor and E_s is the total kinetic energy lowered or the eigenstate energy of the whole ensemble of super carriers. It is a remarkable phenomenon of nature that says about 10^{21} super carriers that are confined into localized states exactly the same way, where each confinement lowers the kinetic energy by the same amount, Δ. Let's now estimate Δ.

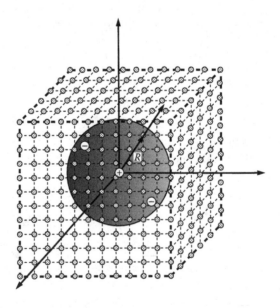

FIGURE 7.4 Motion of superconducting carriers about a lattice site with charge −2|e|. The total angular momentum is L=0 and spin angular momentum S=0.

The Hamiltonian is simply

$$H = -\frac{\hbar^2}{2m_q R^2} \frac{\partial^2}{\partial\theta^2},$$

since

$$P_\theta \rightarrow \hbar i \frac{\partial}{\partial\theta}.$$

Operating H by a wave function $\psi(\theta)$ such that it obeys the mathematical condition $\psi(\theta) = \psi(\theta + 2\pi)$ yields the eigenvalue E_n. The function is of the form

$$\psi(\theta) = e^{jn\theta}.$$

Thus,

$$H\psi(\theta) = E_n\psi(\theta) = \frac{\hbar^2 n^2}{2m_q R^2}\psi(\theta),$$

with

$$E_n = \frac{\hbar^2 n^2}{2m_q R^2}; \, n = 1, 2, 3, \ldots$$

The ground-state excitation is $n = 1$ and $\Delta = E_1$. There are no other bound states. We can put Δ in terms of the Fermi velocity, v_F, by realizing that the orbital motion is quantized in units of $\hbar$ or

$$\hbar = m_q R v_F.$$

Strictly speaking, the velocity of the super carriers is slightly lower than the Fermi velocity.

Thus, we approximate

$$\Delta \approx \frac{\hbar v_F}{2R}.$$

The radius R represents a measure of carrier confinement, and as such, it may be identified with the coherence length parameter, ξ, or $2R = \xi$. The ground-state energy of the Cooper pairs, Δ, is also identified with the energy band gap parameter of a superconductor, E_g, or $2\Delta = E_g$. The above equation may be then rewritten in terms of recognizable parameters.

$$E_g = \frac{2\hbar v_F}{\xi}.$$

Let's estimate ξ for a type I superconductor where typically $v_F \approx 1.5 \times 10^6$ m/s and $E_g \approx 3.2 \times 10^{-22}$ J yielding $\xi \approx 0.6 \times 10^{-6}$ m $\equiv 0.6\,\mu$m. For type II superconductors, oxide superconductors, or high T_c superconductors, E_g is much higher (as much as a factor of 100) (Figure 7.5).

ξ is much smaller than $1\,\mu$m experimentally. T_c, the transition temperature, may be calculated empirically from $T_c \approx E_g/4k$. More accurate prediction of T_c may be obtained from the BCS theory. For $E_g = 3.2 \times 10^{-22}$ J, $T_c \approx 5K$.

The confinement of the super carriers implies microscopic circular loops of current—similar to the classical picture in Chapter 2. According to Faraday's law (sometimes referred to as Lenz's law), the application of an external magnetic field density, B, induces current flow opposite to the current in the loop, which gives rise to the magnetic moment of that loop (see Chapter 2). Let's examine quantum mechanically the application of B to a superconductor. The Hamiltonian for this case becomes

$$H = -\frac{\hbar^2}{2m}\left(\frac{1}{R}\frac{\partial}{\partial\theta} - \frac{jqA_\theta}{\hbar}\right)^2.$$

where the magnetic vector potential component A_θ may be expressed in terms of B or the magnetic flux, Φ.

$$\vec{A} = \frac{1}{2}\left(\vec{B}\times\vec{r}\right) = \frac{1}{2}\left(\frac{\Phi}{\pi R^2}\vec{a}_z \times R\vec{a}_\rho\right).$$

Cylindrical coordinate system is assumed in the above expression. Thus, the Hamiltonian may be simplified as follows:

$$H = -\Delta\left(\frac{\partial}{\partial\theta} - j\frac{\Phi}{\Phi_0}\right)^2,$$

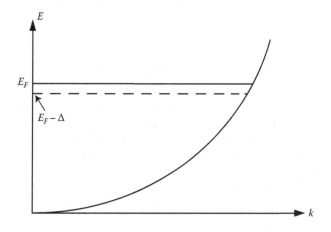

FIGURE 7.5   Energy, E, versus wave constant k. E_F is the Fermi energy.

where $\Phi_0 = 2\pi\hbar/q = 2.07 \times 10^{-15}$ Wb $= 2.07 \times 10^{-7}$ Oe cm^2. Φ_0 is defined as the quantum fluxoid.

As in the previous arguments, the wave function may be assumed to be $e^{jn\theta}$, and therefore, the eigenvalues of H as

$$E_n = \Delta\left(n - \frac{\Phi}{\Phi_0}\right)^2.$$

The ground-state energy is then

$$E_1 = \Delta\left(1 - \frac{\Phi}{\Phi_0}\right)^2.$$

As expected in the limit that $\Phi \to 0$ ($B=0$), the ground-state energy is Δ. The total current flow in this hypothetical microscopic Cooper loop is

$$i = -\frac{dE_1}{d\Phi} = \frac{2\Delta}{\Phi_0}\left(1 - \frac{\Phi}{\Phi_0}\right).$$

Indeed, the current flow due to Φ (B) opposes (the second term in parenthesis) the supercurrent within the loop (the first term in parenthesis). In Figure 7.6, a plot of i versus Φ/Φ_0 is shown. In the limit $\Phi \to \Phi_0$, $i \to 0$ and superconductivity is quenched and there are no supercurrents.

The magnetic field at which superconductivity is quenched is

$$H_c = \frac{\Phi_0}{\mu_0 \pi R^2} = \frac{4\Phi_0}{\mu_0 \pi \xi^2}. \tag{7.9a}$$

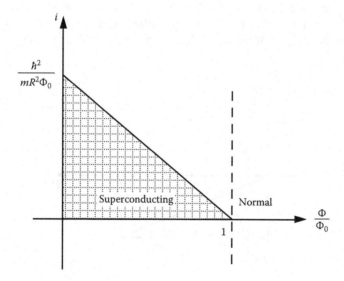

FIGURE 7.6 Induced current for single-loop Cooper carrier.

The induced magnetization due to the application of an external B field is by definition

$$M = n_s m = -n_s \left(i_{ind} \times \pi R^2 \right),$$

where n_s = charge density of Cooper pairs

$$i_{ind} = \text{current induced} = -2 \frac{\Delta \Phi}{\Phi_0^2}.$$

It is simple to show that

$$M = \frac{-2\pi n_s \Delta R^2 \Phi}{\Phi_0^2} = -\frac{2 n_s \Delta \mu_0 H}{\Phi_0^2}.$$

In Figure 7.7, we plot M versus H, the external applied magnetic field for $0 \le H \le H_c$.

Meissner postulated that in a superconductor, the magnetic field H is expelled such that the condition $B=0$ is obeyed within the superconductor (diamagnetic condition). This also means that $M=-H$ in a superconductor or slope of $M/H=-1$ for $H \le H_c$ (see Figure 7.7). Thus, n_s may be calculated simply from the above equation in order to maintain diamagnetism:

$$n_s \cong \frac{4m\Phi_0^2}{\pi^2 \hbar^2 \mu_0 \xi^2} \approx \frac{1}{3} \times 10^{21} \text{ carriers/cm}^3. \tag{7.9b}$$

In summary, in order to obtain complete diamagnetism, a large number of Cooper pair carriers are needed to accomplish this. In Figure 7.8, we show how all of these loops are involved in canceling the external field inside of a superconductor.

We see that the net current is zero inside the superconductor, but the resultant current at the surface is finite but sufficient in magnitude to screen the field inside the superconductor. The alternate point of view is that the screening currents flowing at the edge of the superconductor cancel the external field inside the superconductor.

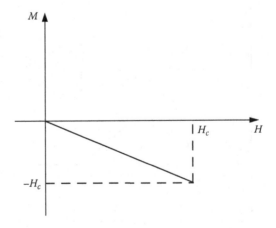

FIGURE 7.7   Magnetization versus external field, H, in a type I superconducting material.

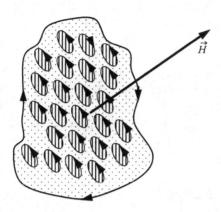

FIGURE 7.8   Internal supercurrents.

MAGNETIC SUSCEPTIBILITY OF SUPERCONDUCTORS: TYPE I

$B=0$ in a superconductor (Meissner effect). This means that $B=\mu_0(H+M)=0$. Since $M=-H$, we have that

$$\chi = \frac{M}{H} = -1.$$

The Meissner effect holds for a field H up to a critical field H_c. For $H \geq H_c$ in which superconductivity is quenched and the superconductor becomes a normal conductor, the B field is simply $B=\mu_0 H$ and $M=0$. Temperature and magnetic field have similar effect on superconductors in that both can quench superconductivity, as shown in Figure 7.9. For temperatures above T_c, the material is in a normal conducting state where only single carriers are involved in the conduction.

LONDON'S PENETRATION DEPTH

Consider a surface boundary that separates a superconductor and free space. In free space, we have

$$B_0 = \mu_0 H,$$

where H is the applied magnetic field. In the superconductor, we have (due to the Meissner effect)

$$B_s = \mu_0 (H + M) = 0.$$

This implies that at the surface $B_0 \neq B_s$. This contradicts one of the fundamental tenets of electromagnetic theory, which states that B normal to a surface is continuous. One way of satisfying the continuity of B at the surface and obeying the Meissner effect is to postulate that B attenuates to zero in some distance λ_L away from the surface. This is the postulate of London. This means that the current in each Cooper "loop"

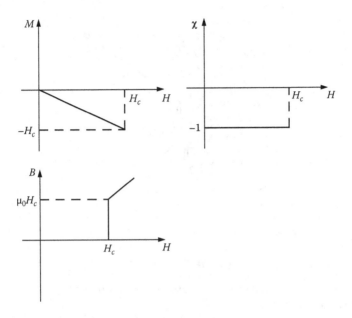

FIGURE 7.9 Magnetic properties of type I superconductors.

as measured from the surface must be different, since a B field can be allowed to exist near the surface. Thus, the decaying length is longer than ξ and is known as the London penetration depth. Physically, one can picture this as the current in the loops varying with distance away from the surface. This means that it is not necessary that each loop carry an identical amount of current. They can carry variable amounts of current as long as it is less than $2\Delta/\Phi_0$ in order to be in a superconducting state. Let's now calculate the decay rate of the screening current.

From Maxwell's equations, we have

$$\vec{\nabla} \times \vec{E} = -\dot{\vec{B}}_s \tag{7.10a}$$

and

$$\vec{\nabla} \times \vec{B} = \mu_0 \vec{J}_s. \tag{7.10b}$$

Displacement currents are negligible in superconductors. Using Equations 7.7b, 7.10a, and 7.10b, we derive the following relationship:

$$\dot{\vec{B}} = -\frac{m}{n_s q^2 \mu_0} \vec{\nabla} \times \vec{\nabla} \times \dot{\vec{B}}. \tag{7.11}$$

Without loss of generality, the above equation simplifies in one dimension to

$$\frac{\partial^2}{\partial x^2} \dot{\vec{B}} = \frac{\dot{\vec{B}}}{\lambda_L^2}, \tag{7.12}$$

where

$$\lambda_L = \frac{1}{q}\sqrt{\frac{m_q}{n_s\mu_0}}.$$

Assuming $n_s \approx (1/2) \times 10^{21}$ carriers/cm^3, we obtain $\lambda_L \approx 0.15\,\mu\text{m}$.

The parameter λ_L is identified as the London's penetration depth, and the solution to Equation 7.12 is

$$\dot{\vec{B}} = \dot{\vec{B}}_0 e^{-x/\lambda_L}. \tag{7.13}$$

$\dot{\vec{B}}$ is the time derivative of the flux density as a function of x inside the superconductor. $\dot{\vec{B}}_0$ is the value of $\dot{\vec{B}}$ at $x=0$, at the surface. This means that changes in flux density do not penetrate far below the surface. At distances far away from the surface, $\dot{\vec{B}} \to 0$, which implies a constant value of $B(x)$ contrary to the Meissner effect.

Now, we are back to the drawing board, H. London suggested that the magnetic behavior of a superconducting metal might be correctly described if it applied not only to $\dot{\vec{B}}$ but also to B itself. That is, the solution for B should take the same form as London assumed

$$B(x) = B_0 e^{-x/\lambda_L}.$$

This implies that the London's equations may be summarized as follows:

$$\vec{J} = \frac{\vec{E}}{\mu_0 \lambda_L^2}. \tag{7.14}$$

Equation 7.14 is very similar to Equation 7.7b. The other London's equation may be derived from taking the curl of Equation 7.14 and asserting as London did that $\dot{\vec{B}} \to \vec{B}$ for physical solution for the attenuation of B in a superconductor. This yields the relation

$$\vec{\nabla} \times \vec{J}_s = -\frac{\vec{B}}{\mu_0 \lambda_L^2} = -\frac{\vec{H}}{\lambda_L^2}. \tag{7.15}$$

The London equations do not replace Maxwell's equations, but are additional conditions obeyed by supercurrents.

TYPE II SUPERCONDUCTORS

There are two critical magnetic fields associated with type II superconductors, H_{c1} and H_{c2}. H_{c1} is defined as the field below which the superconductor obeys the Meissner effect. For fields between H_{c1} and H_{c2}, the sample is made up of superconducting and normal regions. For fields above H_{c2}, the material is normal. For fields

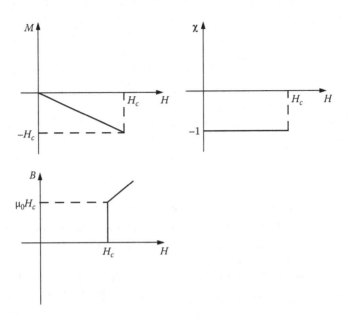

FIGURE 7.10   Type II superconductors.

H greater than H_{c2}, B is simply related to H as $B = \mu_0 H$. The slope of the dotted line shown in Figure 7.10 is equal to μ_0.

Let's now estimate both H_{c1} and H_{c2}. We assume that the magnetic field penetrates the sample in units of Φ_0. Assume further that a magnetic flux strength of the order of Φ_0 is excited within a column of diameter 2ξ, where ξ is defined as the coherence length (Figure 7.11).

Since the region inside this fluxoidal tube is normal, the free energy is that of a normal metal:

$$F_n(J) = E_s + n_s \Delta \pi \xi^2 t - \frac{1}{2} \mu_0 H^2 \pi \xi^2 t, \text{ MKS}, \tag{7.16}$$

where the first term E_s is the energy state of the superconductor, the second term is the total amount of energy required to quench superconductivity so that the medium is in a normal state, and the third term is the magnetic potential energy upon the presence of an H field in the fluxoidal region.

The sum of the first and second terms is simply the Fermi energy of a single carrier times the number of carriers. The thickness of the superconducting film is t and the cross area of the fluxoidal region is $\pi \xi^2$. For simplicity, let's take the reference point as $E_s = 0$. Outside of the fluxoidal region, the magnetic field is decaying within a distance λ_L. The free energy consists only of the magnetization energy, since the free energies of both the normal and superconducting states are referred with respect to the superconducting state. Thus, outside of the fluxoidal region (superconducting region), we have

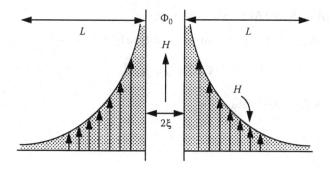

FIGURE 7.11   Fields near fluxoidal region.

$$F_s \cong -\frac{1}{2}\mu_0 H^2 \pi \left(\lambda_L^2 - \xi^2 \right) t. \tag{7.17}$$

We have overestimated the magnetic potential energy in this region by assuming H to be uniform over the length scale of λ_L. It is noted that normal state contribution to the free energy in Equation 7.17 is subtracted from the free energy of the pure superconducting state. The total free energy is the sum of the two free energies or

$$F_T \cong \pi t \left(\xi^2 n_s \Delta - \frac{1}{2}\mu_0 H^2 \lambda_L^2 \right). \tag{7.18}$$

H_{c1} may be determined by setting $F_T = 0$ yielding

$$H_{c1} = \frac{\xi}{\lambda_L} \sqrt{\frac{2 n_s \Delta}{\mu_0}} = \frac{\xi}{\lambda_L} H_c, \tag{7.19}$$

where $H_c = \sqrt{2 n_s \Delta / \mu_0}$ is the critical field for a type I superconductor. Let's prove that this is the case. We have that

$$n_s = \frac{4 m \Phi_0^2}{\pi^2 \hbar^2 \xi^2 \mu_0} = \frac{m \mu_0 \xi^2 H_c^2}{4 \hbar^2} = \frac{\mu_0 H_c^2}{2 \Delta}, \tag{7.20}$$

where
$$H_c = 4\Phi_0 / \mu_0 \pi \xi^2$$
$$\Delta = 2\hbar^2 / m \xi^2$$

Solving for H_c from the above set of equations yields the same result as above.
    H_{c2} was calculated by Abrikosov to be approximately equal to

$$H_{c2} \cong \sqrt{2} \left(\frac{\lambda_L}{\xi} \right) H_c.$$

Recall that for type II superconductors, $\xi \leq \lambda_L$, and typically, $\xi / \lambda_L \approx 2$–3 (see Kittel, 1975). H_c is in the order of (assuming $n_s \approx 0.3 \times 10^{27}$ charges/m^3, and $\Delta = 2 \times 10^{-22}$ J) 10^5 A/m or about 1,000 Oe.

MICROWAVE SURFACE IMPEDANCE

The microwave properties of this mixed state (type II) may be summarized as follows:

$$\vec{\nabla} \times \vec{E} = \vec{J} + j\omega\varepsilon\vec{E},$$

where now in the mixed state

$$\vec{J} = \vec{J}_n + \vec{J}_s,$$ (7.21)

where

$\vec{J}_n$ is the normal current density ($\sigma\vec{E}$)
$\vec{J}_s$ is the superconducting current density

Taking the curl of the above equation yields the following:

$$\vec{\nabla} \times \vec{\nabla} \times \vec{H} = \sigma\vec{\nabla} \times \vec{E} + \vec{\nabla} \times \vec{J}_s + j\omega\varepsilon\vec{\nabla} \times \vec{E}.$$

With the condition that $\vec{\nabla} \cdot \vec{B} = 0$, the above equation may be rewritten as

$$\nabla^2\vec{H} = \left(j\omega\mu_0\sigma + \frac{1}{\lambda_L^2} - \omega^2\mu_0\varepsilon \right)\vec{H} = -\omega^2\mu_0\varepsilon_{eff}\vec{H},$$ (7.22)

where

$$\varepsilon_{eff} = \varepsilon - j\frac{\sigma}{\omega} - \frac{1}{\omega^2\mu_0\lambda_L^2}.$$

The dielectric term is negligible compared to the other terms. It is instructive to describe the microwave properties in terms of ε_{eff} and μ_0. For example, the electromagnetic characteristic impedance may be obtained from

$$Z_0 = \sqrt{\frac{\mu_0}{\varepsilon_{eff}}} = \frac{j\omega\mu_0\lambda_L}{\sqrt{1 + 2j(\lambda_L/\delta)^2}}.$$ (7.23)

Usually, $\lambda_L < \delta$ so that

$$Z_0 \cong j\omega\mu_0\lambda_L\left(1 - j\frac{\lambda_L^2}{\delta^2} \right),$$ (7.24)

where δ is the classical skin depth and is equal to $\sqrt{2/\omega\mu_0\sigma}$. Clearly, the characteristic impedance is inductive as well as resistive. The electromagnetic propagation constant, k, may be calculated from

$$k = \omega\sqrt{\mu_0\varepsilon_{eff}} = -j\frac{1}{\lambda_L}\sqrt{1 + 2j\left(\frac{\lambda_L}{\delta} \right)^2} = \beta - j\alpha.$$

The real part of k, β, is the phase propagation constant and α the attenuation constant. For $\lambda_L < \delta$, we may approximate as follows:

$$\sigma \cong \frac{1}{\lambda_L}\sqrt{1+\left(\frac{\lambda_L}{\delta}\right)^4}$$

and

$$\beta \cong \frac{1}{\lambda_L}\left(\frac{\lambda_L}{\delta}\right)^2.$$

CONDUCTION THROUGH A NON-SUPERCONDUCTING CONSTRICTION

In Figure 7.12, we show a device consisting of two superconducting layers and one normal conducting layer.

The thickness of the thin normal conducting layer is in the order of λ_L or ξ. The normal material can be an insulator or a normal metal. Let's designate the superconductors on the left and right sides of the thin layer (junction) by their respective wave functions, ψ_1 and ψ_2:

$$\psi_1 = \sqrt{n_{s1}}\,e^{i\theta_1} \tag{7.25}$$

and

$$\psi_2 = \sqrt{n_{s2}}\,e^{i\theta_2}, \tag{7.26}$$

where
n_{si} = density of Cooper pairs at site i (1,2)
θ_i = phase angle of the wave

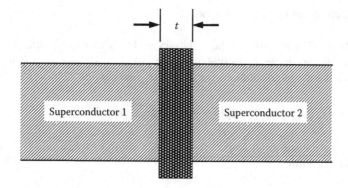

FIGURE 7.12   Conduction through a non-superconducting layer.

We have followed Feynman's formulation in Equations 7.25 and 7.26. The time evolution of the wave functions in the two regions without the junction is

$$ i\hbar \frac{\partial \psi_i}{\partial t} = E_{si}\psi_i, $$

where E_{si} is the eigenstate energy of the ensemble of superconducting carriers in the two regions ($i = 1,2$). With the introduction of the constriction layer, the equations of motion are modified as follows:

$$ i\hbar \frac{\partial \psi_1}{\partial t} = E_{s1}\psi_1 + C\psi_2 \tag{7.27} $$

and

$$ i\hbar \frac{\partial \psi_2}{\partial t} = E_{s2}\psi_2 + C\psi_1. \tag{7.28} $$

The coupling constant C is in units of joules and is a measure of how many carriers have crossed the junction from one region to the other. Putting a battery across the junction promotes conduction across the junction, and as such, the above equations are modified as follows:

$$ i\hbar \frac{\partial \psi_1}{\partial t} = \frac{qV}{2}\psi_1 + C\psi_2 $$

and

$$ i\hbar \frac{\partial \psi_2}{\partial t} = -\frac{qV}{2}\psi_2 + C\psi_1, $$

where
  V is the voltage applied across the junction
  $E_{s1} - E_{s2} = qV$
  $E_{s1} + E_{s2} = 0$
  q is the charge of the Cooper carrier

The condition $E_{s1} + E_{s2} = 0$ is rather arbitrary and it is chosen in order to simplify the algebra. For example, $E_{s1} = qV/2$ and $E_{s2} = -qV/2$ satisfy our assumptions. After substituting the assumed solutions for ψ_1 and ψ_2, one obtains

$$ i\frac{\dot{n}_{s1}}{2} - n_{s1}\dot{\theta}_1 = \frac{qV}{2\hbar} + C\sqrt{n_{s1}n_{s2}}\,e^{i(\theta_2 - \theta_1)}. $$

Interchanging the subscript from 1 to 2 yields a similar expression. This results in the following solutions:

$$ \dot{n}_{s1} = \frac{2C}{\hbar}\sqrt{n_{s1}n_{s2}}\,\sin(\theta_2 - \theta_1), $$

$$\dot{n}_{s2} = -\frac{2C}{\hbar}\sqrt{n_{s1}n_{s2}}\sin(\theta_2 - \theta_1),$$

$$\theta_1 = -\frac{C}{\hbar}\sqrt{\frac{n_{s2}}{n_{s1}}}\cos(\theta_2 - \theta_1) - \frac{qV}{2\hbar},$$

and

$$\theta_2 = -\frac{C}{\hbar}\sqrt{\frac{n_{1s}}{n_{s2}}}\cos(\theta_2 - \theta_1) + \frac{qV}{2\hbar}.$$

The current density flow is simply

$$J = q\ell\dot{n}_{s1} = \frac{2qC\ell}{\hbar}\sqrt{n_{s1}n_{s2}}\sin(\theta_2 - \theta_1).$$

The length scale ℓ is included only to make the units of J correct, A/m^2. Let $n_{s1} = n_{s2} = n_s$ so that

$$J = J_0 \sin(\theta_2 - \theta_1), \tag{7.29}$$

where

$$J_0 = \frac{2qC\ell n_s}{\hbar}.$$

The above relation is often referred to as one of the Josephson relationships, and the other Josephson relationship may be obtained from the condition that

$$\dot{\theta}_2 - \dot{\theta}_1 = \frac{qV}{\hbar}$$

resulting in

$$\delta\theta = \delta\theta_0 + \frac{q}{\hbar}\int V dt. \tag{7.30}$$

$\delta\theta_0$ is the phase angle difference between ψ_2 and ψ_1 at $t=0$. Combining the two Josephson equations, we have the final result:

$$J = J_0 \sin\left(\delta\theta_0 + \frac{q}{\hbar}\int V dt\right). \tag{7.31}$$

ISOTOPIC SPIN REPRESENTATION OF FEYNMAN EQUATIONS

We have implied that both n_{s1} and n_{s2} are large numbers. However, in some practical devices, they may be finite or in some cases relatively small. For small n_{s1} and n_{s2}, Josephson equations need to be modified. But how? The basis of Feynman equations is that as the supercharges move through a thin, normal region, the wave function of the carrier maintains its integrity. The wave function ψ_i describes the state of a collection of n_{si} particles with the same energy. Hence, the superconductivity wave function is a boson wave function, since there can be more than one carrier in the same energy state. Thus, let's designate ψ_i as a boson operator rather than a wave function. The operator is special, since it represents the motion of transporting one carrier across the junction. They are like the raising and lowering operators often cited in the literature in quantum mechanics. They raise or lower the number of carriers on one or the other side of the junction by one carrier. The operators ψ_1 and ψ_2 obey the following algebraic properties:

$$\psi_1 | n_1, n_2 = \sqrt{n_1} | n_1 - 1, n_2 \text{ and } \psi_1^* | n_1, n_2 = \sqrt{n_1 + 1} | n_1 + 1, n_2, \qquad (7.32)$$

where $| n_1, n_2 \rangle$ is the eigenstate of the boson operators. ψ and ψ^* are the lowering and raising boson operators, respectively. We hypothesize that both ψ_1 and ψ_2 obey identically the same equation of motion as that of Feynman derived in the "Conduction through a Non-Superconducting Constriction" section. As operators, the equation of motion may be derived from the commutation relations:

$$i\hbar \frac{\partial \psi_1}{\partial t} = [\psi_1, H] \qquad (7.33a)$$

and

$$i\hbar \frac{\partial \psi_2}{\partial t} = [\psi_2, H]. \qquad (7.33b)$$

In order to put the equations of motion in the same form as that of Feynman, we require that the Hamiltonian be of this form:

$$H = -\frac{qV}{2}\left(\psi_1^*\psi_1 - \psi_2^*\psi_2\right) + C\left(\psi_1^*\psi_2 + \psi_2^*\psi_1\right). \qquad (7.34)$$

Let's apply the commutation relation of Equation 7.33a by considering the first term in Equation 7.34, $\psi_1^*\psi_1$. Thus, write

$$i\hbar \frac{\partial \psi_1}{\partial t} = -\frac{qV}{2}\left[\psi_1, \psi_1^*\psi_1\right].$$

Utilizing the commutation identity for quantum operators,

$$[D, AB] = A[D, B] + [D, A]B,$$

gives the result of the commutation as

$$i\hbar\frac{\partial\psi_1}{\partial t} = \psi_1^*[\psi_1,\psi_1] + [\psi_1,\psi_1^*]\psi_1 = (\psi_1\psi_1^* - \psi_1^*\psi_1)\psi_1,$$

where

$$(\psi_1\psi_1^* - \psi_1^*\psi_1)|n_1,n_2 = (n_1 + 1 - n_1)|n_1,n_2.$$

Indeed,

$$i\hbar\frac{\partial\psi_1}{\partial t} = -\frac{qV}{2}[\psi_1,\psi_1^*\psi_1] = -\frac{qV}{2}\psi_1.$$

By repeating the same procedure outlined above, it is rather straightforward to show that the proposed Hamiltonian is the correct one giving rise to the same equation of motion for the operators ψ_1 and ψ_2, as in Feynman equations (Equations 7.27 and 7.28).

Let's designate the combination of operators in Equation 7.34 with new operators (isotopic spins):

$$T_z = \frac{1}{2}(\psi_1^*\psi_1 - \psi_2^*\psi_2) \tag{7.35}$$

and

$$T_x = \frac{1}{2}(\psi_1^*\psi_2 + \psi_2^*\psi_1) \tag{7.36}$$

so that the Hamiltonian expressed in Equation 7.34 may be expressed as

$$H = -qVT_z + 2CT_x. \tag{7.37}$$

T_x, T_y, and T_z may be derived from the spin matrix σ_α, $\alpha = x$, y, and z.

$$T_\alpha = \frac{1}{2}\psi^\dagger\sigma_\alpha\psi, \tag{7.38}$$

where

$$\psi = \begin{pmatrix} \psi_1 \\ \psi_2 \end{pmatrix} \text{ and } \psi^\dagger = (\psi_1^*\psi_2^*)$$

and

$$\sigma_x = \frac{1}{2}\begin{pmatrix} 0 & 1 \\ 1 & 0 \end{pmatrix}; \sigma_y = \frac{1}{2}\begin{pmatrix} 0 & -i \\ i & 0 \end{pmatrix}; \sigma_z = \frac{1}{2}\begin{pmatrix} 1 & 0 \\ 0 & -1 \end{pmatrix} - \text{spin matrices.}$$

The eigenvalue of the operator T_z is simply

$$\langle n_1, n_2 | T_z | n_1, n_2 \rangle = \left\langle \ldots \left| \frac{1}{2} \left(\psi_1^* \psi_1 - \psi_2^* \psi_2 \right) \right| \ldots \right\rangle = \frac{1}{2}(n_1 - n_2). \tag{7.39}$$

Clearly, $\frac{1}{2}(n_1 - n_2)$ represents the maximum value of the expected value of T_z and is designated as T much like in the spin representation whereby the expected value must range from $-T$ to T in increments of the integer m. For $n_1 > n_2$, the maximum expected value of T_z is T, and for $n_1 < n_2$, the minimum expected value is then $-T$. As in the real spin case, the eigenstate representation of T_z may simply be written as

$$T_z | m, T \rangle = m | m, T \rangle.$$

As such, this implies that $m, \langle T | T_x^2 + T_y^2 + T_z^2 | m, T \rangle = T(T+1)$.

We define T_x, T_y, and T_z as isotopic spin operators obeying the same algebraic commutation relationships as the real spin operators do. For example,

$$iT_x = \left[T_y, T_z \right],$$

$$iT_y = \left[T_z, T_x \right],$$

and

$$iT_z = \left[T_x, T_y \right].$$

The commutation relations are very much analogous to a single spin (real spin) under the influence of an external static magnetic field, H_a, and microwave drive field, $h(t)$.

The Hamiltonian in such case is of the form (CGS)

$$H = -g\mu_B H_a S_z - g\mu_B h(t) S_x. \tag{7.40}$$

While S_x couples the spin states $\left| \frac{1}{2}.S \right\rangle$ and $\left| -\frac{1}{2}.S \right\rangle$, T_x couples various states of $| m, T \rangle$. The transition between various m states implies a transition between regions 1 and 2 or conduction between regions 1 and 2, for example. As it is well known, the transition frequency between the two spin states is

$$\omega = \gamma H_a.$$

This is recognized as the Larmor precessional frequency of the spin motion. This implies that there must also exist a "natural" frequency associated with the transition between m states in the case of the isotopic spins. Let's determine this "natural" frequency. As in the case of the spin, the Larmor frequency may be calculated from the equation of motion of S_x. Thus, let's consider the equation of motion for T_α. Thus,

$$i\hbar \frac{\partial T_\alpha}{\partial t} = [T_\alpha, H]$$

resulting in

$$\hbar \dot{T}_z = 2CT_y,$$

$$\hbar \dot{T}_y = -qVT_x - 2CT_z,$$

and

$$\hbar \dot{T}_x = qVT_y.$$

The uncoupled equation for T_y is

$$\ddot{T}_y + \left(\frac{q^2V^2}{\hbar^2} + \frac{4C^2}{\hbar^2} \right) T_y = 0,$$

and the eigenvalue is

$$\hbar \omega = \sqrt{q^2V^2 + 4C^2}$$

and

$$T_y = A \sin(\omega t).$$

If the junction constricts the current flow, $qV \gg 4C$,

$$\omega = \frac{d\theta}{dt} = \frac{qV}{\hbar},$$

which leads to the first Josephson's equation. The other Josephson's equation may be obtained from

$$J = q\dot{T}_z \left(\frac{\ell}{V} \right) = q \frac{2C}{\hbar} \left(\frac{\ell}{V} \right) T_y = q \frac{2C}{\hbar} \left(\frac{\ell}{V} \right) A \sin(\omega t).$$

The factor ℓ/V is to make J dimensionally correct and V is the volume. A may be determined from Figure 7.13:

$$A = \sqrt{T(T+1) - m^2} \cong T \sqrt{1 - \frac{m^2}{T(T+1)}} = n \sqrt{1 - \frac{m^2}{T(T+1)}},$$

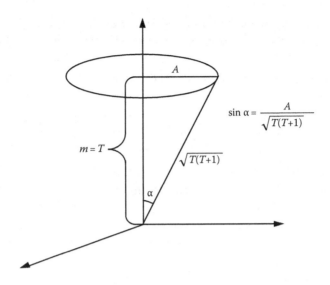

FIGURE 7.13   Determination of sin α.

where

$$n = \frac{1}{2}(n_1 - n_2).$$

Finally, we have the remaining Josephson equation as

$$J = J_0 \sin\left(\theta_0 + \frac{q}{\hbar}\int V dt\right),$$

where

$$J_0 = \frac{2qC(\ell/V)n}{\hbar}\sqrt{1 - \frac{m^2}{T(T+1)}}.$$

PROBLEMS

7.1   Show that the third-order expansion of $L(y)$ about $y=0$ is

$$L(y) = \frac{y}{3} - \frac{y^3}{45} + ...$$

7.2   The Langevin theory applies to dipoles, which may assume any orientation in space. This is not the case in crystal lattices. For a 3D cubic lattice, the dipoles can be oriented only parallel or antiparallel to the E-field. Calculate $\langle \cos \theta \rangle$ as a function of y about $y=0$ for the field parallel to one of the lattice directions. The dipoles may have six possible orientations.

7.3   Calculate ω_0 so that $\varepsilon_R = 10$ at 9 GHz. You may assume the loss factor $\beta = 0$.

7.4 (a) Calculate the dielectric tensor from the general free energy (Equation 7.4). (b) In a magneto-dielectric system, write the total free energy.

7.5 Determine the magnetic field value at which the weight of a superconductor ring sphere of 0.1 mm diameter and density $= 5$ g/cm^3 equals the force exerted on the sphere by a nonuniform magnetic field, as shown in Figure P7.5a and b.

a. $H_i = H_{ext} + H_D = H_{ext} - \dfrac{4\pi M}{3}$; $B_i = H_i + 4\pi M = 0$,

$$M = -\dfrac{3}{2}\left(\dfrac{H_{ext}}{4\pi}\right).$$

b. $F = m\dfrac{\partial B}{\partial x}$ (force equation).

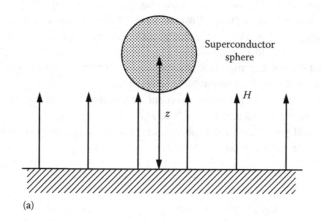

(a)

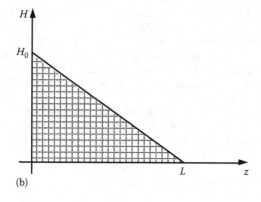

(b)

FIGURE P7.5 A spherical superconducting sample suspended above surface in the presence of a H field.

7.6 Refer to the literature for the values of ξ, n_s, and Δ of well-known super-conductors and calculate H_{c1} and H_{c2} of type II superconductors.

7.7 Write down the equation for the current in a circuit consisting of a resistor R connected in series with a superconducting Josephson junction and voltage V in the circuit.

7.8 A superconductor fills the half-space $x > 0$. In the vacuum ($x < 0$), there is a uniform magnetic field H parallel to y. Find the magnetic field and current distributions in the superconductor in the static case.

7.9 Find the force per unit area on the surface of the superconductor. What is the direction of that force?

7.10 A magnetic field is applied to a superconducting plane coil of self-inductance L through which a current J flows. Determine the current J' which flows in the coil. The cross-sectional area of the coil is S, and the normal to the coil plane makes an angle θ with the direction of H.

7.11 A superconducting coil of self-inductance L is in the normal state in an external magnetic field (the magnetic flux through the coil equals Φ_0). The temperature is lowered so that the coil is in a superconducting state. What is the current that will flow through the coil if the external magnetic field is removed?

7.12 Determine the magnetic moment of a superconducting sphere with radius $R \ll \lambda_L$ in a magnetic field H.

7.13 Determine the magnetic moment of a superconducting sphere with radius $R \gg \lambda_L$ in a magnetic field H.

7.14 Would you expect the Gilbert damping parameter to be isotropic? No solution is provided.

BIBLIOGRAPHY

P.W. Anderson, *Phys. Rev.*, **112**, 1900, 1958.

S. Chikazumi, *Physics of Ferromagnetism*, Clarendon Press, Oxford, U.K., 1997.

R. Coelho, *Physics of Dielectrics*, Elsevier Scientific, Amsterdam, 1979.

E. Fatuzzo and W.J. Merz, *Ferroelectricity*, Interscience, John Wiley & Sons, New York, 1967.

T.L. Gilbert, *IEEE Trans. Mag.*, **40**, 3443, 2004.

J. Hinken, *Superconductor Electronics*, Springer-Verlag, Berlin, 1990.

C. Kittel, *Introduction to Solid State Physics*, John Wiley & Sons, New York, 1975.

C.G. Kuper, *Theory of Superconductivity*, Clarendon Press, Oxford, U.K., 1968.

T.P. Orlando and K.A. Delin, *Foundations of Applied Superconductivity*, Addison-Wesley, Reading, MA, 1991, pp. 575–577.

C.H. Slichter, *Principles of Magnetic Resonance*, Harper & Row, New York, 1963.

M. Tinkham, *Introduction to Superconductivity*, McGraw-Hill, New York, 1975.

A. Widom, Y.N. Srivastava, and C. Vittoria, *Phys. Rev. B*, **46**, 13964, 1992.

SOLUTIONS

7.1

$$L(y) = \coth(y) - \frac{1}{y}.$$

The Laurent series of coth(y) is given by

$$\coth(y) = \frac{1}{y} + \frac{1}{3}y - \frac{1}{45}y^3 + \frac{2}{945}y^5 - \dots$$

So, we get

$$L(y) = \frac{y}{3} - \frac{y^3}{45} + \dots$$

7.2 When dipoles are parallel to the E-field,

$$\cos\theta = 1 \text{ and energy } W = -pE.$$

When dipoles are antiparallel to the E-field,

$$\cos\theta = -1 \text{ and energy } W = pE.$$

The average of cos θ can be calculated as follows:

$$\cos\theta = \frac{1 \times e^{pE/KT} + (-1) \times e^{-pE/KT}}{e^{pE/KT} + e^{-pE/KT}}.$$

Let $y = pE/KT$, then

$$\cos\theta = \frac{e^y - e^{-y}}{e^y + e^{-y}} = \tanh y = y - \frac{1}{3}y^3 - \frac{2}{15}y^5 - \dots$$

7.3 From Equation 7.3,

$$\varepsilon = \varepsilon_0 + \frac{(q^2/m)N}{\omega_0^2 - \omega^2 + j\omega\beta}.$$

If we rearrange the above equation and let β = 0, we get

$$\frac{\varepsilon}{\varepsilon_0} = 1 + \frac{(q^2/m) \cdot (N/\varepsilon_0)}{\omega_0^2 - \omega^2}.$$

From the above equation, it's easy to show

$$\omega_0 = \sqrt{\omega^2 + \frac{\left(q^2 / m\right)\cdot\left(N / \varepsilon_0\right)}{\left(\varepsilon / \varepsilon_0\right) - 1}}.$$

Assuming that the lattice constant is approximately $8\,\text{Å}$, the concentration

$$N \approx \frac{1}{\left(8 \times 10^{-10}\right)^3} = 1.953 \times 10^{21} \text{ cm}^{-3},$$

$$\omega_0 \approx 8.3 \times 10^{14} \text{ rad} / \text{s},$$

$$f_0 = \frac{\omega_0}{2\pi} = 1.32 \times 10^{14} \text{ Hz}.$$

7.4

a. The equation of motion

$$m \frac{d^2 P_i}{dt^2} = -\vec{\nabla}_P F,$$

where

F is the free energy (Equation 7.4)

$$\vec{\nabla}_P = \left(\frac{\partial}{\partial P_i} - \sum_j \frac{\partial}{\partial x_j} \frac{\partial}{\partial E_{ij}} \right) \vec{a}_i, \quad i = 1, 2, \text{ and } 3 \text{ or } x, y, \text{ and } z.$$

First, we consider the x-component of P and expand the right side of equation of motion with the operators in $\vec{\nabla}_P$:

$$-\frac{\partial}{\partial P_x} F = E_x - 2AP_x - 4BP_x^3 - 2C\left(P_y^2 P_x + P_z^2 P_x\right),$$

$$\frac{\partial}{\partial x} \frac{\partial}{\partial E_{xx}} F = D_{11} \frac{\partial^2 P_x}{\partial x^2} + D_{12}\left(\frac{\partial^2 P_y}{\partial x \partial y} + \frac{\partial^2 P_z}{\partial x \partial y} \right),$$

$$\frac{\partial}{\partial y} \frac{\partial}{\partial E_{xy}} F = D_{44}\left(\frac{\partial^2 P_x}{\partial y^2} + \frac{\partial^2 P_y}{\partial x \partial y} \right),$$

$$\frac{\partial}{\partial z} \frac{\partial}{\partial E_{xz}} F = D_{44}\left(\frac{\partial^2 P_x}{\partial z^2} + \frac{\partial^2 P_z}{\partial x \partial z} \right).$$

We rewrite equation of motion for the x-component of P as

$$m\frac{\partial^2 P_x}{\partial t^2} = E_x - 2AP_x - 4BP_x^3 - 2C\left(P_y^2 P_x + P_z^2 P_x\right) + D_{11}\frac{\partial^2 P_x}{\partial x^2} + D_{44}\frac{\partial^2 P_x}{\partial y^2}$$

$$+ D_{44}\frac{\partial^2 P_x}{\partial z^2} + (D_{12} + D_{44})\left(\frac{\partial^2 P_y}{\partial x \partial y} + \frac{\partial^2 P_z}{\partial x \partial z}\right).$$

Assuming $E_x = E_x^{dc} + e_x^{ac}$ and $P_x = P_x^{dc} + p_x^{ac}$, DC parts of E_x and P_x are constants in time and space domain, AC parts of E_x and P_x are small variables and proportional to $e^{j\omega t - j\vec{k}\cdot\vec{r}}$, and only the linear term of P is taken into consideration.

For DC part,

$$0 = E_x - 2AP_x,$$

$$P_x = \frac{1}{2A}E_x.$$

Same for the y- and z-component,

$$P_y = \frac{1}{2A}E_y, \; P_z = \frac{1}{2A}E_z.$$

$$\begin{pmatrix} P_x \\ P_y \\ P_z \end{pmatrix} = \begin{pmatrix} 1/2A & 0 & 0 \\ 0 & 1/2A & 0 \\ 0 & 0 & 1/2A \end{pmatrix}\begin{pmatrix} E_x \\ E_y \\ E_z \end{pmatrix}.$$

For AC part,

$$-m\omega^2 p_x = e_x - 2Ap_x - \left(D_{11}k_x^2 + D_{44}k_y^2 + D_{44}k_z^2\right)p_x - (D_{12} + D_{44})k_x k_y p_y$$

$$- (D_{12} + D_{44})k_x k_z p_z,$$

$$\left(2A + D_{11}k_x^2 + D_{44}k_y^2 + D_{44}k_z^2 - m\omega^2\right)p_x + (D_{12} + D_{44})k_x k_y p_y + (D_{12} + D_{44})k_x k_z p_z = e_x.$$

With similar approach for the y- and z-components, we get

$$(D_{12} + D_{44})k_y k_x p_x + \left(2A + D_{11}k_y^2 + D_{44}k_x^2 + D_{44}k_z^2 - m\omega^2\right)p_y + (D_{12} + D_{44})k_y k_z p_z = e_y,$$

$$(D_{12} + D_{44})k_z k_y p_y + \left(2A + D_{11}k_z^2 + D_{44}k_y^2 + D_{44}k_x^2 - m\omega^2\right)p_z = e_z.$$

Rewrite the above three linear functions in matrix form:

$$[G] \cdot \begin{bmatrix} p_x \\ p_y \\ p_z \end{bmatrix} = \begin{bmatrix} e_x \\ e_y \\ e_z \end{bmatrix},$$

where the 3×3 matrix

$$[G] = \begin{bmatrix} \begin{pmatrix} 2A + D_{11}k_x^2 + D_{44}k_y^2 \\ +D_{44}^2 k_z^2 - m\omega^2 \end{pmatrix} & (D_{12} + D_{44})k_x k_y & (D_{12} + D_{44})k_x k_z \\ (D_{12} + D_{44})k_y k_x & \begin{pmatrix} 2A + D_{11}k_y^2 + D_{44}k_x^2 \\ +D_{44}^2 k_z^2 - m\omega^2 \end{pmatrix} & (D_{12} + D_{44})k_y k_z \\ (D_{12} + D_{44})k_z k_x & (D_{12} + D_{44})k_z k_y & \begin{pmatrix} 2A + D_{11}k_z^2 + D_{44}k_y^2 \\ +D_{44}^2 k_x^2 - m\omega^2 \end{pmatrix} \end{bmatrix}.$$

Dielectric tensor $[\chi] = [G]^{-1}$.

b. The total free energy of a magneto-dielectric system is the sum of the magnetic and electrical free energies:

$$F_{Total} = F_m^{[\cdot]} + F_e + F_{e,m} + F_{\sigma,m},$$

where

$$F_m = -\vec{M} \cdot \vec{H} + (1/2)\left(N_x M_x^2 + N_y M_y^2 + N_z M_z^2\right) - \left(A/M^2\right)\vec{M} \cdot \nabla^2 \vec{M}$$

$$+ K_1\left(\alpha_1^2 \alpha_2^2 + \alpha_1^2 \alpha_3^2 + \alpha_2^2 \alpha_3^2\right) + K_u \cos^2 \psi,$$

$$F_e = -Nq\vec{E} \cdot \vec{U} + (1/2)C_{11}\left(e_{11}^2 + e_{22}^2 + e_{33}^2\right)$$

$$+ C_{12}\left(e_{11}e_{22} + e_{22}e_{33} + e_{11}e_{33}\right) + (1/2)C_{44}\left(e_{12}^2 + e_{23}^2 + e_{31}^2\right)$$

$$+ A\left(P_x^2 + P_y^2 + P_z^2\right) + B\left(P_x^4 + P_y^4 + P_z^4\right) + C\left(P_y^2 P_z^2 + +P_z^2 P_x^2 + P_x^2 P_y^2\right).$$

$F_{e,m} =$ Magneto-dielectric coupling term

$$F_{\sigma,m} = B_1 \sum_i e_{ii}\alpha_i^2 + B_2 \sum_{i \neq j} e_{ij}\alpha_i\alpha_j,$$

where
$B_1 = (3/2)\lambda_{100}(C_{12} - C_{11})$
$B_2 = -3\lambda_{111}C_{44}$
$\lambda_{100}, \lambda_{111} = $ magnetostriction constant

7.5 The sum of the gravity and magnetic force on the superconductor sphere is equal to zero:

$$F_g + F_m = -\rho v g + \frac{\partial}{\partial z}\left(-\vec{m}\cdot\vec{B}\right) = 0,$$

$$-Mv\frac{\partial B}{\partial z} = \rho v g,$$

where v is the volume of the sphere.

From hints, $M = -(3/2)H_{ext}$, so we get the following:

$$\frac{3}{2}H\mu_0\frac{\partial H}{\partial z} = \rho g.$$

We rewrite the above equation in the differential equation forms:

$$H\cdot dH = \frac{2\rho g}{3\mu_0}\cdot dz.$$

Assuming $H = H_0$ at $z = 0$, we get the solution

$$\frac{1}{2}H^2 = \frac{2\rho g}{3\mu_0}z + \frac{1}{2}H_0^2,$$

$$H = \sqrt{H_0^2 + \frac{4\rho g}{3\mu_0}z}.$$

7.6 For type II superconductors,

$$H_{c1} = \frac{\xi}{\lambda}H_c \ln\left(\frac{\lambda}{\xi} - 0.27\right),$$

$$H_{c2} = \frac{\lambda}{\xi}\sqrt{2}H_c,$$

where $H_c = \sqrt{n_s\Delta 8\pi}$.

TABLE S.7.6

Superconducting Parameters

Material	λ (nm)	ξ (nm)	Δ (meV)	n_s (nm⁻³)	H_{c1} (T)	H_{c2} (T)
Pb–In	150	30	1.2	0.017	0.009	0.2
Pb–Bi	200	20	1.7	0.018	0.008	0.5
Nb–Ti	300	4	1.5	0.25	0.007	13
Nb–N	200	5	2.4	0.73	0.024	15
PbMo₆S₈	200	2	2.4	1.9	0.02	61
V₃Ga	90	2.5	2.3	2.2	0.045	23
V₃Si	60	3	2.3	5.4	0.1	20
Nb₃Sn	65	3	3.4	4.1	0.1	23
Nb₃Ge	90	3	3.7	5.4	0.1	38

Ref.: T.P. Orlando and K.A. Delin, *Foundations of Applied Superconductivity*, Addison-Wesley, Reading, MA, 1991, pp. 575–577.

7.7 For a Josephson junction (Figure S7.7), we have the current-phase relation $J = J_c \sin \phi$ and voltage-phase relation:

$$\frac{\partial \phi}{\partial t} = \frac{2|e|V}{\hbar}.$$

The total voltage V_0 is the sum of the voltages on the resistor and Josephson junction, that is,

$$V_0 = V_R + V_J = RI_c \sin\phi + \frac{\hbar}{2|e|}\frac{\partial \phi}{\partial t}.$$

The equation is

$$\frac{\hbar}{2|e|}\frac{\partial \phi}{\partial t} = V_0 - RI_c \sin\phi.$$

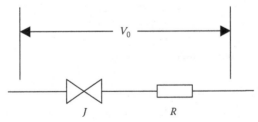

FIGURE S7.5 A Josephson junction and a resistor are connected in series.

7.8 Magnetic field distribution for superconductor in 1D is

$$\vec{H}(x) = H_0 e^{-x/\lambda}\vec{a}_y,$$

where $\lambda = (1/q)\sqrt{m/n_x\mu_0}$.
Current distribution is

$$\vec{J}_s(x) = \vec{\nabla} \times \vec{H} = \frac{\partial}{\partial x}B(x)\vec{a}_z = -\frac{H_0}{\lambda}e^{-x/\lambda}\vec{a}_z.$$

7.9 The force per unit area is

$$f = \int_0^\infty \vec{J}_s \times \vec{B}dx = \vec{a}_x \frac{B_0^2}{\mu_0\lambda}\int_0^\infty e^{-2x/\lambda}dx = \frac{B_0^2}{2\mu_0}\vec{a}_x.$$

7.10 In this case, when the static magnetic field is applied to the superconductor, the induced current will generate a magnetic field, which shields the coil from the applied external magnetic field. The total flux through the coil stays the same.
The induced current is

$$I_{ind} = \frac{\Delta\phi}{L} = \frac{B\pi r^2\cos\theta}{L} = \frac{\mu_0\pi r^2}{L}H_0\cos\theta.$$

The total current is

$$I' = I - I_{ind} = I - \frac{\mu_0\pi r^2}{L}H_0\cos\theta.$$

7.11 The current and flux in the coil will stay the same:

$$I = \frac{\phi_0}{L}.$$

7.12 The magnetic moment $M \rightarrow 0$, when radius $R/\lambda \rightarrow 0$.
7.13 The internal field is the sum of external field and demagnetizing field:

$$H_i = H_0 = H_D = H_0 - \frac{1}{3}M.$$

The total magnetic flux in superconductors is equal to zero:

$$B = \mu_0(H_i + M) = 0.$$

We solve M from the two above equations:

$$M = -\frac{3}{2}H_0.$$

8 Kramers–Kronig Equations

Let's consider the time response of a magnetic system, and in particular, the time dependence of the magnetic field $H(t)$. $H(t)$ may be related to $H(\omega)$ via the Fourier transform

$$H(t) = \int_{-\infty}^{\infty} H(\omega) e^{j\omega t} d\omega \tag{8.1}$$

and

$$H(\omega) = \frac{1}{2\pi} \int_{-\infty}^{\infty} H(t) e^{-j\omega t} dt. \tag{8.2}$$

Since $H(t)$ is a measurable quantity,

$$H(t) = H^*(t), \tag{8.3}$$

which implies

$$H(t) = \int_{-8}^{\infty} H(\omega) e^{j\omega t} d\omega = \left[\int_{-\infty}^{\infty} H(\omega) e^{j\omega t} d\omega \right]^* = \int_{-\infty}^{\infty} H^*(\omega) e^{-j\omega t} d\omega. \tag{8.4}$$

Let $z = -\omega$

$$\int_{-\infty}^{\infty} H^*(\omega) e^{-j\omega t} d\omega = -\int_{\infty}^{-\infty} H^*(z) e^{jzt} dz = \int_{-\infty}^{\infty} H^*(-z) e^{jzt}. \tag{8.5}$$

Let $\omega = z$

$$\int_{-\infty}^{\infty} H^*(-z) e^{jzt} dz = \int_{-\infty}^{\infty} \underline{H^*(-\omega) e^{j\omega t} d\omega} \equiv \int_{-\infty}^{\infty} \underline{H^*(\omega) e^{j\omega t} d\omega} = \int_{-\infty}^{\infty} H(\omega) e^{j\omega t} d\omega. \tag{8.6}$$

The two underlined integrals in Equation 8.6 are equal only if

$$H^*(-\omega) = H(\omega). \tag{8.7}$$

DOI: 10.1201/9781003431244-8

The magnetic field is related to the magnetic flux density, $B(\omega)$, by

$$B(\omega) = \mu(\omega)H(\omega) \tag{8.8}$$

and

$$B^*(\omega) = \mu^*(\omega)H^*(\omega). \tag{8.9}$$

Since $B(\omega)$ is also a measurable quantity, we require that

$$B(\omega) = B^*(-\omega), \tag{8.10}$$

which implies

$$\mu(\omega)H(\omega) = \mu^*(-\omega)H^*(-\omega). \tag{8.11}$$

Use the property

$$H(\omega) = H^*(-\omega), \tag{8.12}$$

so that

$$\left[\mu(\omega) - \mu^*(-\omega)\right]H(\omega) = 0, \tag{8.13}$$

where (in MKS)

$$\mu(\omega) = 1 + \chi(\omega),$$
$$\mu^*(-\omega) = 1 + \chi^*(-\omega). \tag{8.14}$$

Equation 8.13 also implies that

$$\left[\chi(\omega) - \chi^*(-\omega)\right]H(\omega) = 0. \tag{8.15}$$

Since $H(\omega)$ is a measurable quantity, i.e., $H(\omega) \neq 0$, we have

$$\chi(\omega) = \chi^*(-\omega), \tag{8.16}$$

where, for example,

$$\chi(\omega) = \chi'^{(\omega)} - j\chi''(\omega). \tag{8.17}$$

Substituting the above relation into Equation 8.16, we have two subsequent relations

$$\chi'(\omega) - j\chi''(\omega) = \chi'(-\omega) + j\chi''(-\omega). \tag{8.18}$$

resulting in

$$\chi'(\omega) = \chi'(-\omega), \qquad \text{even function of } \omega$$
$$\chi''(\omega) = -\chi''(-\omega), \quad \text{odd function of } \omega. \tag{8.19}$$

Let's relate $\chi'(\omega)$ to $\chi''(\omega)$ by considering the integral

$$I = \oint_C \frac{\omega\chi(\omega)d\omega}{\omega^2 - \omega_0^2}. \tag{8.20}$$

The contour of integration is defined in Figure 8.1 to avoid residues.

Near the mathematical poles, use the following transformation (Figure 8.2):

$$\omega - \omega_0 = \rho e^{j\theta}. \tag{8.21}$$

Thus,

$$d\omega = \rho e^{j\theta} jd\theta = (\omega - \omega_0) jd\theta. \tag{8.22}$$

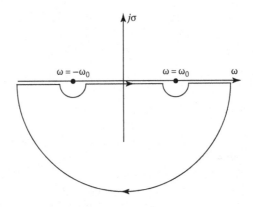

FIGURE 8.1   Contour of integration for Equation 8.20.

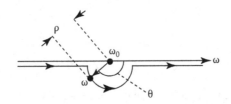

FIGURE 8.2   Transformation in the vicinity of the mathematical poles.

Similarly, near the $-\omega_0$ pole,

$$d\omega = \rho e^{j\theta} jd\theta = (\omega + \omega_0) jd\theta. \qquad (8.23)$$

The integral becomes

$$I = P\int_{-\infty}^{0} \frac{\omega\chi(\omega)d\omega}{\omega^2 - \omega_0^2} + \int_{\pi}^{0} \frac{\omega_0\chi(\omega_0)(\omega - \omega_0) jd\theta}{2\omega_0(\omega - \omega_0)} + \int_{\pi}^{0} \frac{-\omega_0\chi(-\omega_0)(\omega + \omega_0) jd\theta}{-2\omega_0(\omega + \omega_0)} = 0, \qquad (8.24)$$

where we've used the approximation

$$\omega^2 - \omega_0^2 = (\omega + \omega_0)(\omega - \omega_0)|_{\omega \equiv \omega_0} \cong 2\omega_0(\omega - \omega_0) \quad \text{and}$$

$$\omega^2 - \omega_0^2 = (\omega + \omega_0)(\omega - \omega_0)|_{\omega \equiv -\omega_0} \cong -2\omega_0(\omega + \omega_0). \qquad (8.25)$$

Thus, the integral simplifies to the following:

$$I = P\int_{-\infty}^{\infty} \frac{\omega\chi(\omega)d\omega}{\omega^2 - \omega_0^2} - \frac{\pi j}{2}\big[\chi(\omega_0) + \chi(-\omega_0)\big] = 0, \qquad (8.26)$$

yielding

$$P\int_{-\infty}^{\infty} \frac{\omega\chi(\omega)d\omega}{\omega^2 - \omega_0^2} = \frac{\pi j}{2}\big[2\chi'(\omega_0)\big]. \qquad (8.27)$$

Using Equation 8.19,

$$P\int_{-\infty}^{\infty} \frac{\omega(\chi'(\omega) - j\chi''(\omega))d\omega}{\omega^2 - \omega_0^2} = \pi j\chi'(\omega_0). \qquad (8.28)$$

Since $\chi'(\omega)$ is an even function of ω (see Equation 8.19) and $\omega\chi'(\omega)$ is odd,

$$\int_{-\infty}^{\infty} \frac{\omega\chi'(\omega)d\omega}{\omega^2 - \omega_0^2} = 0. \qquad (8.29)$$

However, $\chi''(\omega)$ is an odd function of ω (see Equation 8.19) and $\omega\chi''(\omega)$ is even

$$\int_{-\infty}^{\infty} \frac{-j\omega\chi''(\omega)d\omega}{\omega^2 - \omega_0^2} = -2j\int_{0}^{\infty} \frac{\omega\chi''(\omega)d\omega}{\omega^2 - \omega_0^2}. \qquad (8.30)$$

Combining the above result with Equation (8.28), we obtain

$$\chi'(\omega_0) = -\frac{2}{\pi} P \int_0^\infty \frac{\omega \chi''(\omega) d\omega}{\omega^2 - \omega_0^2}.$$
(8.31)

There is an inverse relation for $\chi''(\omega)$. For this relation, we need to consider the integral

$$I = \oint_C \frac{\chi(\omega) d\omega}{\omega^2 - \omega_0^2}.$$
(8.32)

After applying the same integration contour procedure as before, we obtain

$$I = P \int_{-\infty}^\infty \frac{\chi(\omega) d\omega}{\omega^2 - \omega_0^2} - \frac{\pi j}{2} \left[\frac{\chi(\omega_0)}{\omega_0} - \frac{\chi(-\omega_0)}{\omega_0} \right] = 0,$$
(8.33)

yielding the result that

$$P \int_{-\infty}^\infty \frac{\chi(\omega) d\omega}{\omega^2 - \omega_0^2} = \frac{\pi \chi''(\omega_0)}{\omega_0}.$$
(8.34)

Since $\chi'(\omega)$ is an even function of ω (see Equation 8.19),

$$\int_{-\infty}^\infty \frac{\chi'(\omega) d\omega}{\omega^2 - \omega_0^2} = 2 \int_0^\infty \frac{\chi'(\omega) d\omega}{\omega^2 - \omega_0^2}.$$
(8.35)

However, $\chi''(\omega)$ is an odd function of ω (see Equation 8.19):

$$\int_{-\infty}^\infty \frac{\chi''(\omega) d\omega}{\omega^2 - \omega_0^2} = 0.$$
(8.36)

Finally, we come to the conclusion that

$$\chi''(\omega_0) = \frac{2\omega_0}{\pi} P \int_{-\infty}^\infty \frac{\chi'(\omega) d\omega}{\omega^2 - \omega_0^2}.$$
(8.37)

Equations 8.31 and 8.37 are referred to as the Kramers–Kronig relations. These relationships are valid for diagonal as well as off-diagonal elements of the susceptibility tensor.

The mathematical procedure to calculate the susceptibility is to vary ω_0 over the frequency range of interest. Usually $\chi'(\omega)$ or $\chi''(\omega)$ is measured. Thus, the other quantity may be determined using Kramers–Kronig relations. MATLAB® code is included in the appendix to calculate susceptibilities.

PROBLEMS

8.1   Start with

$$\chi''(\omega) = \frac{M_0(\Delta\omega/\gamma)}{\left(H - (\omega/\gamma)\right)^2 + (\Delta\omega/\gamma)^2},$$

where

$$M_0 = 400 \text{ G}$$

$$\Delta\omega/\gamma = 200 \text{ Oe}$$

$$H = 3,000 \text{ Oe}$$

$$\gamma = 2\pi \times 2.8 \times 10^6 \text{ Hz/Oe}$$

$$\chi'(\omega) = \frac{M_0\left(H - (\omega/\gamma)\right)}{\left(H - (\omega/\gamma)\right)^2 + (\Delta\omega/\gamma)^2}.$$

Calculate $\chi'(\omega)$ using Kramers–Kronig relations near 8–9 GHz and compare with the exact expressions.

8.2   The Kramers–Kronig relations developed in Chapter 7 were considered for scalar expressions of χ' and χ''. Using the relations

$$\int \frac{\omega\chi_{ik}(\omega)}{\omega^2 - \omega_0^2} d\omega = 0,$$

$$\int \frac{\chi_{ik}(\omega)}{\omega^2 - \omega_0^2} d\omega = 0,$$

and

$$\chi'_{ik}(\omega) = \chi'_{ik}(-\omega); \quad \chi''_{ik}(\omega) = -\chi''_{ik}(-\omega)$$

proves that the same Kramers–Kronig relations may be used for $\chi'_{ik}(\omega)$ and $\chi''_{ik}(\omega)$.

Hint: Use contour integration where the contour is around the poles $+\omega_0$ and $-\omega_0$.

8.3   What are some possible sources of error that may influence the results of a typical numerical Kramers–Kronig integration?

8.4   Describe the relevance of $\chi'_{ik}(\omega) = \chi'_{ik}(-\omega)$ and $\chi''_{ik}(\omega) = -\chi''_{ik}(-\omega)$ in the development of the Kramers–Kronig relation. Why must these conditions hold?

8.5   For the special case of wave propagation where the propagation direction is perpendicular to the static magnetization, the effective permeability is defined as follows (see Chapter 6 or 9):

$$\mu_{eff} = \mu - \frac{\kappa^2}{\mu},$$

where both μ and κ obey Kramers–Kronig relationships. Does μ_{eff} also obey Kramers–Kronig relationships?
No solution is provided.

BIBLIOGRAPHY

B.S. Gourary, *J. Appl. Phys.* **28**, 283, 1957.
H.A. Kramers, *Atti Long. Int. Fisici Como* **2**, 545, 1927.
R.L. Kronig, *Physica* **3**, 1009, 1936.
E.T. Whitaker and G.N. Watson, *Modern Analysis*, The MacMillan Co., New York, 1943.

SOLUTIONS

8.1   Kramers–Kronig relations are used to calculate $\chi'(\omega)$ and $\chi''(\omega)$ (Figure S.8.1a and b). We use MATLAB® for the calculation.

```
function x1 = my_x1 (f0, d, sp,
M0, H, r, dH)
S1 = [d :d :sp];
S2 = [-sp :d : -d];
S1 = f0 + S1;
S2 = f0 + S2;
L = [S2, S1];
nL = length(L);
X2 = zeros(1, nL);
for nn = 1: nL
X2(nn) = M0*dH/
((H-L(nn)/r)^2 + (dH)^2);
end
s = 0;
for nn = 1 : nL
s = s + X2(nn)*L(nn)/
(L(nn) - f0)/(L(nn) + f0);
end
x1 = 2/pi * s * d;
function x2 = my_x2(f0, d, sp,
M0, H, r, dH)
S1 = [d :d :sp];
S2 = [-sp :d : -d];
S1 = f0 + S1;
S2 = f0 + S2;
L = [S2, S1];
```

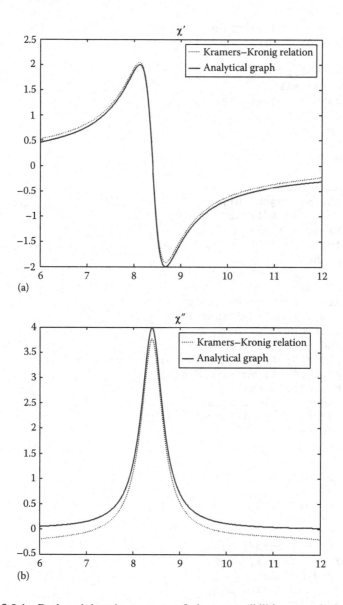

FIGURE S.8.1 Real and imaginary parts of the susceptibilities as calculated using Equations 8.31 and 8.37.

```
nL = length(L);
x1 = zeros(1, nL);
for nn = 1 : nL
X1(nn) = (M0*(H-L(nn)/r)/
((H-L(nn)/r)^2 + (dH)^2));
end
s = 0;
for nn = 1 : nL
```

```
s = s + X1(nn)*L(nn)/
(L(nn) - f0)/(L(nn) + f0);
end
x2 = 2/pi * s * d;
clear all
f0 = [6: 0.01: 12];
d = 0.01;
sp = 4;
M0 = 0.4;
H = 3;
r = 2.8;
dH = 0.1;
nf0 = length(f0);
X1 = zeros(1, nf0);
X1_an = X1;
for nnf0 = 1 : nf0
X1(nnf0) = my_x1
(f0(nnf0), d, sp, M0, H,
r, dH);
X1_an(nnf0) = M0*(H-f0
(nnf0)/r)/((H-f0(nnf0)/
r)^2 + dH^2);
end
plot(f0, X1,':'); hold on;
plot(f0, X1_an); hold off;
title('X-prime');
legend('Kramer-Kronig
relation','Analytical graph')
clear all
f0 = [6: 0.01: 12];
d = 0.01;
sp = 4;
M0 = 0.4;
H = 3;
r = 2.8;
dH = 0.1;
nf0 = length(f0);
X2 = zeros(1, nf0);
X2_an = X2;
for nnf0 = 1 : nf0
X2(nnf0) = my_x2
(f0(nnf0), d, sp, M0,
H, r, dH);
X2_an(nnf0) = M0*
(H - f0(nnf0)/r)/
((H-f0(nnf0)/r)^2 + dH^2);
end
plot(f0, X2,':'); hold on;
plot(f0, X2_an); hold off;
title('X-double prime');
legend('Kramer-Kronig
relation','Analytical graph')
```

8.2   We define two tensors, I_{ik}^a and I_{ik}^b as

$$I_{ik}^a = \int_C \frac{\omega \chi_{ik}(\omega)}{\omega^2 - \omega_0^2}\, d\omega$$

and

$$I_{ik}^b = \int_C \frac{\chi_{ik}(\omega)}{\omega^2 - \omega_0^2}\, d\omega.$$

The two integrals are analytic in C. Therefore, those are obviously zero. With two conditions given in this problem and same technique of contour integrals for every tensor element, Kramers–Kronig relations can be obtained.

8.3   Kramers–Kronig relations work only for a single domain. Therefore, there is an error at low frequencies if one uses these relations where one usually has domains.

Second, when we measure $\chi'(\omega)$ ($\chi''(\omega)$) experimentally and calculate $\chi''(\omega)$ ($\chi'(\omega)$) by Kramers–Kronig relations, those relations require $\chi'(\omega)$ ($\chi''(\omega)$) to have infinite integration over frequency. Experimentally, one has only access to a range of frequencies.

8.4   For simplicity, write

$$k \cong k_0\left(1 - j\frac{\mu'}{2\mu''}\right) = \beta - j\alpha \quad \text{where } \alpha = \frac{\mu'}{2\mu''}k_0 \quad \text{and} \quad \beta = k_0.$$

A wave is represented by $e^{-j(\omega t - kx)} = e^{-j\omega t}e^{-j\beta x}e^{\mp \alpha x}$. Irrespective of wave propagation direction, the wave must be attenuated either way. Mathematically, it means changing the sign of μ'', as ω changes sign.

9 Electromagnetic Wave Propagation in Anisotropic Magneto-Dielectric Media

Spin wave dispersions of magnetically ordered materials have been calculated by various authors. The reader is referred to seminal publications by Holstein–Primakoff, Dyson, Keffer, Akhiezer, etc., on the subject matter. Spin wave excitations in magnetic materials as well as superconductivity materials involve the collective excitations or interactions of ~10^{22} particles in a macroscopic body. Second quantization calculational methods have been formulated to approximate the interaction energy of 10^{22} particles. These theoretical formulations are beyond the scope of this book. We have adopted a much simpler approach whereby interactions of particles at a local site are represented by a molecular field, which is of the same form from site to site.

Our calculational method utilizes a semiclassical approach where the magnetic moment of an ordered magnetic material is represented by a classical vector $\vec{m}$ rather than by a quantum spin operator. The magnetic moment is assumed to be uniform in amplitude over a small region of the material. The object of our calculations is to determine the spatial variation of $\vec{m}$, dynamic magnetization vector under the influence of a molecular exchange field. We assume that within the crystal unit cell, $\vec{m}$ is uniform, although there may be many magnetic sublattices within the crystal unit cell. For example, in garnet ferrite materials, there may be as many as three magnetic sublattices in the crystal unit cell. Clearly, spin wave excitations whose wavelengths are greater than the lattice constant of a unit cell are considered semiclassical for the dispersion of spin waves, since $\vec{m}$ is uniform over one unit cell. For wavelengths smaller than the lattice constant, one may no longer assume $\vec{m}$ to be constant over the unit cell, especially if it contains magnetic sublattices. For these cases, spin wave excitations occur at much higher frequency, usually in the optical frequency regime. For the former case, where $\vec{m}$ may be assumed to be uniform over the crystal unit cell, the spin wave dispersion is often referred to as the acoustic magnon dispersion, since it is the lowest frequency excitation. It is well known that in the semiclassical approach, there are two uniform dynamic fields besides static internal fields: the dynamic exchange and the volume demagnetizing fields. The exchange dynamic field is a molecular field representing exchange coupling between spins. The volume demagnetizing field can be approximated to be constant over one-half wavelength. Clearly, $\vec{m}$ changes sign over that distance. This means that the dynamic magnetization vector, $\vec{m}$, "sees" a uniform demagnetizing field in its motion, since $\vec{m}$ is uniform over the unit cell, which is much less than the wavelength of the spin wave. As such,

DOI: 10.1201/9781003431244-9

it is meaningful to treat the motion of $\vec{m}$ within the unit cell in a classical way, where all the fields acting on $\vec{m}$ are uniform.

The exchange field is defined as follows:

$$\vec{h}_{exc.} = \frac{2A}{M^2} \nabla^2 \vec{M},$$

where A = exchange stiffness constant (erg/cm) ~10^{-6} erg/cm.

Assume that $\vec{M} = \vec{M}_0 + \vec{m}$, where $\vec{M}_0$ is the static magnetization and is constant throughout the magnetic sample. Then,

$$\vec{h}_{exc.} = -\frac{2A}{M^2} k^2 \vec{m}, \quad \vec{m} \propto e^{-j\vec{k}\cdot\vec{r}}.$$

It is too cumbersome to carry the subscript "0" on M in the denominator of the above equation.

There may be internal demagnetizing fields as a result of spin wave excitations. For example, assume the following spin wave configuration (Figure 9.1a).

Each site is represented by a classical vector $\vec{m}$ under the influence of an exchange molecular field, $\vec{h}_{exc.}$, and demagnetizing field. The site may be thought of as a point in space as classical theory implies, but in actuality, $\vec{m}$ is averaged over a unit cell, as shown in Figure 9.1b.

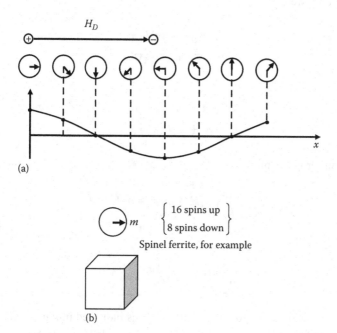

(a)

$\vec{m}$ $\left\{ \begin{array}{l} 16 \text{ spins up} \\ 8 \text{ spins down} \end{array} \right\}$

Spinel ferrite, for example

(b)

FIGURE 9.1 Comparisons of length scales.

Spin wave propagation is in the $+x$-direction, and h_D is the volume demagnetizing field, also in the x-direction. We have picked the x-direction arbitrarily.

Let's calculate this demagnetizing field, h_D. Gauss law states that

$$\vec{\nabla} \cdot \vec{b} = 0.$$

Rewrite to include h_D

$$\vec{\nabla} \cdot \left(\vec{h}_D + 4\pi\vec{m} \right) = 0,$$

$$\vec{\nabla} \cdot \vec{h}_D = -4\pi\vec{\nabla} \cdot \vec{m}.$$

The term on the right side can be thought of as a magnetic "charge," analogous to the electric case. Note the polarity in Figure 9.1a.

Since $\vec{h}_D$ is not related to ohmic or displacement current, we must have that

$$\vec{\nabla} \times \vec{h}_D = 0$$

which implies that

$$\vec{h}_D = -\vec{\nabla}\Psi,$$

where $\psi \equiv$ a scalar magnetic potential.

In summary, we have

$$\vec{\nabla} \cdot \vec{h}_D = -4\pi\vec{\nabla} \cdot \vec{m},$$

$$-\vec{\nabla} \cdot \left(\vec{\nabla}\Psi \right) = -4\pi\vec{\nabla} \cdot \vec{m},$$

or

$$\vec{\nabla}^2 \Psi = 4\pi\vec{\nabla} \cdot \vec{m}.$$

The solution for ψ is simply

$$\Psi = j\frac{4\pi}{k^2}\left(m_x k_x + m_y k_y + m_z k_z \right)e^{-j\vec{k}\cdot\vec{r}},$$

which finally yields

$$\vec{h}_D = -\vec{\nabla}\Psi = -\frac{4\pi}{k^2}\left(\vec{m} \cdot \vec{k} \right)\vec{k}.$$

Let's assume that the direction of $\vec{k}$ is arbitrary, as indicated in Figure 9.2.

We define the z-axis to be along the static internal field, $\vec{H}_0$, or static magnetization, $\vec{M}_0$. In the linear approximation, both dynamic components of the magnetization

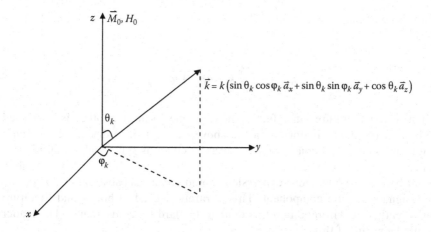

FIGURE 9.2   The direction of the propagation constant, k, relative to internal field directions.

are in the x–y plane. The spin wave vector, $\vec{k}$, may be assumed at arbitrary direction with polar angle, θ_k, and azimuth angle, φ_k, as shown in Figure 9.2. Note that one can always assume $\vec{k}$ to be in the x–y plane or the x-axis, for example. Thus, based on these approximations and assumptions, the dynamic exchange and demagnetizing fields may be assumed to be in the x–y plane.

Now we are in a position to calculate the effects of spin wave excitations on the FMR condition, for example. As before (see Chapter 6), we have from the equation of motion that

$$j\frac{\omega}{\gamma}\vec{m} = \vec{M}_0 \times \vec{h} + \vec{m} \times \vec{H}_0,\tag{9.1}$$

where

$$\vec{h} = \vec{h}_{exc.} + \vec{h}_D$$

$H_0 \equiv$ internal static magnetic field

$$\vec{H}_0 = H_0 \vec{a}_z$$

The demagnetizing field $\vec{h}_D$ is defined as

$$\vec{h}_D = -4\pi m_x \sin^2 \theta_k \vec{a}_x, \qquad \varphi_k = 0.$$

There is no demagnetizing field for $\bar{k}$ along $\bar{H}_0$ or $\bar{a}_z$ in the linear approximation. Also,

$$\bar{h}_{exc.} = -\frac{2A}{M^2}k^2\left(m_x\bar{a}_x + m_y\bar{a}_y\right).$$

Often in the literature, an effective field of one form or another is expressed with M in the denominator, as in the above expression, to designate the static magnetization. By all conventions, M usually contains static and dynamic components of the magnetization. So, by this convention as well as the one described in our book, M in the above expression should contain a subscript of "0" or "S" to designate a static component. This is rarely obeyed in books and literature consistently. We will maintain this double standard for convenience. The reader should be aware of this.

Substituting into the equations of motion, we have that

$$j\frac{\omega}{\gamma}m_x = \left(H_0 + \frac{2A}{M}k^2\right)m_y, \tag{9.1a}$$

$$j\frac{\omega}{\gamma}m_y = -\left(H_0 + \frac{2A}{M}k^2 + 4\pi M\sin^2\theta_k\right)m_x. \tag{9.1b}$$

The FMR condition is simply

$$\left(\frac{\omega}{\gamma}\right)^2 = \left(H_0 + \frac{2A}{M}k^2 + 4\pi M\sin^2\theta_k\right)\left(H_0 + \frac{2A}{M}k^2\right), \tag{9.2}$$

For $\theta_k = 0$, we have

$$\frac{\omega}{\gamma} = H_0 + \frac{2A}{M}k^2.$$

SPIN WAVE DISPERSIONS FOR SEMI-INFINITE MEDIUM

Pure spin wave dispersions in semi-infinite media for long wavelengths have been calculated since the 1940s using second quantization calculation method. Our semi-classical approach derived here gives rise to exactly the same dispersion relations. For propagation of spin waves in finite media, the calculations become rather complex mathematically however the approach. Nevertheless, we will navigate through these mathematical complexities to obtain approximate spin wave dispersions for finite mediums in the subsequent sections.

In Figure 9.3, ω/γ is plotted as a function of k for various values of θ_k. We are assuming no magnetic anisotropy energy.

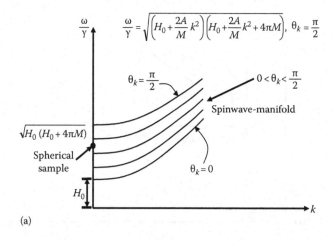

(a)

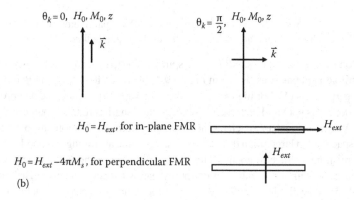

(b)

FIGURE 9.3 Spin wave dispersions for semi-infinite medium. This film is assumed to be semi-infinite in extent.

SPIN WAVE DISPERSION AT HIGH k-VALUES

In the absence of a semiclassical theory for spin wave dispersions at high k-values or short wavelengths, we ad hoc introduce an empirical expression for dispersion. For k-values near the Brillouin zone ($k = \pi/a$), the spin wave dispersion is modified according to Kittel as follows (see reference):

$$\frac{\omega}{\gamma} = \sqrt{\left(H_0 + \frac{2A}{M}\left(\frac{4}{a^2}\right)\sin^2\left(\frac{ka}{2}\right) \right)\left(H_0 + \frac{2A}{M}\left(\frac{4}{a^2}\right)\sin^2\left(\frac{ka}{2}\right) + 4\pi M \sin^2\theta_k \right)},$$

where a is the lattice constant of unit cell comprising of all magnetic sublattices.

As one would expect, in the limit of small k-values or long wavelengths (where semiclassical theory is applicable), the above dispersion converges to that of Equation 9.2. In particular, we have, for example,

$$\frac{\omega}{\gamma} = H_0 + \frac{2A}{M}k^2, \quad \text{for } \theta_k = 0.$$

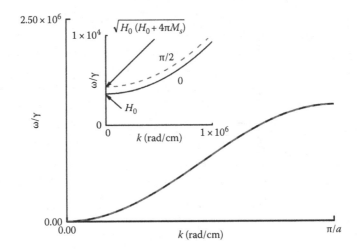

FIGURE 9.4 Spin wave dispersion at high k-values. YIG parameters were assumed and $H_0 = 3,500$ Oe. The resolution at $k = 0$ is poor to resolve a gap in field.

Thus, at long wavelengths or small k's, the spin wave energy scales as k^2. Also, the spin wave manifold is in place, as shown in Figure 9.4. However, for k near the Brillouin zone ($k = \pi/a$), (ω/γ) levels off. For high k-values, the continuous or semiclassical approach that we have adopted is not valid or correct, since the wavelengths are in the order of the lattice constant. A semiclassical approach implies averaging of magnetic moments over a domain region in space much larger than the lattice constant, while averaging at small length scales implies vanishing average moments. This also implies vanishing volume demagnetizing fields. Hence, no spin wave manifold or energy gap is expected near the Brillouin zone regime. Near the Brillouin zone, one must treat theoretically the spin variable as a quantum operator rather than as classical vectors. We are not about to do that here.

In the literature, this is often referred to as the "acoustic" branch of the magnon or spin wave dispersion. The optical branches are at higher frequencies and they involve interaction between the various magnetic sublattices of the crystal, for example.

THE $k = 0$ SPIN WAVE LIMIT

At this point, we superimpose the FMR conditions as calculated in Chapter 6 onto the dispersion plots of the previous sections. Clearly, there is a disconnection here. The FMR conditions apply for finite size samples, whereas the dispersions of pure spin waves of the previous sections apply for semi-infinite media. In order to get around this disconnect, it is argued that FMR excitations occur at $k = 0$, which is a spin wave excitation of infinite wavelength. Although it is cute, there are no cigars. It is well known that one has other types of wave propagation in finite size media. The problem then becomes how does one transition from a purely spin wave to another type of wave in finite media. We will address this problem later. For now, let's examine the disconnect.

SPHERE

For $k = 0$, the FMR condition is given as $\omega/\gamma = H_{ext}$.

The internal field is defined as

$$H_0 = H_{ext} - \frac{4\pi M_s}{3} = \frac{\omega}{\gamma} - \frac{4\pi M_s}{3}$$

or

$$\frac{\omega}{\gamma} = H_0 + \frac{4\pi M_s}{3}.$$

We need to express ω/γ in terms of H_0, since the dispersions plots of the previous section only recognize the internal field in a semi-infinite medium. Now, let's place this excitation relative to the previous plots.

This is approximately 2/3 of the way on the spin wave manifold, as shown in Figure 9.5. Let's calculate (θ_k) for a sphere, see Figure 9.5. Write that

$$H_0 + \frac{4\pi M_s}{3} = \sqrt{H_0\left(H_0 + 4\pi M_s \sin^2 \theta_k\right)}.$$

Assume for simplicity that $H_0 > 4\pi M_s$

$$H_0 + \frac{4\pi M_s}{3} \approx H_0 + 2\pi M_s \sin^2 \theta_k,$$

$$\sin^2 \theta_k = \frac{2}{3},$$

$$\theta_k \approx 54°.$$

Spin waves propagate in all directions in the sphere. However, only spin waves with φ_k and $54° \leq \theta_k \geq 90°$ are energetically degenerate with the FMR mode excitation. This is an important consideration in formulating coupling schemes between FMR and spin waves, for example.

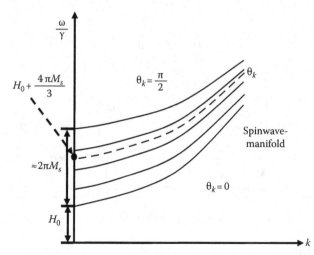

FIGURE 9.5 Relative position of the FMR excitation (sphere) in the dispersion plot.

THIN FILMS

Perpendicular FMR configuration (Figure 9.6)

For $k = 0$, the FMR condition (see Chapter 6) is given as

$$\frac{\omega}{\gamma} = H_0 = H_{ext} - 4\pi M_s,$$

$$\therefore \theta_k = 0.$$

In this case, the FMR mode excitation is not energetically degenerate with any spin wave mode excitations.

For $k \neq 0$, the dispersion is the same as before:

$$\frac{\omega}{\gamma} = H_0 + \frac{2A}{m} k^2 = H_{ext} - 4\pi M_s + \frac{2A}{M} k^2.$$

Could there be $\theta_k = \pi/2$ excitations? Yes.

This dispersion would be

$$\frac{\omega}{\gamma} = \sqrt{\left(H_{ext} - 4\pi M_s + \frac{2A}{M} k^2 \right)\left(H_{ext} - 4\pi M_s + \frac{2A}{M} k^2 + 4\pi M_s \right)}.$$

PARALLEL FMR CONFIGURATION (FIGURE 9.7)

For $k = 0$, the FMR resonance condition is from Chapter 6:

$$\frac{\omega}{\gamma} = \sqrt{(H_{ext})(H_{ext} + 4\pi M_s)} \equiv \sqrt{H_0 (H_0 + 4\pi M_s)},$$

where

$$H_0 = H_{ext} \text{ and}$$

$$\therefore \theta_k = \frac{\pi}{2}.$$

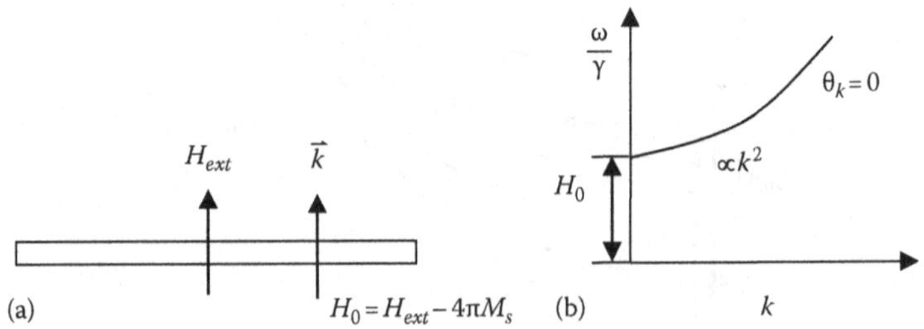

FIGURE 9.6 Relative position of the FMR excitation (thin film) in the dispersion.

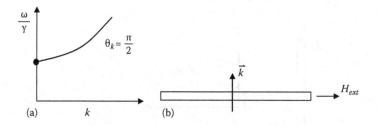

FIGURE 9.7   For $k \neq 0$, the dispersion is as before.

In this case, all spin wave states are energetically degenerate with the FMR mode excitation.

For $k \neq 0$, the dispersion is as before:

$$\frac{\omega}{\gamma} = \sqrt{\left(H_{ext} + \frac{2A}{M} k^2 \right)\left(H_{ext} + \frac{2A}{M} k^2 + 4\pi M_s \right)} \text{ or}$$

$$= \sqrt{\left(H_0 + \frac{2A}{M} k^2 \right)\left(H_0 + \frac{2A}{M} k^2 + 4\pi M_s \right)}, \ \theta_k = \pi/2.$$

For $\vec{k}\vec{H}_0$ and $\theta_k = 0$, we have that

$$\frac{\omega}{\gamma} = H_{ext} + \frac{2A}{M} k^2 = H_0 + \frac{2A}{M} k^2, \ \theta_k = 0.$$

NEEDLE

The FMR condition ($k = 0$) is (Figure 9.8)

$$\frac{\omega}{\gamma} = H_{ext} + 2\pi M_s, \text{ FMR.}$$

where $H_{ext} = H_0$. Then, $\omega/\gamma = H_0 + 2\pi M_s$. The general spin wave dispersion relation is

$$\frac{\omega}{\gamma} = \sqrt{\left(H_{ext} + \frac{2A}{M} k^2 \right)\left(H_{ext} + \frac{2A}{M} k^2 + 4\pi M_s \sin^2 \theta_k \right)}.$$

For $\theta_k = 0$, we have

$$\frac{\omega}{\gamma} = H_{ext} + \frac{2A}{M} k^2.$$

The upper branch of the spin wave manifold, $\theta_k = \pi/2$, is then

$$\frac{\omega}{\gamma} = \sqrt{\left(H_{ext} + \frac{2A}{M} k^2 \right)\left(H_{ext} + \frac{2A}{M} k^2 + 4\pi M_s \right)}.$$

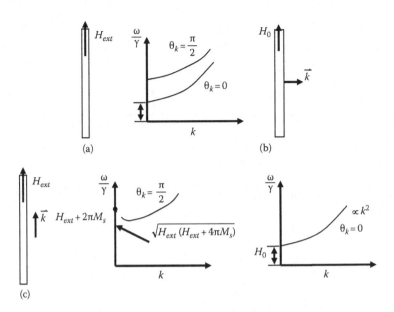

FIGURE 9.8 Spin wave excitations in magnetic cylinders.

For $k = 0$, FMR is slightly above the FMR manifold. This presents an interesting dilemma. In the limit that $k \to 0$, the spin wave dispersion must converge to FMR ($k = 0$). The details of how it may occur are not provided in this semiclassical theory. Nevertheless, the FMR mode is still degenerate to all spin wave states.

Clearly, FMR ($k = 0$) may exist above or at the onset of the upper branch of the spin wave dispersion depending on the value of H relative to the saturation magnetization.

Applying H perpendicular to the needle axis is not practical because one needs to overcome the demagnetizing field.

SURFACE OR LOCALIZED SPIN WAVE EXCITATIONS

The trend in modern technology is toward miniaturization and efficiency. By necessity, this means the use of nanomaterials, thin-film composites, and equivalent. Small particles or thin films imply that surface and/or interfacial effects are important to modern applications. For now, let's consider magnetic interfacial effects between a magnetic film and its substrate. Films are routinely grown at high temperatures on a substrate, and as such, there is no such thing as a "sharp" transition between the two dissimilar materials. The chemistry would simply not allow it. We can safely assume that the magnetic properties or parameters are not uniform near or at the interface.

For the case of uniform M_0, the spin wave dispersion for $\theta_k = 0$ is

$$\frac{\omega}{\gamma} = H_0 + \frac{2A}{M} k^2. \tag{9.3}$$

We may rewrite Equation 9.3 as (for uniform M_0)

$$\left(\frac{\omega}{\gamma} - H_0 + \frac{2A}{M}\frac{\partial^2}{\partial z^2}\right)m = 0$$

The above equation was derived from Equations 9.1a and 9.1b by combining the two to form an excitation for circular mode:

$$m = m_x - jm_y.$$

Also, we assume one-dimensional propagation so that

$$\nabla^2 \to \frac{\partial^2}{\partial z^2}$$

in the exchange field operator.

Equation 9.3 may be modified for *nonuniform* M_0 by redefining H_0. The exchange field is as before:

$$\bar{H}_{exc.} = \frac{2A}{M^2}\nabla^2\bar{M} = \frac{2A}{M^2}\frac{\partial^2}{\partial z^2}(M_0\bar{a}_z + \bar{m}).$$

This means that the internal static field is redefined as follows in a nonuniform M_0:

$$H_0' = H_0 + \frac{2A}{M^2}\frac{\partial^2}{\partial z^2}M_0. \tag{9.4}$$

Thus, Equation 9.3 may simply be rewritten as follows for nonuniform M_0:

$$\left(\frac{\omega}{\gamma} + \frac{2A}{M}\frac{\partial^2}{\partial z^2} - H_0 - \frac{2A}{M^2}\frac{\partial^2}{\partial z^2}M_0\right)m = 0. \tag{9.5}$$

Putting Equation 9.5 in the following form, we have a recognizable equation

$$\frac{2A}{M}\frac{\partial^2}{\partial z^2}m + \left(\frac{\omega}{\gamma} - H_0 - \frac{2A}{M^2}\frac{\partial^2}{\partial z^2}M_0\right)m = 0.$$

Comparing the above equation to Schrodinger's equation below

$$\frac{\hbar^2}{2\rho}\frac{\partial^2\Psi}{\partial z^2} + (E - V)\Psi = 0, \tag{9.6}$$

we identify

$$\Psi \to m,$$

$$E \to \frac{\omega}{\gamma},$$

$$\frac{\hbar^2}{2\rho} \to \frac{2A}{M},$$

and

$$V \to H_0 + \frac{2A}{M^2} \frac{\partial^2}{\partial z^2} M_0.$$

Let's examine typical interfacial region of a film.

As shown in Figure 9.9, the second derivative of M gives rise to a maximum and a minimum in the potential V. The minimum in V is identified with a potential well. As in any other potential well problem, the stability or the existence of a localized state, in this case a surface spin wave state, depends on the depth (energy) and the width of the well. Physically, this means that for a proper field excitation, it is possible to excite microwave magnetization components located only at the interface, but not in the film itself! Since the excitation is localized, it is a non-propagation mode excitation.

For perpendicular FMR, $H_0 = H_{ext} - 4\pi M_s$, and of course, $4\pi M_s$ may vary as well at the interface. Thus, for the perpendicular field configuration, the addition of the magnetization factor may "wash" out the potential well and the bound surface state.

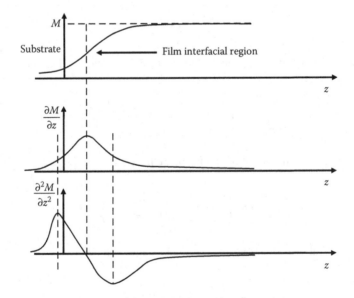

FIGURE 9.9 Variation of static magnetization near interface between magnetic film and substrate.

For the parallel field case, there is no extra magnetization factor to deal with, and indeed, this type of analysis gives rise to a surface bound state. For more details, the reader is referred to Vittoria (1998).

PURE ELECTROMAGNETIC MODES OF PROPAGATION: SEMI-INFINITE MEDIUM

Besides spin waves, there may be pure electromagnetic wave propagation in a medium.

Wave propagation in a semi-infinite medium characterized by ε and μ_0 is given as

$$k^2 = \omega^2 \varepsilon \mu_0.$$

Let's plot ω versus Re(k) and $I(k)$ for (a) $\varepsilon = \varepsilon'$ (insulator) and (b) $\varepsilon = -j(\sigma/\omega)$ (metal) (Figures 9.10 and 9.11).

Case (a) insulator:

$$\varepsilon = \varepsilon',$$
$$k^2 = \omega^2 \varepsilon' \mu' \text{ (no losses).}$$

Case (b) metal:

$$\varepsilon = -j\frac{\sigma}{\omega}, \text{ metal}$$

$$k^2 = \omega^2 \left(-j\frac{\sigma}{\omega} \right) \mu_0 = -j\omega\sigma\mu_0,$$

$$\beta - j\alpha = \text{Re}(k) - j\,\text{Im}(k) = \frac{1}{\sqrt{2}}(1-j)\sqrt{\sigma\mu_0\omega}.$$

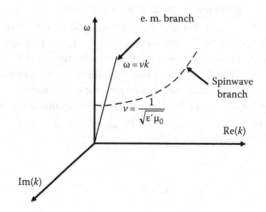

FIGURE 9.10   Electromagnetic wave dispersion in an insulator.

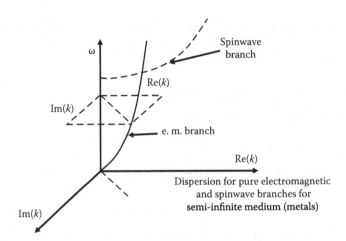

FIGURE 9.11 Electromagnetic wave dispersion for a metal.

Dispersion for pure electromagnetic and spin wave branches for semi-infinite medium (dielectric case):

$$\beta = \mathrm{Re}(k) = \sqrt{\frac{\omega\mu_0\sigma}{2}}; \ \mathrm{Im}(k) = \sqrt{\frac{\omega\mu_0\sigma}{2}} = \alpha.$$

We identify α as the attenuation constant and β the phase constant.

Clearly, with no coupling between the equation of motion and Maxwell's equations, one can only excite pure electromagnetic or spin wave modes. Let's now couple the two sets of equations.

COUPLING OF THE EQUATION OF MOTION AND MAXWELL'S EQUATIONS

Let's now couple the equation of motion with Maxwell's equation. In general, we may assume $\vec{M}$ in the z-direction and an electromagnetic wave propagating in the $\vec{r}$ direction. Also, transverse electromagnetic (TEM) wave of propagation is assumed. The microwave properties of the magnetic film may be represented by a permeability tensor and the permittivity by a scalar quantity. This is not necessary. The permittivity can also be tensorial and our theoretical approach would still be the same. However, it would introduce a lot more terms in our formulation. The electromagnetic magnetic field, $\vec{h}$, may be expressed in terms of a propagation vector $\vec{k}$, along the $\vec{r}$ direction as follows (Figure 9.12):

$$\vec{h} = \left(h_x\vec{a}_x + h_y\vec{a}_y + h_z\vec{a}_z\right)e^{j\left(\omega t - \vec{k}\cdot\vec{r}\right)}.$$

Forward propagation is along the $+\vec{r}$ direction or radially out.

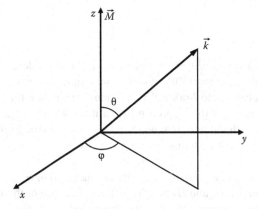

FIGURE 9.12 Wave propagation direction, $\vec{k}$, relative to $\vec{M}$.

We may summarize the equation of motion by a tensor permeability (see Chapter 6). The permeability tensor was derived from the equation of motion for the magnetization. The equation of motion contains internal and external fields deduced from the free energy. Effectively, the free energy has replaced the role of the constitutive relations used in the past to complement Maxwell's equations. The same methodology may be applied in obtaining a permittivity tensor. In essence, the equations of motion are coupled to Maxwell's equations via the free energy:

$$[\mu] = \mu_0 \begin{bmatrix} \mu_{xx} & j\kappa & 0 \\ -j\kappa & \mu_{yy} & 0 \\ 0 & 0 & 1 \end{bmatrix}, \text{ MKS,}$$

where

$$\mu_{xx} = 1 + \frac{4\pi M H_1}{\Omega^2},$$

$$\mu_{yy} = 1 + \frac{4\pi M H_2}{\Omega^2},$$

$$\kappa = \frac{4\pi M \dfrac{\omega}{\gamma}}{\Omega^2},$$

$$H_1 = H_{01} + \frac{2A}{M} k^2, \text{ and}$$

$$H_2 = H_{02} + \frac{2A}{M} k^2.$$

H_{01} and H_{02} are internal magnetic fields to be defined later. A is the exchange stiffness constant.

$$\Omega^2 = H_1 H_2 - \left(\frac{\omega}{\gamma}\right)^2.$$

It is noted that the permeability tensor elements are k dependent in contrast to conventional definition of permeability. As such, we can represent magnetic wave excitations to be substituted into Maxwell's equations rather than the reverse way as in the past. The advantage of this approach is that it allows (1) to identify terms in the general dispersion with the pure magnetic and electromagnetic excitations and (2) to identify the nature of the coupling between the two sets of equations: equations of motion and Maxwell's equations. More discussions on this matter will ensue.

Volume dynamic demagnetizing fields are not included in H_1 or H_2, when coupled to Maxwell's equations. H_1 and H_2 may be different because of the internal magnetic anisotropy fields, for example. Damping may be included as follows by writing ω/γ as

$$\frac{\omega}{\gamma} \to \frac{\omega}{\gamma}(1 - j\alpha)$$

assuming Gilbert form for magnetic damping, where α is the Gilbert damping parameter. If one assumes Landau–Lifshitz form for damping, ω/γ is modified as follows:

$$\frac{\omega}{\gamma} \to \frac{\omega}{\gamma}\left(\frac{1}{1 + j(\lambda / \gamma M_s)}\right),$$

where λ is the Landau–Lifshitz damping parameter. For small λ,

$$\alpha \simeq \frac{\lambda}{\gamma M_s}$$

It is interesting to point out that the above inclusions for damping are a result of the fact that the permeability for the case of the external magnetic field applied perpendicular to the film plane is defined as follows (depending on the form for damping):

$$\mu = 1 + \frac{4\pi M}{H_0 - (\omega/\gamma)(1 - j\alpha)},$$

assuming Gilbert relaxation model. However,

$$\mu = 1 + \frac{4\pi M}{H_0 - \left(\frac{\omega}{\gamma}\right)\left(\frac{1}{1 + j(\lambda/\gamma M)}\right)},$$

assuming Landau–Lifshitz damping form. H_0 is applied perpendicular to the film plane.

Maxwell's equations are coupled to permeability via Faraday's law

$$\vec{\nabla} \times \vec{e} = -j\omega \vec{b} = -j\omega \bar{\bar{\mu}} \cdot \vec{h}, \text{ and} \tag{9.7}$$

$$\vec{\nabla} \times \vec{h} = j\omega\varepsilon_{eff}\vec{e}, \tag{9.8}$$

where μ is a dyadic operator and

$$j\omega\varepsilon_{eff} = j\omega\varepsilon + \alpha = j\omega\left(\varepsilon + \frac{\alpha}{j\omega}\right).$$

Thus,

$$\varepsilon_{eff} = \varepsilon - j\frac{\sigma}{\omega},$$

where

$$\varepsilon_{eff} = \varepsilon, \text{ dielectric,}$$
$$\varepsilon_{eff} = -j(\sigma/\omega), \text{ metal.}$$

Henceforth, we will drop "*eff*" subscript and Maxwell's equations are combined to give

$$\vec{\nabla} \times \left(\vec{\nabla} \times \vec{h}\right) = j\omega\varepsilon\vec{\nabla} \times \vec{e} = j\omega\varepsilon\left[-j\omega\ddot{\mu} \cdot \vec{h}\right],$$

$$\vec{\nabla}\left(\vec{\nabla} \cdot \vec{h}\right) - \nabla^2\vec{h} = \omega^2\varepsilon\mu_0\left[\left(\mu_{xx}h_x + j\kappa h_y\right)\vec{a}_x + \left(-j\kappa h_x + \mu_{yy}h_y\right)\vec{a}_y + h_z\vec{a}_z\right],$$

where $\vec{h} = \left(h_x\vec{a}_x + h_y\vec{a}_y + h_z\vec{a}_z\right)e^{-j\vec{k}\cdot\vec{r}}$. Thus, we are in a position to address each term in Maxwell's equations:

$$\vec{\nabla}\left(\vec{\nabla} \cdot \vec{h}\right)$$

$$= h_x e^{-j\vec{k}\cdot\vec{r}}\left[\vec{a}_x\left(-j\right)k\sin\theta\cos\varphi + \vec{a}_y\left(-j\right)k\sin\theta\sin\varphi + \vec{a}_z\left(-j\right)k\cos\theta\right]$$

$$\times\left[-jk\sin\theta\cos\varphi\right]$$

$$+ h_y e^{-j\vec{k}\cdot\vec{r}}\left[\vec{a}_x\left(-j\right)k\sin\theta\cos\varphi + \vec{a}_y\left(-j\right)k\sin\theta\sin\varphi + \vec{a}_z\left(-j\right)k\cos\theta\right]$$

$$\times\left[-jk\sin\theta\sin\varphi\right]$$

$$+ h_z e^{-j\vec{k}\cdot\vec{r}}\left[\vec{a}_x\left(-j\right)k\sin\theta\cos\varphi + \vec{a}_y\left(-j\right)k\sin\theta\sin\varphi + \vec{a}_z\left(-j\right)k\cos\theta\right]$$

$$\times\left[-jk\cos\theta\right],$$

$$\nabla^2\vec{h} = -k^2 e^{-j\vec{k}\cdot\vec{r}}\left[\vec{a}_x h_x + \vec{a}_y h_y + \vec{a}_z h_z\right], \text{ and}$$

$$\bar{\nabla} \times \bar{\nabla} \times \bar{h} = \left[\left(k^2 - k^2 \sin^2\theta_k \cos^2\varphi_k\right)h_x + \left(-k^2 \sin^2\theta_k \sin\varphi_k \cos\varphi_k\right)h_y\right.$$
$$\left. + \left(-k^2 \sin\theta_k \cos\theta_k \cos\theta_k\right)h_z\right]\bar{a}_x e^{-j\bar{k}\cdot\bar{r}}$$
$$+ \left[\left(-k^2 \sin^2\theta_k \cos\varphi_k \sin\varphi_k\right)h_x + \left(k^2 - k^2 \sin^2\theta_k \sin^2\varphi_k\right)h_y\right.$$
$$\left. + \left(-k^2 \sin\theta_k \sin\varphi_k \cos\theta_k\right)h_z\right]\bar{a}_y e^{-j\bar{k}\cdot\bar{r}}$$
$$+ \left[\left(-k^2 \sin^2\theta_k \cos\varphi_k \cos\theta_k\right)h_x + \left(-k^2 \sin\theta_k \sin\varphi_k \cos\theta_k\right)h_y\right.$$
$$\left. + \left(k^2 - k^2 \cos^2\theta_k\right)h_z\right]\bar{a}_y e^{-j\bar{k}\cdot\bar{r}}.$$

Dropping $e^{-j\bar{k}\cdot\bar{r}}$ and subscript "k" and combining with Equation 9.8, we obtain the following:

$$0 = \left(k^2 - k^2 \sin^2\theta\cos^2\varphi - \theta^2\varepsilon\mu_0\mu_{xx}\right)h_x + \left(-k^2 \sin^2\theta\sin\varphi\cos\varphi - \omega^2\,\varepsilon\mu_0 j\kappa\right)h_y$$
$$+ \left(-k^2 \sin\theta\cos\varphi\cos\varphi\right)h_z,$$

(9.8a)

$$0 = \left(-k^2 \sin^2\theta\cos\varphi\sin\varphi + \omega^2\varepsilon\mu_0 j\kappa\right)h_x + \left(k^2 - k^2 \sin^2\theta\sin^2\varphi - \omega^2\varepsilon\mu_0\mu_{yy}\right)h_y$$
$$+ \left(-k^2 \sin\theta\sin\varphi\cos\theta\right)h_z,$$

(9.8b)

$$0 = \left(-k^2 \sin\theta\cos\varphi\cos\theta\right)h_x + \left(-k^2 \sin\theta\sin\varphi\cos\theta\right)h_y + \left(k^2 \sin^2\theta - \omega^2\varepsilon\mu_0\right)h_z.$$

(9.8c)

The above set of equations is of the form

$$[A][h] = 0,$$

where

$$[h] = \begin{pmatrix} h_x \\ h_y \\ h_z \end{pmatrix}.$$

Assuming a tensor property for the permittivity would have not changed the form of the secular equation. The general dispersion would have contained more terms.

However, a tensorial permittivity implies another set of equations of motion from which the tensor could be derived from, much like the permeability tensor. The point is that both tensors may be derivable from the general free energy rather than from ad hoc means as in a set of constitutive relationships.

The dispersion relation may be obtained by setting det(A) = 0. After expanding the determinant, we obtain

$$k^4 \left[1 + \sin^2 \theta \left(\mu_{xx} \cos^2 \phi + \mu_{yy} \sin^2 \phi - 1 \right) \right] -$$
$$-k^2 \omega^2 \varepsilon \mu_0 \left[\mu_{xx} + \mu_{yy} + \sin^2 \theta \left(\mu_{xx} \mu_{yy} - k^2 - \mu_{xx} \sin^2 \phi - \mu_{yy} \cos^2 \phi \right) \right] + \qquad (9.9)$$
$$+ \omega^4 \varepsilon^2 \mu_0^2 \left(\mu_{xx} \mu_{yy} - \kappa^2 \right) = 0.$$

The general dispersion relationship, Equation 9.9, appears to be deceptively simple, but it is rather complex. It contains many degrees of freedom: the angle between the static magnetization direction and wave propagation is chosen to be arbitrary; the permeability tensor elements are k-dependent and internal anisotropy fields, insulators versus metals, etc. It is not practical to cover every case of interest here. We have limited our discussions to two cases of interest: (A) $\theta = 0$ and (B) $\theta = 90°$.

A. $\theta = 0$, $\vec{M}\vec{k}$, the wave dispersion becomes

$$k^4 - k^2 \omega^2 \varepsilon \mu_0 \left(\mu_{xx} + \mu_{yy} \right) + \omega^4 \varepsilon^2 \mu_0^2 \left(\mu_{xx} \mu_{yy} - \kappa^2 \right) = 0. \qquad (9.10)$$

Rewrite Equation 9.10 in the following factor form:

$$\left(k^2 - \omega^2 \varepsilon \mu_0 \mu_{xx} \right) \left(k^2 - \omega^2 \varepsilon \mu_0 \mu_{yy} \right) = \omega^4 \varepsilon^2 \mu_0^2 \kappa^2, \qquad (9.10a)$$

where

$$\mu_{xx} = \mu_{yy} = \mu.$$

The above relationship implies that the internal magnetic anisotropy fields due to cubic and hexagonal anisotropy energies are assumed to be zero.

As such, Equation 9.10a yields

$$k^2 - \omega^2 \varepsilon \mu_0 \mu = \pm \omega^2 \varepsilon \mu_0 \kappa \qquad (9.10b)$$

or

$$k^2 = \omega^2 \varepsilon \mu_0 \left(\mu \pm \kappa \right), \qquad (9.10c)$$

where

$$\mu = 1 + \frac{4\pi MH}{\Omega^2},$$

$$H = H_0 + \frac{2A}{M}k^2,$$

$$\mu = \frac{4\pi M(\omega/\gamma)}{\Omega^2},$$

$$\Omega^2 = H^2 - \frac{\omega^2}{\gamma^2}, \text{ and}$$

$$\frac{\omega}{\gamma} \rightarrow \frac{\omega}{\gamma}(1 - j\alpha).$$

Note here that $H_1 = H_2 = H = H_{0+}(2A/M)k^2$. There is no volume demagnetizing field included here explicitly. However, H_1 and H_2 could include magnetic anisotropy fields yielding $H_1 \neq H_2$, and therefore, $\mu_{xx} \neq \mu_{yy}$.

Let's assume the thin-film geometrical configuration in Figure 9.13 where the propagation direction is perpendicular to the film plane. For a magnetic metal, we have that $\varepsilon_{eff} = \varepsilon - j(\sigma/\omega)$. Finally, we have

$$\mu \pm \kappa = 1 + \frac{4\pi M_s H}{\Omega^2} \pm \frac{4\pi M_s(\omega/\gamma)}{\Omega^2},$$

$$\mu \pm \kappa = 1 + \frac{4\pi M_s(H \pm (\omega/\gamma))}{(H + (\omega/\gamma))(H - (\omega/\gamma))},$$

$$+\text{Sign}: \mu + \kappa = 1 + \frac{4\pi M_s}{H - (\omega/\gamma)}, \text{resonant mode, and}$$

$$-\text{Sign}: \mu - \kappa = 1 + \frac{4\pi M_s}{H + \left(\dfrac{\omega}{\gamma}\right)}, \text{nonresonant mode.}$$

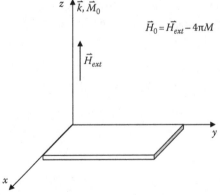

FIGURE 9.13 Wave propagation direction relative to thin-film geometry.

For now, consider only resonant mode

$$k^2 = \omega^2 \varepsilon \mu_0 \left(1 + \frac{4\pi M_s}{\left(H - (\omega/\gamma) \right)} \right) \tag{9.11}$$

or

$$\left(\frac{k^2}{\omega^2 \varepsilon \mu_0} - 1 \right) \left(H - \frac{\omega}{\gamma} \right) = 4\pi M_s$$

or

$$H - \frac{\omega}{\gamma} = \frac{4\pi M_s}{\left(k^2 / \omega^2 \varepsilon \mu_0 \right) - 1}.$$

Putting the above expression into a recognizable form,

$$\frac{\omega}{\gamma} = H - \frac{4\pi M_s}{\left(k^2 / \omega^2 \varepsilon \mu_0 \right) - 1}$$

or

$$\frac{\omega}{\gamma} = H_0 + \frac{2A}{M} k^2 - \frac{4\pi M_s}{\left(k^2 / \omega^2 \varepsilon \mu_0 \right) - 1}. \tag{9.11a}$$

This is a convenient form to note, since it allows us to compare with the pure case of spin wave excitations. It is observed that the coupling to Maxwell's equations has introduced the third term in the RHS in the dispersion relation (see Equation 9.11a). The first term on the RHS of Equation 9.11a is the FMR condition, as calculated in Chapter 6 for the static internal field applied perpendicular to the film plane. The first two terms on the RHS represent the "pure" spin wave dispersion resulting from the equation of motion without coupling to Maxwell's equations. The third term on the RHS represents the coupling factor to Maxwell equations. Setting the denominator of this third term to zero determines the "pure" electromagnetic branches of k. The resultant equation (all three terms) is an admixture of spin waves and electromagnetic waves. Thus, the calculated k-value is not exactly a pure spin wave or electromagnetic wave. In order to make the discussion more transparent, Equation 9.11a is rewritten in a factored form as follows:

$$\left[\frac{\omega}{\gamma} - \left(H_0 + \frac{2A}{M} k^2 \right) \right] \left[\left(\frac{k^2}{\omega^2 \varepsilon \mu_0} \right) - 1 \right] = -4\pi M_s. \tag{9.11b}$$

By setting the RHS of Equation 9.11b equal to zero, the two "pure" branches can be easily obtained. This is equivalent to decoupling the equations of motion from Maxwell's equations. Equation 9.11b presents some interesting opportunities. For example, acoustic wave dispersions may be included in Equation 9.11a or 9.11b via the permittivity, ε, and the form of Equation 9.11b will not be altered. However, ε would be k dependent (sixth order). The author suspects (not proven) that the form of Equation 9.11b to be the same even if tensorial ε is included in the analysis. The second factor in Equation 9.11b would contain more terms involving off-diagonal elements of tensor ε. It is straightforward to include metallic effects in Equation 9.11b via ε (see the presentation below).

The non-resonant mode dispersion may be obtained by putting $(\omega/\gamma) \to -(\omega/\gamma)$.

METALS

Consider only the term $k^2/\omega^2\, \varepsilon\mu_0$ contained in the third term of Equation 9.11a. For a metal, we may write

$$\varepsilon = -j\frac{\sigma}{\omega}.$$

Thus,

$$\omega^2\varepsilon\mu_0 = -j\omega^2\frac{\sigma}{\omega}\mu_0 = -j\omega\sigma\mu_0,$$

$$\delta_0^2 = (\text{skindepth})^2 = \frac{2}{\omega\sigma\mu_0},$$

$$\omega\sigma\mu_0 = \frac{2}{\delta_0^2},$$

$$\therefore\ \frac{k^2}{\omega^2\varepsilon\mu_0} = \frac{k^2}{-j\omega^2\sigma\mu_0} = j\frac{k^2}{2/\delta_0^2} = \frac{j}{2}(k\delta_0)^2.$$

Thus, the dispersion for magnetic metals becomes

$$\frac{\omega}{\gamma} = H_0 + \frac{2A}{M}k^2 - \frac{4\pi M_s}{\dfrac{j}{2}(k\delta_0)^2 - 1} \qquad (9.11c)$$

or

$$\frac{\omega}{\gamma} = H_0 + \frac{2A}{M}k^2 - \frac{4\pi M_s}{\left(\dfrac{k^2}{k_0^2} - 1\right)},$$

where

$$\omega^2 \varepsilon \mu_0 = k_0^2 = -\frac{2}{\delta_0^2} j.$$

DIELECTRICS

$$\varepsilon = \varepsilon' - j\varepsilon'', \quad \varepsilon'' \ll \varepsilon',$$
$$\omega^2 \varepsilon \mu_0 = \omega^2 \mu_0 (\varepsilon' - j\varepsilon'') = k_0^2,$$

and

$$\frac{k^2}{k_0^2} = \frac{k^2}{\omega^2 \mu_0 \varepsilon' \left(1 - j\dfrac{\varepsilon''}{\varepsilon'} \right)}.$$

The dispersion then becomes

$$\frac{\omega}{\gamma} = H_0 + \frac{2A}{M} k^2 - \frac{4\pi M_s}{\left(\dfrac{k^2}{k_0^2} - 1 \right)},$$

where $k_0^2 = \omega^2 \varepsilon' \mu_0 \left(1 - j\dfrac{\varepsilon''}{\varepsilon'} \right)$.

SOLUTION FOR k^2

We have in general that

$$\frac{\omega}{\gamma} = H_0 + \frac{2A}{M} k^2 - \frac{4\pi M_s}{\left(k^2/k_0^2 \right) - 1}. \tag{9.11d}$$

Rewrite

$$\left(\frac{\omega}{\gamma} - H_0 \right) \left(\frac{k^2}{k_0^2} - 1 \right) = \frac{2A}{M} k^2 \left(\frac{k^2}{k_0^2} - 1 \right) - 4\pi M_s,$$

$$\frac{2A}{M} \frac{k^4}{k_0^2} - k^2 \left[\frac{2A}{M} + \frac{(\omega/\gamma) - H_0}{k_0^2} \right] - 4\pi M_s \left(\frac{\omega}{\gamma} - H_0 \right) = 0, \tag{9.12}$$

$$k^4 - k^2 \left[k_0^2 + \frac{M}{2A} \left(\frac{\omega}{\gamma} - H_0 \right) \right] + \left(\frac{\omega}{\gamma} - H_0 - 4\pi M_s \right) k_0^2 \frac{M}{2A} = 0.$$

Let

$$\frac{(\omega/\gamma) - H_0}{2(A/M)} = k_s^2$$

and

$$k_s^2 - \frac{4\pi M_s}{2(A/M)} = \Delta k_s^2,$$

where

k_s is the pure spin wave propagation constant

$\Delta k_s^2 \equiv$ change from pure spin wave constant squared

As such, Equation 9.12 may simply be rewritten as

$$k^4 - k^2\left(k_0^2 + k_s^2\right) + \left(\Delta k_s^2\right)k_0^2 = 0,$$

where $k_0^2 = \omega^2 \varepsilon \mu_0 \equiv$ pure electromagnetic mode.

For propagation of the non-resonant mode, we write

$$\frac{\omega}{\gamma} \rightarrow -\frac{\omega}{\gamma}$$

Thus, the propagation constants for non-resonant modes can be obtained by changing sign on ω. We have obtained wave dispersion for the following parameters:

a. $g = 2.10$, $4\pi M_s = 10,000$ (G), $H_0 = 3,500$ (Oe), $\Delta\omega/\gamma = 30$ (Oe), $\sigma = 0.7 \times 10^5$ (mho/cm), $A = 1.14 \times 10^{-6}$ (erg/cm), $\theta_k = 0$ ($H_0 = 3,500$ Oe), and $\theta_k = (\pi/2)$ ($H_0 = 1,103.3$ Oe).

These parameters simulate permalloy material (Figures 9.14 and 9.15).

b. $g = 2.005$, $4\pi M_s = 1,750$(G), $H_0 = 3,500$(Oe), $\varepsilon_{R'} = 14.5$, $\varepsilon''/\varepsilon' = 0.001$, $\Delta\omega/\gamma = 1$ (Oe), $A = 0.4 \times 10^{-6}$ (erg/cm), $\theta_k = 0$ ($H_0 = 3,500$ Oe), and $\theta_k = (\pi/2)$ ($H_0 = 2,732.72$ Oe).

The above parameters simulate YIG ferrite (Figures 9.16 and 9.17).

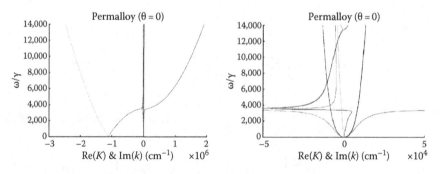

FIGURE 9.14 Wave dispersions for H perpendicular to film plane (permalloy).

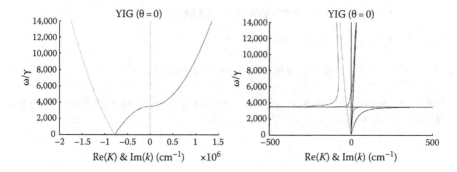

FIGURE 9.15 Dispersions for H perpendicular to film plane (YIG).

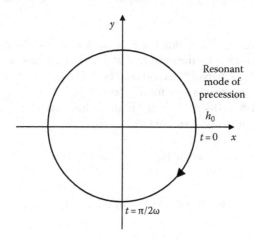

FIGURE 9.16 Normal mode motion of spin wave mode.

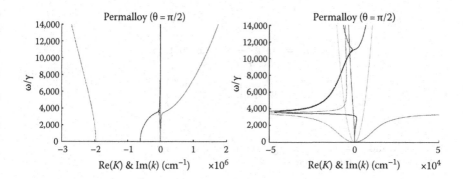

FIGURE 9.17 Dispersions for H in the plane (permalloy).

NORMAL MODES OF SPIN WAVE EXCITATIONS

For $\theta = 0$ and $\mu_{xx} = \mu_{yy} = \mu$, Equation 9.8b reduces to

$$\omega^2 \varepsilon \mu_0 j \kappa h_x + \left(k^2 - \omega^2 \varepsilon \mu_0 \mu\right) h_y = 0. \tag{9.13}$$

The dispersion for the resonant spin wave mode is $k^2 = \omega^2 \varepsilon \mu_0 (\mu + \kappa)$. Substituting the dispersion relation into Equation 9.13 yields the result that

$$j h_x = h_y.$$

Assume $h_x(t) = h_0 \cos(\omega t)$, $h_x = h_0$ (in complex notation), then

$$h_y(t) = \text{Re}\left[j h_0 e^{j\omega t} \right] = \text{Re}\left[h_0 e^{j\left(\omega t + \frac{\pi}{2}\right)} \right] = -h_0 \sin(\omega t).$$

The time dependence of the fields traces a clockwise circular motion (see Figure 9.16). The sense of rotation is the same as that found for the case of the normal modes in FMR calculated in Chapter 5, as it should be.

There are two cases to consider for $\theta_k = \pi/2$.

A. $\theta_k = \pi/2$ and $\phi_k = 0$—propagation in the x-direction:

From Equations 9.8a and 9.8b, we have

$$-\omega^2 \varepsilon \mu_0 \mu_{xx} h_x - \omega^2 \varepsilon \mu_0 j \kappa h_y = 0, \tag{9.13a}$$

$$\omega^2 \varepsilon \mu_0 j \kappa h_x + \left(k^2 - \omega^2 \varepsilon \mu_0 \mu_{yy}\right) h_y = 0. \tag{9.13b}$$

The dispersion is simply

$$-\mu_{xx}\left(k^2 - k^2 - \omega^2 \varepsilon \mu_0 \mu_{yy}\right) - \omega^2 \varepsilon \mu_0 \kappa^2 = 0$$

or

$$k^2 = -\omega^2 \varepsilon \frac{\mu_0}{\mu_{xx}} \kappa^2 + \omega^2 \varepsilon \mu_0 \mu_{yy},$$

$$k^2 = \omega^2 \varepsilon \mu_0 \left(\frac{\mu_{xx}\mu_{yy} - \kappa^2}{\mu_{xx}}\right). \tag{9.14}$$

B. $\theta_k = \pi/2$ and $\phi_k = \pi/2$—propagation in the y-direction:

From Equation 9.8, we have

$$\left(k^2 - \omega^2 \varepsilon \mu_0 \mu_{xx}\right) h_x - \omega^2 \varepsilon \mu_0 j \kappa h_y = 0, \tag{9.13c}$$

$$\omega^2 \varepsilon \mu_0 j \kappa h_x - \omega^2 \varepsilon \mu_0 \mu_{yy} h_y = 0. \tag{9.13d}$$

The dispersion is then

$$-\mu_{yy}\left(k^2 - \omega^2 \varepsilon \mu_0 \mu_{xx}\right) - \omega^2 \varepsilon \mu_0 \kappa^2 = 0,$$

$$k^2 = \omega^2 \varepsilon \mu_0 \left(\frac{\mu_{xx}\mu_{yy} - \kappa^2}{\mu_{xx}}\right). \tag{9.15}$$

The other solution or dispersion is

$$k^2 = \omega^2 \varepsilon \mu_0,$$

which is the pure electromagnetic mode of propagation.

Notice the subtle difference between Equations 9.14 and 9.15. Thus, if the internal or volume demagnetizing fields are zero, cases B and C are identical as they should be. Thus, for anisotropic internal fields as reflected in μ_{xx} and μ_{yy} will affect k depending on the direction of propagation. If $\mu_{xx} = \mu_{yy}$, the two cases are exactly the same.

For the isotropic case, we have

$$\mu_{xx} = \mu_{yy} = \mu = 1 + \frac{4\pi MH}{\Omega^2}$$

and

$$\kappa = \frac{4\pi M\left(\omega/\gamma\right)}{\Omega^2}; \ \Omega^2 = H^2 - \frac{\omega^2}{\gamma^2}; \ H = H_0 + \frac{2A}{M}k^2; \ \frac{\omega}{\gamma} \to \frac{\omega}{\gamma}\left(1 - j\alpha\right),$$

and

$$\mu_{\mathit{eff}} = \frac{\mu^2 - \kappa^2}{\mu} = \frac{\left(1 - \dfrac{4\pi MH}{\Omega^2}\right)\left(1 + \dfrac{4\pi MH}{\Omega^2}\right) - \dfrac{\left(4\pi M\left(\omega/\gamma\right)\right)^2}{\Omega^4}}{1 + \dfrac{4\pi MH}{\Omega^2}},$$

$$\mu_{\mathit{eff}} = 1 + \frac{4\pi M\left(H + 4\pi M\right)}{\Omega^2 + 4\pi MH}.$$

Let

$$\omega^2 \varepsilon \mu_0 = k_0^2,$$

then

$$\frac{k^2}{k_0^2} - \mu_{\mathit{eff}} = 0 \ \left(\text{see Equation 9.15}\right).$$

The dispersion relation becomes

$$\frac{k^2}{k_0^2} - 1 - \frac{4\pi M (H + 4\pi M)}{\Omega^2 + 4\pi MH} = 0.$$

Rewriting,

$$\Omega^2 + 4\pi MH = \frac{4\pi M}{\left(k^2/k_0^2 \right) - 1}(H + 4\pi M). \tag{9.16}$$

Again note that the coupling to Maxwell's equations has produced the right-hand side term in Equation 9.16. Setting the right-hand-side term to zero is equivalent to decoupling Maxwell's equations from the equation of motions or

$$\Omega^2 + 4\pi MH = 0.$$

We recognize the dispersion relation of Equation 9.2 with no coupling to Maxwell's equation as

$$H^2 - \frac{\omega^2}{\gamma^2} + 4\pi MH = 0.$$

Putting the above expression more into a recognizable form, we have

$$\frac{\omega^2}{\gamma^2} = H(H + 4\pi M) = \left(H_0 + \frac{2A}{M}k^2 \right)\left(H_0 + \frac{2A}{M}k^2 + 4\pi M \right), \ \theta_k = \frac{\pi}{2},$$

which is indeed the same result as in Equation 9.2 obtained from only the equation of motion.

    Let's obtain the dispersion relation from Equation 9.16 including the coupling term. Collecting terms and expanding, one obtains

$$\left(\frac{k^2}{k_0^2} - 1 \right)\left[\left(H_0 + \frac{2A}{M}k^2 \right)\left(H_0 + \frac{2A}{M}k^2 + 4\pi M \right) - \frac{\omega^2}{\gamma^2} \right] = 4\pi M\left[H_0 + \frac{2A}{M}k^2 + 4\pi M \right].$$

Putting the above expression in ascending process of k, we obtain

$$\left(\frac{k^2}{k_0^2} - 1 \right)\left[H_0(H_0 + 4\pi M) + k^2\left(\frac{2A}{M}H_0 + \frac{2A}{M}H_0 + \frac{2A}{M}4\pi M \right) + \frac{4A^2}{M^2}k^4 - \frac{\omega^2}{\gamma^2} \right]$$

$$= 4\pi M\left[H_0 + \frac{2A}{M}k^2 + 4\pi M \right],$$

and re-arranging the above equation becomes

$$\left(\frac{k^2}{k_0^2}-1\right)\left[H_0\left(H_0+4\pi M\right)-\frac{\omega^2}{\gamma^2}+k^2\left(\frac{2A}{M}H_0+\frac{2A}{M}H_0+\frac{2A}{M}4\pi M\right)+\frac{4A^2}{M^2}k^4\right]$$

$$=4\pi M\left[H_0+\frac{2A}{M}k^2+4\pi M\right].$$

Multiplying all the factors, one obtains

$$0=\frac{k^2}{k_0^2}\left(H_0\left(H_0+4\pi M\right)-\frac{\omega^2}{\gamma^2}\right)+\frac{2A}{M}\frac{k^4}{k_0^2}\left(2H_0+4\pi M\right)+\frac{4A^2}{M^2}\frac{k^6}{k_0^2}-\frac{2A}{M}k^2$$

$$\left(2H_0+4\pi M\right)-\frac{4A^2}{M^2}k^4+\frac{\omega^2}{\gamma^2}-H_0\left(H_0+4\pi M\right)-\frac{2A}{M}4\pi Mk^2-4\pi M\left(H_0+4\pi M\right).$$

We note that the full dispersion is of the form

$$\left(k^6+ak^4+bk^2+c\right)\left(k^2-k_0^2\right)=0.$$

The dispersion contains a pure electromagnetic mode and "mixed" magneto-dielectric modes cubic in k^2. Let's identify the arbitrary constants a, b, and c in the above equation by focusing only on the cubic equation.

$$\frac{4A^2}{M^2}\frac{k^6}{k_0^2}+k^4\left(\frac{2A}{M}\frac{1}{k_0^2}\left(2H_0+4\pi M\right)-\frac{4A^2}{M^2}\right)$$

$$+k^2\left[\frac{H_0\left(H_0+4\pi M\right)-\left(\omega^2/\gamma^2\right)}{k_0^2}-\frac{2A}{M}\left(2H_0+8\pi M\right)\right]+\frac{\omega^2}{\gamma^2}-\left(H_0+4\pi M\right)^2=0.$$

Putting the above equation into a simpler form, we have that

$$k^6+k^4\left[\underbrace{\frac{2H_0+4\pi M}{2(A/M)}}_{a}-k_0^2\right]+k^2\left[\underbrace{\frac{H_0\left(H_0+4\pi M\right)-\left(\omega^2/\gamma^2\right)}{4\left(A^2/M^2\right)}-\frac{2\left(H_0+4\pi M\right)}{2(A/M)}k_0^2}_{b}\right]$$

$$+\left[\underbrace{\frac{\left(\omega^2/\gamma^2\right)-\left(H_0+4\pi M\right)^2}{4\left(A^2/M^2\right)}}_{c}\right]k_0^2=0.$$

$$(9.17)$$

The solution to the above cubic equation can be approximated by perturbation method, for example. The solutions are approximately as follows:

$$k_1^2 = k_{x1}^2 + \Delta 1,$$
$$k_2^2 = k_{x2}^2 + \Delta 2,$$

and

$$k_3^2 = k_{x3}^2 + \Delta 3,$$

where k_{x1}^2, k_{x2}^2, and k_{x3}^2 are the solutions to the cubic equation in the absence of the coupling to Maxwell's equation. The values of k_{x1}^2 and k_{x2}^2 may be estimated by approximating the cubic equation as a quadratic equation and dropping the constant term. This approximation is valid for high k-values. A simpler way is to use Equation 9.2 to solve for the two estimates, as shown below:

$$\frac{2A}{M} k_{x1}^2 = -(H_0 + 2\pi M) - \sqrt{\frac{\omega^2}{\gamma^2} + (2\pi M)^2} \text{ and}$$

$$\frac{2A}{M} k_{x2}^2 = -(H_0 + 2\pi M) + \sqrt{\frac{\omega^2}{\gamma^2} + (2\pi M)^2}.$$

The estimate for k_{x3}^2 may be obtained from the cubic equation in the limit of low k-values, yielding

$$k_{x3}^2 = -\frac{c}{b}.$$

Using perturbation methods, we approximate the Δ's as follows:

$$\Delta_1 = \frac{k_0^2 \Delta k_{x1}^4}{\left(k_{x1}^2 - k_{x2}^2\right)\left(k_{x1}^2 - k_0^2\right)},$$

$$\Delta_2 = \frac{k_0^2 \Delta k_{x2}^4}{\left(k_{x2}^2 - k_{x1}^2\right)\left(k_{x2}^2 - k_0^2\right)},$$

$$\Delta_3 = \frac{k_0^2 \Delta k_{x3}^4}{\left(k_0^2 - k_{x1}^2\right)\left(k_0^2 - k_{x2}^2\right)}.$$

This root of the cubic equation has no counterpart in the semiclassical approach of the previous sections. For example, Equation 9.2 does not contain any hints about this root. In the literature, this root is often referred to as a surface spin wave mode, since spin wave fields attenuate rapidly. Hence, these modes are located near the surface. The reader is referred to the literature (as a start, see Vittoria, 1998).

Besides the above dispersion, we have a pure electromagnetic branch where $k^2 = \omega^2 \varepsilon \mu_0$.

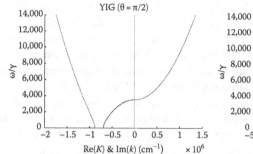 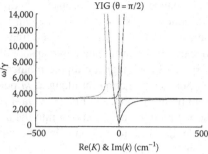

FIGURE 9.18 Dispersions for H in the plane (YIG).

There are now computer codes that can solve cubic equations readily. Refer to MATLAB® code in Appendix 9.A.

In Figures 9.17 and 9.18, wave dispersion plots are presented for permalloy and YIG for $\theta_k = 90°$. Parameters for permalloy and YIG are provided in the previous section.

MAGNETOSTATIC WAVE EXCITATIONS

So far, we have discussed wave propagation in a magneto-dielectric medium semi-infinite in extent. We learned from Figures 9.14, 9.15, 9.17, and 9.18 that pure spin wave branches do not intersect the frequency axis as k approaches zero. Near $k = 0$, the k-values consist of mixed branches of spin waves and electromagnetic waves even for semi-infinite media. For finite size media, there are magnetostatic waves to contend with for k near zero. It is beyond the scope of this book to include all three types of waves in one analysis. Here, we consider the effect of magnetostatic excitations on the pure spin wave branch or dispersion and ignore the electromagnetic branches. We present qualitative arguments for quantitative calculations that are again beyond the scope of this book. It is helpful to visualize magnetostatic waves as being spin waves with low k-values.

Our qualitative approach follows along these lines of thought: (1) at high k-values, the branch is pure spin wave; (2) the pure spin wave branch must converge to the FMR condition at exactly $k = 0$ point; and (3) at low k-values, the pure spin wave branch may be modified because either the volume or surface demagnetizing energies have changed from the pure spin wave branch situation. The regime of small k-values where the pure spin wave branch is modified is indicated by dotted lines. The geometry of interest is a finite size magnetic film.

$\bar{M}$ PERPENDICULAR TO FILM PLANE

a. $\theta_k = 0$:

For linear excitations, there are two microwave components of the magnetization, m_x and m_y, transverse to the static magnetization and are both in the film plane. The static magnetization and propagation directions are perpendicular to the film plane, z-direction. Let's imagine that we can subdivide the film into many flat portions, as indicated in Figure 9.19a as dotted lines along the x-axis. Let's further imagine that we can take an instantaneous snapshot of the film from its edge and take note of the position of the magnetic dipole moment, m, in those portions. For example, in the top portion at each point along x, m points in the same direction or m is in phase from point to point, see Figure 9.19a. In the next portion, m again is in phase along the x-direction but is pointing into the paper; this is indicated as a cross mark in the figure. In the third portion below the surface, now m points in the $-x$-direction and so on. The point here is that the phase of m is changing along the z-direction, because the wave is propagating along that direction.

Let's translate this picture into a dispersion modification if any. At high k-values, the branch is pure spin wave, as indicated in Figure 9.19b as a line. At some low arbitrary, low k-value, introduce a gap in the spin wave branch. The gap extends from the FMR point on the vertical axis (designated as H_0 and $k = 0$ point, point 1 in Figure 9.19) to an arbitrary point on the pure spin wave branch (point 2). What is certain about those two points is that they will be there even after the modifications of the pure spin wave branch in that gap. The fundamental question is the following: as $k \to 0$ and as we move away from point 2 to 1, has the volume or the surface demagnetizing energies changed from point 2? To put this question in a different vein, let's ask the following question: as the number of "artificial" subdivisions of the film (see

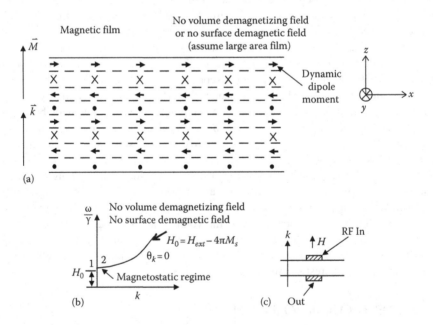

(a)

(b)

(c)   Out

FIGURE 9.19 (a) Instantaneous display of the magnetic moment as seen from the edge of the film. The static magnetization and wave propagation are normal to the film plane. (b) Unmarked line indicates pure spin wave branch and dotted line the magnetostatic wave branch. (c) Delay line device with input and output microstrip lines connections.

Figure 9.19a with horizontal dotted lines) reduce to a single one (that of the film itself), has the volume and surface demagnetizing energies changed from point 2? If the answer is that there are no changes, then the pure spin wave branch will proceed from point 2 to 1 in its natural way or the intended way or as calculated in the previous sections by semiclassical ways (shown in Figure 9.19b as dotted line). Otherwise, one needs to estimate the qualitative changes. This is what we are attempting to do here. In this case or example, there are no changes in volume demagnetizing energy, because m is uniform along the x-direction regardless of whether there are one or many subdivisions. The surface demagnetizing energy is small for any case, since the film is infinite in distance along the x-direction.

Spin wave dispersions can be measured for k-values up to the Brillouin zone using neutron and/or optical scattering experiments. We limit our discussions here for long wavelength spin wave excitations at microwave frequencies (~GHz). For example, spin wave dispersions are often measured with the SWR (spin wave resonance) measurement technique at ~10 GHz. Standing spin wave modes are observed as absorption peaks in an EPR type of setup. For a special frequency and static magnetic field combination, standing modes are excited obeying the following relationship, for example,

$$\frac{\omega}{\gamma} = H_0 + \frac{2A}{M} k_n^2,$$

where

$H_0 \equiv$ static internal field
$k_n = \pi n/t$, $n = 0, 1, 2, 3, \ldots.$
$t \equiv$ thickness of the film

The k-value is quantized by the film thickness and we assume in the above equation that the static magnetization is perpendicular to the film plane. Thus, by fixing frequency and varying the static field, each standing spin wave mode may be observed by EPR technique. For more details, the reader is referred to the literature under the names of Vittoria, Wigen, and others.

Another microwave technique that utilizes the spin wave dispersion especially in the magnetostatic wave region (gap region between points 1 and 2 in Figure 9.19b) is the launching or propagation of microwave signals through a magnetic film in order to fabricate delay lines (see Figure 9.19c). However, in this case, it may be rather difficult to shield the output sensor from the input microwave field, since the separation between input and output may be rather small for film or even thick films or slabs of ferrites (Figure 9.19c). The dispersion now is continuous or not discrete as shown below:

$$\frac{\omega}{\gamma} = H_0 + \frac{2A}{M} k^2 \text{ or}$$

$$\omega = \gamma H_0 + \gamma \frac{2A}{M} k^2.$$

Taking derivative of the above equation, we have

$$\frac{d\omega}{dk} = v = \gamma \frac{2A}{M} 2k \approx 2 \times 10^7 \times 2 \times \frac{0.8 \times 10^{-6}}{140} \times 200.$$

Finally, we have that

$$v \cong 40 \left(\frac{cm}{s} \right) - \text{rather low!}$$

$$\tau = \frac{t}{v} = \frac{100}{40} \times 10^{-4} = 0.25 \text{ ms—not practical device (no local isolation).}$$

Now, let's maintain the same static field bias conditions as in the previous case, but assume the spin wave propagates in the film plane rather than normal to the plane.

b. $\theta_k = \pi/2$:

Since the wave propagates in the x-direction, the phase of m also changes along the same direction. In Figure 9.20a, the phase of m in each portion (region inside two dotted vertical lines) is constant. The interpretation of • and × in Figure 9.20a is simply that m is out of the paper and into the paper directions, respectively. Clearly, now there is volume demagnetizing energy when the subdivisions or portions (high k's) are small. Each subdivision represents about a quarter wavelength. At every half wavelength, the moments oppose each other and give rise to volume demagnetizing fields. Hence, the spin wave branch (point 2 and higher) is at the top of the spin wave manifold (see the previous sections about manifold). Point 1 representing FMR

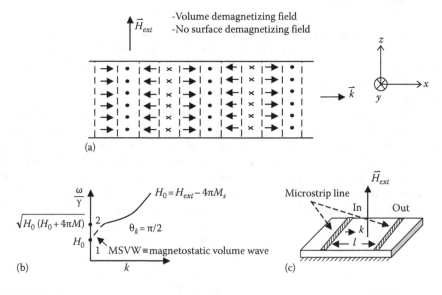

FIGURE 9.20 (a) Magnetic moment distribution for magnetostatic wave (MSW) propagation in the film plane; (b) approximate magnetostatic wave dispersion; and (c) delay line device.

($k = 0$) has not changed position on the frequency scale from the previous case, which was at the bottom of the manifold. Now, we have a dilemma: point 2 is at the top of the manifold and point 1 is at the bottom of manifold. How to reconcile this difference as $k \to 0$? The solution or the argument is rather simple. As the subdivisions increase in size, the separation between opposing moments also increases. Hence, the demagnetizing field or energy must decrease since this energy scales inversely with distance. This is equivalent to having point 2 traverse to the bottom of the manifold as the subdivision coalesces to one single subdivision from one edge of the film to the other as $k \to 0$. At this condition, there is zero volume demagnetizing energy and zero surface demagnetizing energy. In conclusion, there has been a full reduction in volume demagnetizing energy in going from point 2 to 1 (see Figure 9.20b). We have extrapolated this reduction to be linear. Quantitative calculations show a linear dependence at very low k's (see the references list). Point 2 has been calculated by others to be in the order of $k \sim 1,000$/cm. Let's now refine or quantify approximately the conjecture of Figure 9.20b. Again, qualitative arguments will be used in order to estimate the approximate linear dispersion of magnetostatic wave of Figure 9.20b. Let's assume for simplicity that the magnetic moment, m, varies in the z-direction rather than being constant, as shown in Figure 9.20a. Thus, there are two components of $\bar{k}$, k_x, and k_z. The component k_x clearly applies for a wave propagating in the x-direction and k_z in the z-direction. The value of k_z is discrete and equal to π/t, where t is the thickness of the film. For the case, $k_x = k_z = \pi/t$, $k = \sqrt{2}(\pi/t)$, and $\theta_k = 45°$. Let's show where in the spin wave manifold we are situated (Figure 9.21).

Let's calculate approximately the slope $\Delta\omega/\Delta k = v = $ velocity of MSW (magnetostatic wave):

$$\left(H_0^2 + H_0 2\pi M_s\right) - H_0^2 = \frac{\omega_{45°}^2}{\gamma^2} - \frac{\omega_0^2}{\gamma^2} \cong 2\frac{\omega}{\gamma}\frac{\Delta\omega}{\gamma},$$

$$\therefore \Delta\omega \cong \frac{\gamma H_0 (2\pi M_s)}{2H_0} \cong \gamma\pi M_s.$$

(9.18)

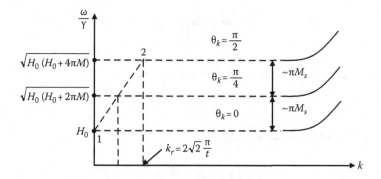

FIGURE 9.21 Magnification of the magnetostatic wave (MSW) region between points 1 and 2 (see Figure 9.20b).

Let's estimate point 2 on the k-axis or as defined in Figure 9.21 as k_r. According to Figure 9.21 or the previous estimates, we have

$$k_r = 2\sqrt{2}\frac{\pi}{t} \approx \frac{10}{t} = 1{,}000 \text{ cm}^{-1}, t = 100 \text{ μm}.$$

This estimate is remarkably in agreement with the current literature. Now let's turn our attention to the delay line device in Figure 9.20c. From the linear slope exhibited in Figure 9.21, it is simple to calculate the MSW velocity

$$v \cong \frac{\gamma(\pi M_s)}{\sqrt{2}\left(\dfrac{\pi}{t}\right)} \cong \frac{(\pi M_s)t\gamma}{\sqrt{2\pi}}. \tag{9.19}$$

Let's estimate v for a YIG film of $10\,\mu m$ thick:

$$v \cong 400\times10^{-3}\frac{2\pi}{\sqrt{2\pi}} g \times 1.4\times10^6 (\text{cm/s}) = 1.74\times10^6 (\text{cm/s}).$$

The delay time, τ, for a 1 cm slab of YIG would then be

$$\tau = \frac{1}{v} = 0.57\,(\mu s).$$

Let's now estimate approximately the propagation loss over 1 cm slab of YIG. The time dependence of the MSW wave is of the form

$$\propto e^{j\omega t}.$$

If the radial frequency is allowed to be complex,

$$\frac{\omega}{\gamma} \rightarrow \frac{\omega}{\gamma}(1 - j\alpha)$$

yielding an attenuation of the form over a delay time, τ,

$$e^{-\Delta\omega\tau} \tag{9.20}$$

where $\Delta\omega/\gamma = $ FMR linewidth, and

$$\gamma = 2\pi g \times 1.4\times10^6\,(\text{Hz/Oe}).$$

By definition, the amount of loss may be expressed in dB:

$$dB = -8.68 \frac{\Delta \omega}{2\gamma} \cdot (\gamma)\tau = -4.34 \frac{\Delta \omega}{\gamma}(\gamma)\tau,$$

$$dB = -4.34 \frac{\Delta \omega}{\gamma}(2\pi \times 2.8)10^6 \tau, \tag{9.21}$$

Thus, for a YIG slab whose $\Delta \omega/\gamma \sim 1/4$ Oe at X-band and $\tau \sim 0.55$ (μs), we have an insertion loss over 1 cm of slab of YIG:

$$dB = -76.4 \times \frac{1}{4} \times 0.55 \cong -10.$$

Ideally or practically, it is desirable to reduce losses to -3 dB or less. In this case, if the thickness is tripled, losses could be reduced by a factor of 3, for example. These are the type of trade-offs one must do in the design of delay line devices utilizing MSW waves. In practical devices, the frequency for which MSW are nondispersive (linear) is rather small.

Usually, the nondispersive part is in the order of 100–200 MHz, which is much below than $\gamma(2\pi M_s)$ as implied in this simple argument. For detailed calculations of MSW waves, the reader is referred to the extensive literature on this subject matter (see references as a start).

The above discussion has dealt with the fundamental modes of propagation whereby

$$k_z = \frac{\pi}{t}. \tag{9.22}$$

For higher mode excitations of MSW waves

$$k_z = n\frac{\pi}{t}, n \text{ is an integer,}$$

this dispersion would shift to the right of the fundamental mode, as shown in Figure 9.22.

For finite thickness of films or slabs, both k_x and k_z are excited whereby the resultant propagation constant, k, is approximately linearly dependent on frequency. For more precise calculation of the dispersion relations, the reader is referred to the literature. As such, for ferrite, lateral dimensions of a film or slab values of k_x and k_z may be discrete. This is very much analogous to a microwave cavity whereby electromagnetic propagation constants are also discrete to give rise to various modes of oscillations in a microwave cavity. Similarly, magnetostatic standing resonance modes may also be excited in a finite size (thickness and lateral dimensions) film. Ferromagnetic resonance (FMR) is a technique by which magnetostatic resonance modes may be observed. The literature on detailed calculations of magnetostatic

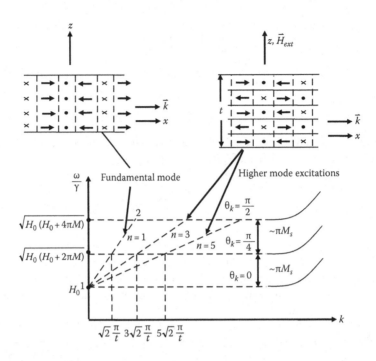

FIGURE 9.22 Higher mode excitations of MSW.

wave dispersions, observation of magnetostatic resonant modes, as well as delay line device applications is very extensive.

$\vec{H}$ IN THE FILM PLANE

A. $\theta_k = 0$:

This is the case when the static magnetization wave propagation direction is parallel to each other and in the film plane.

The surface demagnetizing field, $h_D^{(s)}$, is perpendicular to the microwave component of the magnetization at all points on the surface (Figure 9.23). As in the previous cases, the microwave dipole is transverse to the static magnetization in linear excitation. This means that m is always perpendicular to the surface demagnetizing field (see Figure 9.23). Hence,

$$v \int h_D^{(s)} \cdot d\overline{m} = 0. \tag{9.23}$$

It is noted in Figure 9.23 that the dipole moments never oppose each other along the x-direction to generate a volume demagnetizing field. In summary, at high k-values (spin wave regime), the spin wave branch is at the bottom of the manifold, point 2. However, point 1 (FMR condition or $k = 0$ point) occurs at the top of the manifold. As before, we reconcile this by simply drawing a straight line from point 2 to point 1.

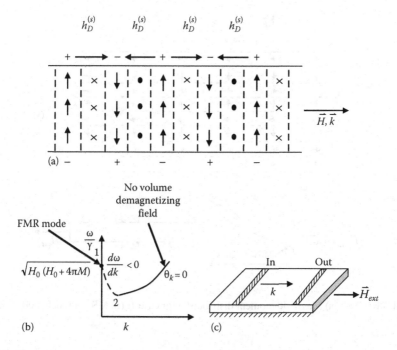

FIGURE 9.23 (a) Magnetic moment distribution for MSW propagation, static magnetization in the film plane; (b) approximate MSW dispersion; and (c) delay line device.

Now, we have a curious dispersion whereby $d\omega/dk \leq 0$. This is sometimes referred to as the magnetostatic backward volume wave (MSBW). The reader may wonder where did the extra energy come from? As k approaches zero, the size of the dipole subdivision (region in between vertical dotted lines in Figure 9.23) takes up the whole film. As such, magnetic poles are generated on both film surfaces to generate a surface demagnetizing energy. The direction of this surface demagnetizing field is perpendicular to h_D^S, as depicted in Figure 9.23.

B. $\theta_k = \pi/2$:

There are two cases to consider for static magnetization in the plane: (1) the wave propagates normal to both the static magnetization and the film plane; (2) the wave propagates normal to the static magnetization but it is also in the plane of the film.

CASE 1

In Figure 9.24, the transverse moments are shown relative to the external static field or magnetization. Clearly, points 1 and 2 are located at the top of the manifold since there is a volume demagnetizing field in the z-direction. It is quite simple to conclude that the MSW branch is the extension in dotted lines shown in Figure 9.24. For example, for t less than half wavelength, the volume demagnetizing field transitions into a surface demagnetizing field since surface magnetic "charges" are induced on the

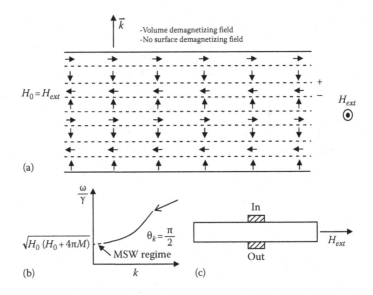

FIGURE 9.24 (a) Magnetic moment distribution for case 1; (b) MSW branch; and (c) delay line device.

two surfaces of the film. This dispersion is very similar to the case where the MSW branch is at the bottom of the manifold (static magnetization and wave propagation are both perpendicular to the film plane). As in that case, MSW devices are not very practical due to electrical isolation between input and output coupling lines.

CASE 2

Figure 9.25 shows static magnetization and wave propagation in the plane of the film but transverse to each other.

We have a unique situation whereby both the volume and surface demagnetizing fields coexist (Figure 9.25). Since these fields are planar fields and parallel to $\vec{m}$, both give rise to demagnetizing energies. A clarification is in order here. The surface field depicted in Figure 9.25a is along the x-direction, but the surface field in Figure 9.24a at $k = 0$ is along the z-direction. They are both surface fields, since surface charges are induced in both cases. The surface fields in Figure 9.25a contribute to surface demagnetizing energy, because the surface field is parallel to m near the surface. This result is to be contrasted with the one in Figure 9.23a, where m is perpendicular to the surface field. Hence, in the configuration of Figure 9.23a, there is no contribution to the surface energy.

The pure spin wave branch is at the top of the manifold since volume demagnetizing fields are induced for short wavelengths (small subdivisions). Recall that each subdivision is about a quarter wavelength. At some critical wavelength or k-value, surface energy begins to contribute to the makeup of the dispersion curve. It is critical because this energy depends on an integral similar to Equation 9.23. For high k-values, the integral is small since V is small (short wavelength). V is defined as the

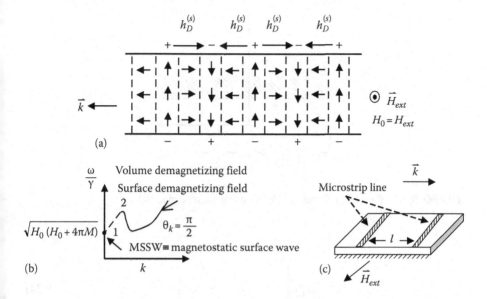

FIGURE 9.25 (a) Magnetic moment distribution for case 2; (b) MSSW branch; and (c) delay line device.

volume of that region whereby there is an interaction between the surface field and the magnetic moment near the surface. V scales roughly with wavelength. For very long wavelengths, the integral is also small since the surface field is small (long distance between surface charges). However, somewhere in between these two limits of wavelengths, V and the surface field are sufficiently large to make a contribution to the dispersion (point 2 in Figure 9.25a). For infinite wavelength ($k = 0$, FMR point 1), the volume demagnetizing field and the surface demagnetizing field along the x-direction vanish to zero, but the surface field normal to the film plane (along the z-direction) is maximum. The region from point 1 to 2 in Figure 9.25a is referred to as the magneto-static surface wave (MSSW) branch. The resume reader is referred to the literature for detailed calculation of the dispersion for MSSW. Clearly, the criticality in k is related to the thickness of the film and the separation between the magnetic film and the electrical ground plane, for example.

In the positive slope region, the MSSW velocity is typically for YIG (Vittoria):

$$v \cong 5 \times 10^6 \, (\text{cm/s}), \text{ and}$$

$$\tau = \frac{1}{5 \times 10^6} = 0.2 \times 10^7 \, (\text{s}) = 200 \, (\text{ns}).$$

FERRITE BOUNDED BY PARALLEL PLATES

We assume propagation along the z-direction, see Figure 9.26. TM and TE modes of propagation are degenerate. For H along the x-axis, the permeability tensor may be written as follows:

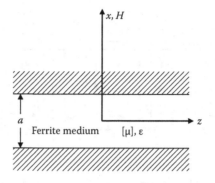

FIGURE 9.26 Parallel plate waveguide enclosing a ferrite material.

$$[\mu] = \begin{bmatrix} 1 & 0 & 0 \\ 0 & \mu & -j\kappa \\ 0 & j\kappa & \mu \end{bmatrix}. \tag{9.24}$$

From Maxwell equations, we have

$$\vec{\nabla} \times \vec{e} = -j\frac{\omega}{c}\ddot{\mu} \cdot \vec{h}$$

and

$$\vec{\nabla} \times \vec{h} = j\frac{\omega}{c}\mu\vec{e}.$$

Assuming field solutions of the form e^{-jkz}, we have from the $\vec{\nabla} \times \vec{e}$ equation the following:

$$ke_y = -\frac{\omega}{c}h_x, \tag{9.25a}$$

$$jke_x + \frac{\partial}{\partial x}e_z = j\frac{\omega}{c}\left(\mu h_y - j\kappa_z\right), \tag{9.25b}$$

and

$$\frac{\partial}{\partial x}e_y = -j\frac{\omega}{c}\left(\mu h_z + j\kappa h_y\right). \tag{9.25c}$$

Note that μ is not k dependent in this example.

The above set of equations simplify further if there is no propagation along the x-direction:

$$ke_y = -\frac{\omega}{c}h_x, \tag{9.26a}$$

$$ke_x = \frac{\omega}{c}\left(\mu h_y - j\kappa h_z\right), \tag{9.26b}$$

and

$$\mu h_z = -j\kappa h_y. \tag{9.26c}$$

From the $\vec{\nabla} \times \vec{h}$ equation, we obtain

$$kh_y = \frac{\omega}{c}\varepsilon e_x \tag{9.27a}$$

and

$$kh_x = -\frac{\omega}{c}\varepsilon e_y. \tag{9.27b}$$

We combine the set of Equations 9.25 and 9.27 to obtain the characteristic equation

$$\left[k^2 - \left(\frac{\omega}{c}\right)^2 \varepsilon\mu_{\mathit{eff}}\right]e_x, e_y = 0, \tag{9.28}$$

where

$$k = (\omega/c)\sqrt{\varepsilon\mu_{\mathit{eff}}}, \text{ CGS}$$

$$\mu_{\mathit{eff}} = (\mu^2 - \kappa^2)/\mu \equiv \mu_0\left(\mu'_R - j\mu_R''\right)$$

$$\varepsilon \equiv \text{dielectric constant} \equiv \varepsilon_0\left(\varepsilon'_R - j\varepsilon_R''\right)$$

$$c = \text{velocity of light}$$

Equation 9.28 is exactly of the same form as Equation 9.14 or 9.15. This is no surprise since wave propagation is perpendicular to static magnetization. Clearly, k is complex and let's put it in the form

$$k = \beta - j\alpha,$$

where β and α are the propagation phase and attenuation constants, respectively. For small values of ε'' and μ'' relative to ε' and μ', we may approximate β and α as

$$\beta \cong \frac{2\pi}{\lambda},$$

$$\lambda = \frac{\lambda_0}{\sqrt{\mu'_R \varepsilon'_R}},$$

$$\lambda_0 = \frac{c}{f},$$

$$\alpha = \frac{\pi}{\lambda}\left[\frac{\varepsilon''}{\varepsilon'} + \frac{\mu''}{\mu'}\right], \text{ and}$$

$$f = \text{frequency}.$$

The attenuation or losses in a magneto-dielectric medium are additive, as shown in the above equation. Losses may be expressed in dB's or

$$dB = 20\log\left(e^{-\lambda z}\right) = -8.68\mu z.$$

It is practical to measure losses relative to the length of a device or

$$\frac{dB}{cm} = -\frac{27.3}{\lambda}\left[\frac{\varepsilon''}{\varepsilon'} + \frac{\mu''}{\mu'}\right].$$

PROBLEMS

9.1   If $\vec{M}$ and $\vec{k}$ were interchanged in directions, would the results generated so far change? Explain.

9.2   Show that in the limit $\mu'' \geq \mu'$, $\beta = 2\pi/\lambda$, but

$$\alpha \cong \frac{\pi}{\lambda}\left[\frac{\varepsilon''}{2\varepsilon'} + 1\right]\sqrt{2\frac{\mu''}{\mu'}}.$$

9.3   Assume a TE_{102} microwave cavity with dimensions $2.5 \times 1.25 \times 5 \, \text{cm}^3$. Calculate the frequency shift of the cavity $\Delta\omega$ for a dielectric sample placed at a position in the cavity where the electric field is maximum. The sample size is $0.1 \times 0.1 \times 0.1 \, \text{cm}^3$ and the relative dielectric constant is $(10 - j1)$.

9.4   For the case of $\theta_k = \pi/2$, show that the following is true:

$$\left(k^6 + C_1 k^4 + C_2 k^2 + C_3\right)\left(k^2 + C_4\right) = 0,$$

and $(k^2 + C_4)$ factor corresponds to the electromagnetic skin depth mode.

9.5   Most, if not all, commercial computer codes are predicated on the assumption that

$$\mu_{xx} = \mu_{yy}.$$

How would you proceed, if this assumption is not correct for some ferrites? No solution is provided.

APPENDIX 9.A

We consider the empty cavity with no magnetic film. We will compare this later (Chapter 10) when we place a magnetic film in a microwave cavity.

EMPTY MICROWAVE CAVITY

Equivalent circuit of Iris hole to a rectangular shape microwave cavity

$$
\begin{pmatrix} V_1 \\ I_1 \end{pmatrix} = \begin{pmatrix} a & 0 \\ 0 & 1/a \end{pmatrix} \begin{pmatrix} V_2 \\ I_2 \end{pmatrix},
$$

$$
\frac{V_1}{I_1} = a^2 \frac{V_2}{I_2}.
$$

AT RESONANCE

The combination of Iris hole and cavity may be represented as follows:

At resonance, $l = \lambda$ so that

$$
\frac{V_2}{I_2} = \frac{1-|\Gamma_\alpha|}{1+|\Gamma_\alpha|} Z_0,
$$

where

$$
|\Gamma_\alpha| = |\Gamma_L| e^{-2\alpha l} = e^{-2\alpha l} = e^{-2\alpha\lambda}, \; |\Gamma_L| = 1
$$

$l \equiv$ length of cavity $\equiv \lambda$ (at resonance)

$$
Z_0 = Z_{00}\left(1 - j\frac{\omega}{\gamma}\right).
$$

Off resonance, we have that

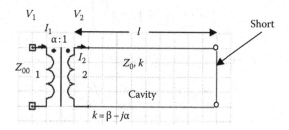

FIGURE 9.A.1 Equivalent circuit of a shorted rectangular microwave cavity (typical of a standard EPR spectrometer).

$$\frac{V_2}{I_2} = \frac{1 - |\Gamma_\alpha|^2 - 2j|\Gamma_\alpha|\sin(2\beta l - \varphi_L)}{1 + |\Gamma_\alpha|^2 - 2|\Gamma_\alpha|\cos(2\beta l - \varphi_L)} Z_0, \text{ where}$$

$$\varphi_L = \pi,$$

$$\beta = \frac{2\pi}{\lambda},$$

$$\lambda = \frac{3}{\sqrt{5}}\lambda_0, \text{ rectangular } TE_{1,0} \text{ mode,}$$

$$\lambda_0 = \frac{c}{f}, \text{ and}$$

$$Z_{00} = \frac{3}{\sqrt{5}}\eta_0, \ \eta_0 = 120\pi.$$

The input impedance, Z_1, at resonance may be written as

$$\frac{V_1}{I_1} = Z_1 = a^2 \frac{1 - e^{-2\alpha l}}{1 + e^{-2\alpha l}} Z_0, \ l = \lambda.$$

For no-reflection condition, we require that

$$Z_1 = Z_0,$$

$$a^2 = \frac{1 + e^{-2\alpha l}}{1 - e^{-2\alpha l}},$$

and

$$Z_0 = \frac{3}{\sqrt{5}}\eta_0\left(1 - j\frac{\alpha}{\beta}\right),$$

where Q of the cavity may be defined as follows:

$$\frac{\alpha}{\beta} = \frac{1}{2Q}, \ Q = \frac{\omega_0}{\Delta\omega} = \frac{f_0}{\Delta f}.$$

Recall from circuit theory that

$$Q = \frac{\omega L}{R} = \frac{\omega L}{2\alpha Z_{00}} = \frac{\omega\sqrt{LC}}{2\alpha} = \frac{\beta}{2\alpha} \Rightarrow \frac{\beta}{\alpha} = 2Q \Rightarrow \frac{\alpha}{\beta} = \frac{1}{2Q}$$

or

$$\alpha = \frac{\beta_{res}}{2Q}, \text{ at resonance,} \qquad (9.A.1)$$

where $\beta_{res} = 2\pi/l$ and $l = \lambda\left(3/\sqrt{5}\right)(c/f_0), f_0 \equiv$ resonance frequency of cavity. With Equation 9.A.1, we can obtain α from the resonance Q of the cavity. However, for frequencies tuned away from resonance, we have that

$$Z_1 = Z_{00}\left(1 - j\frac{f_0/f}{2Q}\right)\left(\frac{1 + e^{-2\alpha\lambda_{res}}}{1 - e^{-2\alpha\lambda_{res}}}\right)\left(\frac{1 - |\Gamma_\alpha|^2 - 2j|\Gamma_\alpha|\sin(2\beta l - \varphi_L)}{1 + |\Gamma_\alpha|^2 - 2|\Gamma_\alpha|\cos(2\beta l - \varphi_L)}\right)$$

$$\beta = \frac{2\pi}{\lambda}, \quad \frac{\beta_{res}}{\beta} = \frac{\lambda}{l} = \frac{f_0}{f}, \quad \lambda = \frac{3}{\sqrt{5}}\frac{c}{f}(\text{any frequency}),$$

$$\text{and } Z_{00} = \frac{3}{\sqrt{5}}\eta_0, \quad _0 = 12\eta 0\pi.$$

In general, one may write $S_{11} = (Z_1 - Z_{00})/(Z_1 + Z_{00})$ as a function of frequency.

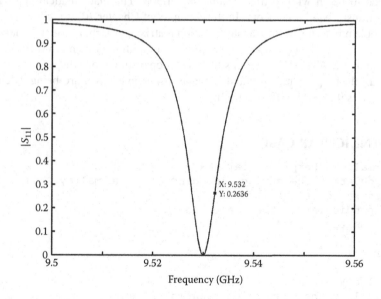

X: 9.532
Y: 0.2636

FIGURE 9.A.2 The magnitude of the reflection coefficient from a microwave cavity tuned at 9.53 GHz is plotted as a function of frequency. The input power to the cavity is assumed to be constant with frequency.

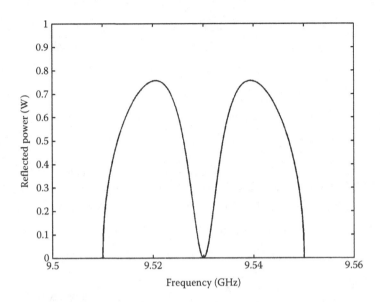

FIGURE 9.A.3 This plot is the same as the one in Figure 9.A.2 except that incident power is assumed to be parabolic function of frequency. This is the case in most EPR spectrometers.

In the above plot, the magnitude of the reflection coefficient is plotted.

In the above plot, some realism is injected in the plot in order to simulate the power generated in a klystron tube, as in an EPR equipment. Q is about what is assumed to begin with in this calculation, ~1,500. The mathematical approach is that of transmission line theory. In the appendix of Chapter 10, we will re-examine this problem in terms of the transfer function matrix approach. As such, it makes the analysis simpler when considering the insertion of a film in the microwave cavity.

Computer MATLAB® programs for wave propagation mode dispersions are provided. Both the perpendicular and in-plane dispersions cases are being calculated (see Figures 9.14, 9.15, 9.17, and 9.18).

PERPENDICULAR CASE

```
% Permalloy, Perpendicular FMR
% Calculating the surface impedance of a permalloy film with
magnetization
% Perpendicular to the plane. Theta_k = 0.
% CGS units.
clc;
clear all;
close all;
global d V
% Calculate the dispersion relation first.
global w mu0 eps Xi1s Xi2s k1s k2s Z1s Z2s h1r h1l h2r h2l
mu0 = 4*pi*10^(-9); % henry/cm
eps0 = 1/(36*pi)*10^(-11); % F/cm
```

```
Z0 = sqrt(mu0/eps0);
sigma = 0.7*10^5; % mho/cm
A = 1.14e-6; % erg/cm
g = 2.10;
M = 10000/(4*pi); % Gauss
Ho = 2500:1:4500; % Oe
L = length(Ho);
gamma = 2*pi*1.4*10^6*g; % Hz/Oe
w = gamma*3500; % GHz
f = w/(2*pi);
dH = 2*4.33e-3*w/gamma; % Oe
omega = w*(1-j*4.33e-3);
w = omega;
eps = -j*sigma/omega; % F/cm
k0_sq= -j*omega*sigma*mu0;
% Resonant mode.
c1 = ones(1,L);
c2 = -(k0_sq+(w/gamma-Ho)*M/(2*A));
c3 = M/(2*A)*k0_sq*(w/gamma-Ho-4*pi*M);
k1_sq = -c2./(2*c1)+sqrt(c2.^2 - 4*c1.*c3)./(2*c1);
k2_sq = -c2./(2*c1) - sqrt(c2.^2 - 4*c1.*c3)./(2*c1);
k1 = sqrt(k1_sq);
k1 = sign(-imag(k1)).*k1; % Pick up the physical mode
(non-growing).
k2 = sqrt(k2_sq);
k2 = sign(-imag(k2)).*k2;
% Check the result by integrating the poynting vector.
Xi1s = Xi1(i); Xi2s = Xi2(i); k1s = k1(i); k2s = k2(i);
Z1s = Z1(i); Z2s = Z2(i);
h1r = h_vec(1,i); h1l = h_vec(2,i); h2r = h_vec(3,i); h2l
= h_vec(4,i);
end
```

IN-PLANE CASE

```
%Permalloy in plane FMR
clc;
clear all;
close all;
global w mu0 Qi1s Qi2s Qi3s k1s k2s k3s h1r h1l h2r h2l h3r
h3l Z1s Z2s Z3s eps
g = 2.1;
fourpiMs = 10000;
dH = 30;
sigma = 0.7e5;
A = 1.14e-6;
mu0 = 4*pi*10^-9;
f = 9.53e9;
e0 = 1/(36*pi)*10^-11;
gamma = g.*0.8805e7;
```

```
omega = 2.*pi*f-j.*(dH.*0.924525e7);
w = omega;
topn = j.*fourpiMs.*w/gamma;
H0 = 10:1:3000;
L = length(H0);
H1 = H0;
H2 = H0+fourpiMs;
e = -j.*sigma./omega;
M = fourpiMs./(4.*pi);
k0 = sqrt(omega.^2.*e.*mu0);
a1 = 1;
a2 = (2.*H0+fourpiMs)./(2.*A./M) -k0.^2;
a3 = (H0.*(H0+fourpiMs) - (omega./gamma).^2)./(4.*A.^2./M.^2)
-2.*(H0+fourpiMs)./(2.*A./M).*k0.^2;
a4 = ((omega./gamma).^2-(H0+fourpiMs).^2)./(4.*A.^2./M.^2).
*k0.^2;
a = a2;
b = a3;
c = a4;
[k1_sq,k2_sq,k3_sq] = solve3order(a,b,c);
k0_sq = -j.*omega.*sigma.*mu0;%k0squared
k1 = -sign(imag(k1_sq)).*sqrt(k1_sq);
k2 = -sign(imag(k2_sq)).*sqrt(k2_sq);
k3 = -sign(imag(k3_sq)).*sqrt(k3_sq);
end
```

BIBLIOGRAPHY

J.D. Adam and S.N. Bajpai, *IEEE Trans. Mag.* **18**, 1598, 1982.

R.W. Damon and J.R. Eshbach, *J. Phys. Chem. Solids* **19**, 308, 1961.

T. Holstein and H. Primakoff, *Phys. Rev.* **58**,1098, 1940.

B. Lax and K. Button, *Microwave Ferrites and Ferrimagnetics,* McGraw-Hill Book Co., Inc., New York, 1962.

M. Sparks, *Ferromagnetic Relaxation Theory*, McGraw-Hill Book Co., New York, 1964.

D.D. Stancil, Theory of Magnetostatic Waves, Springer-Verlag, Berlin, 1993.

D.D. Stancil and A. Prabhakar, *Spin Waves: Theory and Applications*, Springer-Verlag, Berlin, 2009.

C. Vittoria, *Elements of Microwave Networks*, World Scientific Publishing Co., Singapore, 1998.

C. Vittoria and J.H. Schelleng, *Phys. Rev.* B **16**, 4020, 1977.

C. Vittoria and N.D. Wilsey, *J. Appl. Phys.* **45**, 414, 1974.

C. Vittoria, J.N. Craig, and G.C. Bailey, *Phys. Rev.* B **10**, 3945, 1974.

SOLUTIONS

9.1 Let's look at the equations of motion which include microwave exchange field:

$$\frac{j\omega}{\gamma} = \vec{m} = \vec{M}_0 \times \left(\vec{h} + \vec{h}_{ex} \right) + \vec{m} \times \vec{H}_0,$$

where $\vec{h}_{ex} = -(2A/M)k^2\vec{m} - 4\pi\vec{k}(\vec{m} \cdot \vec{k})/k^2$

From Figure S9.1, one can see that the equation of motion will stay the same by interchanging vectors.

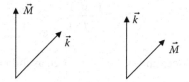

FIGURE S9.1 Directions of $\vec{M}$ relative to $\vec{k}$.

9.2 The usual form of propagation constant of wave can be written as follows:

$$\gamma = \alpha + j\beta = \frac{j2\pi}{\lambda_0}\sqrt{\varepsilon' - j\varepsilon''}\sqrt{\mu' - j\mu''},$$

where λ_0 is the wavelength in air.

$$\gamma = j\frac{2\pi}{\lambda_0}\sqrt{\varepsilon'\mu'}\left(1 - j\frac{\varepsilon''}{\mu'}\right)^{1/2}\left(1 - j\frac{\mu''}{\mu'}\right)^{1/2}$$

$$= j\frac{2\pi}{\lambda}\left(1 - \frac{j}{2}\frac{\varepsilon''}{\varepsilon'}\right)\left(\frac{\mu''}{\mu'}\right)^{1/2}\left(\frac{\mu'}{\mu''} - j\right)^{1/2}$$

$$= j\frac{2\pi}{\lambda}\left(1 - \frac{j}{2}\frac{\varepsilon''}{\varepsilon'}\right)\left(\frac{\mu''}{\mu'}\right)^{1/2}\left(e^{-j(\pi/4)}\right)\left(1 + \frac{j}{2}\frac{\mu'}{\mu''}\right)$$

$$= j\frac{2\pi}{\lambda}\left(\frac{\sqrt{2}}{2} - j\frac{\sqrt{2}}{2}\right)\left(1 - \frac{j}{2}\frac{\varepsilon''}{\varepsilon'}\right)\left(\sqrt{\frac{\mu''}{\mu'}} + \frac{j}{2}\sqrt{\frac{\mu'}{\mu''}}\right).$$

Finally, we have that

$$\alpha \approx \frac{\pi}{\lambda}\sqrt{\frac{2\mu''}{\mu'}}\left(1 + \frac{1}{2}\frac{\varepsilon''}{\varepsilon'}\right).$$

9.3 From references listed below, one can write the dielectric perturbation to a microwave cavity as follows:

$$\frac{1}{2}\left(\frac{f - f_0}{f}\right) = \frac{-\int_v\left[\varepsilon_0\chi_\varepsilon\vec{E}\cdot\vec{E}^*\right]dv}{\int_V\varepsilon_0\vec{E}\cdot\vec{E}^*dV}.$$

Suppose we have a small enough perturbation, it means that the E field can be considered as constant. Then, we are able to write the above equation as follows:

$$\chi_\varepsilon \cong \frac{1}{2} \frac{\Delta f}{f} \frac{V}{v}.$$

The electric susceptibility can be written as follows:

$$\varepsilon' \cong 1 + \frac{1}{2}\left(\frac{\Delta f}{f_{102}}\right)\frac{V}{v},$$

where $f_{102} = 8.4\,\text{GHz}$.

Thus, we have $\Delta f = 2 f_{102} \frac{v}{V}(\varepsilon' - 1) = 2 \times (8.4 \times 6.4 \times 10^{-5}) \times 9 = 9.67$ MHz.

9.4 $Ak^4 + Bk^2 + C = \left(k^6 + C_1 K^4 + C_2 K^2 + C_3\right)\left(k^2 + C_4\right) = 0.$

Suppose $\theta = \pi/2$ and $\varphi = 0$. Then, we have the following for the A, B, and C:

$$A = \mu_{xx} = 1 + \frac{4\pi M\left(H_0 + (2A/M)k^2\right)}{\Omega^2}$$

$$B = -\mu_{xx}\left(1 + \mu_{yy}\right) + \kappa^2$$

$$C = \mu_{xx}\mu_{yy} - \kappa^2$$

$$\mu = 4\pi M \frac{\omega}{\gamma\Omega^2}$$

$$\mu_{yy} = 1 + \frac{4\pi M\left(H_0 + (2A/M)k^2 + 4\pi M\right)}{\Omega^2} = \mu_{xx} + \left(\frac{4\pi M}{\Omega}\right)^2$$

$$\Omega^2 = \left(H_0 + \frac{2A}{M}k^2\right)\left(H_0 + \frac{2A}{M}k^2 + 4\pi M\right) - \frac{\omega^2}{\gamma^2}.$$

Quadratic solution of the first equation would be as follows:

$$k_1^2 = \frac{\omega^2 \mu_0 \mu\left(\mu_{xx}\mu_{yy} - \kappa^2\right)}{\mu_{xx}},$$

$$k_2^2 = \omega^2 \mu_0 \varepsilon.$$

Let $C_4 = -\omega^2\mu_{0e}$; this corresponds to the electromagnetic skin depth mode. The dispersion mode can be written as follows:

$$\mu_{xx}^2\Omega^2 + \left((4\pi M)^2 + j\frac{\Omega^2 k^2}{2\delta^2}\right)\mu_{xx} - \kappa^2\Omega^2 = 0,$$

where $\varepsilon = -j(\sigma/\omega)$ and $\varepsilon = \sqrt{2/\sigma\mu_0\omega}$.

The above equation can be reduced to the following:

$$k^6 + C_1 k^4 + C_2 k^2 + C_3 = 0,$$

where

$$C_1 = 1 + \frac{H_0}{2\pi M} + j\frac{A}{2\pi M\delta^2}$$

$$C_2 = \frac{H_0}{4\pi M} - \left(\frac{\omega}{4\pi M\gamma}\right)^2\left(\frac{H_0}{4\pi M}\right)^2 + j\frac{2A}{\pi M\delta^2}\left(1 + \frac{H_0}{4\pi M}\right)$$

$$C_3 = j\frac{A}{\pi M\delta^2}\left\{\left(1 + \frac{H_0}{4\pi M}\right)^2 - \left(\frac{\omega}{4\pi M\gamma}\right)^2\right\}.$$

10 ATLAD Deposition of Magnetoelectric Hexaferrite Films and Their Properties

BASIC DEFINITIONS OF ORDERED FERROIC MATERIALS

- Antiferromagnetic: The exchange interaction between magnetic moments in different sublattices within a unit cell couples to each other such that their moments oppose each other below the Neel temperature. The net moment is zero. Examples include magnetic transition metal oxides and chlorides, the Rare Earth Chlorides and Rare Earth hydroxides compounds, Chromium oxides, and alloys such as iron manganese (FeMn).
- Ferrimagnetic: Again, the antiferromagnetic exchange coupling dominates, but the net magnetic moment is nonzero. Examples include the garnet, spinel, and hexaferrites. The Neel temperature is well above room temperature.
- Ferromagnetic: All magnetic moments within the unit cell are all exchange coupled pointing in the same direction. Most ferromagnets tend to be metallic.
- Antiferroelectric: The electric dipole moments cancel each other completely within each crystallographic unit cell. This is the analog to antiferromagnetism.
- Ferroelectric: They are characterized as having a spontaneous electric polarization for temperatures below the Curie temperature. It is the analog of ferromagnetism.
- Piezoelectricity: Describes the influence of an applied linear electric field on strain or a change in polarization due to an applied stress.
- Piezomagnetic: Represents a change in strain due to an applied magnetic field or a change in magnetization upon an applied stress.
- Electrostriction: Describes a change in strain due to an applied electric field.
- Magnetostriction: Describes a change in strain due to an applied magnetic field.
- Multiferroic: A material possessing at least two ferroic properties: ferroelectricity and ferrimagnetism.

DOI: 10.1201/9781003431244-10

PARITY AND TIME REVERSAL SYMMETRY IN FERROICS

In ferroelectric material, the electric dipole moment **p** is represented by a positive point charge that lies asymmetrically within a crystallographic unit cell that has no net charge, $\mathbf{p} = qr$, where **p** is the electric dipole moment, q is the electric charge, and r is the distance between the two charges. Inverting r ($r \rightarrow -r$) reverses **p**, since there is no time dependence explicitly. In ferromagnets, the magnetic dipole moment is represented classically simply as $m = \pi r^2 dq/dt$, where $dS = \pi r^2$. A spatial inversion produces no change, since $(r^2) = (-r)^2$, but time reversal reverses the orbital motion, and thus, **m**. In summary, **p** obeys time symmetry but not parity. However, **m** obeys parity but not time symmetry. Multiferroics, such as magnetoelectric hexaferrites, are both ferrimagnetic and ferroelectric breaking time reversal and parity symmetries.

TENSOR PROPERTIES OF THE MAGNETOELECTRIC COUPLING IN HEXAFERRITES

There has been steady progress ever since the first measurement of α, the linear magnetoelectric (ME) coupling. Recent theoretical has been focused on ways to increase α. From these studies, there evolved a hint that the strength of the piezoelectric and the piezomagnetic constants may be important in being able to enhance α. The increase in α as measured recently in hexaferrites is significant because it can impact every facet of modern technologies whereby power consumption, miniaturization, compatibility with other technologies, and computing circuitry are critical or crucial.

The argument presented in various models is that the spin spiral configuration as formulated by the Dzyaloshinskii–Morya interaction equation is essential in exciting ME effects in hexaferrites. Hexaferrites have hexagonal crystal symmetry and the majority of them are ferrimagnetic. For strong exchange coupling, the moments oppose each other and are collinear. A net moment results as the opposite moments do not cancel each other similar to an antiferromagnet. However, as the exchange coupling between opposite spins is reduced upon special ionic substitutions, the opposite spins still oppose each other, but they are both canted with respect to the C-axis of the hexaferrite. Sr ion substitutes in Z-type or M-type hexaferrites straining the bonding of the chemical combination of Fe-O-Fe bonding located near the Sr ion or the 12k site. The change in the bonding angle has two ramifications: (1) It affects the exchange coupling, and therefore, the ground state of the spins globally, giving rise to spin spiral configurations. Thus, changes in strains translate into changes in magnetic moment via the magnetostriction interaction. (2) The application of an electric field, **E**, strains the material and induces electric polarization giving rise to piezoelectric effects as well as changes in magnetic moment, the ME effect. The conduit between the electric and magnetic system is the strain. In the literature, it was referred to as the "slinky helix" spin spiral model; see illustration of the model below.

The linear ME coupling parameter, α, of hexaferrites at room temperature is fairly high in comparison to other types of single-phase materials and laminated composite structures. Potential for new applications in medical sensors, recording media, computer and electronic integrated circuits (IC), and wireless communication networks appears to be promising. In a ME material, the application of a magnetic field, **H**, induces an electric polarization, **P**, and this is referred to as a direct ME effect. In the converse ME effect, the application of an electric field, **E**, induces magnetic polarization or **M**. Landau–Lifshitz proposed in 1957 the possibility of a linear relationship between the electric field, **E**, and the magnetic field, **H**. Dzyaloshinski showed that such linear relationship between **E** and **H** may be possible, if spins at different sites were non-collinear. The notion of non-collinearity of the spins was re-enforced theoretically by Moriya in considering mechanisms of the local magnetic anisotropy and exchange coupling between spin sites extending single ion models for magnetic anisotropy. At about the same time, experimental confirmation of magnetoelectricity was discovered in Cr_2O_3 by Astrov and Folen and Rado.

Since the discovery of magnetoelectricity, various theoretical models have been proposed to explain magnetoelectricity in terms of spin non-collinearity and specifically with spin spiral configurations. Most of these models apply to single-phase ME materials and not pertaining necessarily to hexaferrites. A thermodynamic argument was proposed to model magnetoelectricity in hexaferrites. The model may simply be summarized briefly as follows. The ME coupling parameter, α, in hexaferrites is tensorial and it scales as the product of the piezomagnetic and piezoelectric strain tensors. The thermodynamic argument is of sufficient generality that it may be applicable to materials other than hexaferrites. Dynamic field excitations in ME hexaferrites are being considered in this section, utilizing the M-type hexaferrite magnetic structure as a model for the calculations. The arguments presented here are equally applicable to other hexaferrites as well as other ME materials with different symmetries or composites.

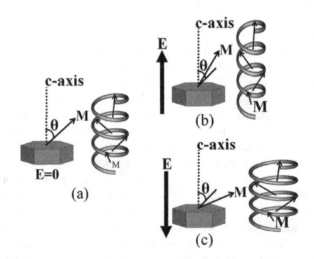

A microscopic theory encompassing the above two points must necessarily be a global theory which includes all of the magnetic ions (including Co ions for example) in the unit cell of a hexaferrite and not sites near the 12k site in order to obtain a quantitative estimate of α. A thermodynamic argument is made that this is a very difficult task and that the parameters d and λ represent macroscopic approach to magnetism.

For M- and Z-type hexaferrites, the crystal structure is hexagonal with space group $P6_3/mmc$. Y-type hexaferrites are not of the same space group, and therefore, the Y-types do exhibit the ME effect. In general, α is a (3×3) tensor and its definition is of the following form:

$$[\alpha] = [d][C][L^M] \text{ (s/m)}. \tag{10.1}$$

The tensor $[d]$ is recognized as the piezoelectric tensor in units of m/v, $[C]$ is the elastic modulus tensor in units of J/m^3, and $[\Lambda^M]$ is the piezomagnetic tensor in units of m/a. Thus, the above relationship shows that, indeed, there is an explicit connection between the piezoelectric and the piezomagnetic tensors via the strain! The product $[d][C]$ yields the polarization $[P]$ in units of coul/m^2, where $[P]$ is a (3×6) tensor. Specifically, for this space group symmetry, $[d]$ and $[C]$ are defined as

$$[d] = \begin{bmatrix} 0 & 0 & 0 & 0 & d_5 & 0 \\ 0 & 0 & 0 & d_4 & 0 & 0 \\ d_1 & d_1 & d_1 & 0 & 0 & 0 \end{bmatrix} \text{ and}$$

$$[C] = \begin{bmatrix} C_{11} & C_{12} & C_{13} & 0 & 0 & 0 \\ C_{12} & C_{11} & C_{13} & 0 & 0 & 0 \\ C_{13} & C_{13} & C_{33} & 0 & 0 & 0 \\ 0 & 0 & 0 & C_{44} & 0 & 0 \\ 0 & 0 & 0 & 0 & C_{44} & 0 \\ 0 & 0 & 0 & 0 & 0 & C_{66} \end{bmatrix}$$

The piezomagnetic tensor is defined as follows:

$[\Lambda^M] = [a][T]$, where

$$[a] = \begin{bmatrix} a_1 & 0 & a_3 & 0 & 0 & 0 \\ 0 & a_1 & a_3 & 0 & 0 & 0 \\ a_3 & a_3 & a_2 & 0 & 0 & 0 \\ 0 & 0 & 0 & a_4 & 0 & 0 \\ 0 & 0 & 0 & 0 & a_4 & 0 \\ 0 & 0 & 0 & 0 & 0 & a_5 \end{bmatrix},$$

with $a_1 = (\lambda_A - \lambda_B)/M^2$, $a_2 = -\lambda_C/M^2$, $a_3 = -\lambda_B/M^2$, $a_4 = 2a_1$, and $a_5 = [4\lambda_D - (\lambda_A + \lambda_C)]/M^2$. The λs are the magnetostriction constants for a hexagonal crystal structure.

M represents either the saturation magnetization in strong magnetic fields or the remanent magnetization in the absence of a magnetic field. The matrix elements of $[a]$ are in units of m^2/a^2 and are a linear function of the magnetostriction constants of the hexaferrites. For cubic crystal structures, see W. P. Mason and/or L. D Landau and E. M. Lifshitz. The matrix $[T]$ is a (6×3) tensor whose elements are a linear product of the magnetic matrix elements of the susceptibility and the components of the magnetization $(M_1, M_2,$ and $M_3)$ and are in units of a/m. Thus, the first row is designated as $T_{1k} = M_1\chi_{1k}$, where $k = 1, 2,$ and 3. For the next two rows, the subscript 1 is replaced by 2 and 3, respectively. In the next three rows, the components of M are admixed.

$$T_{4k} = \left(M_1\chi_{2k} + M_2\chi_{1k} \right)/2,$$

$$T_{5k} = \left(M_3\chi_{1k} + M_1\chi_{3k} \right)/2,$$

$$T_{6k} = \left(M_2\chi_{3k} + M_3\chi_{2k} \right)/2.$$

If the remanence magnetization is along the z-axis (the c-direction) of the hexagonal structure, then the magnetic susceptibility has the approximate diagonal form:

$$[\chi] \cong \begin{bmatrix} \chi_\perp & 0 & 0 \\ 0 & \chi_\perp & 0 \\ 0 & 0 & \chi_\| \end{bmatrix},$$

The subscripts ($\perp$) and ($\|$) are perpendicular and parallel to the C-axis. It is approximated that the diagonal χ's are much higher valued than the off-diagonal elements. However, a diagonal $[\chi]$ does not imply a diagonal $[\alpha]$ tensor. Nevertheless, it is approximated as being diagonal or

$$[\alpha] \cong \begin{bmatrix} \alpha_\perp & 0 & 0 \\ 0 & \alpha_\perp & 0 \\ 0 & 0 & \alpha_\| \end{bmatrix}.$$

where $\alpha_\perp = 2(\lambda_A - \lambda_B)d_5 C_{44}\chi_{11}/M_R$ and $\alpha_\| = -2(\lambda_B C_{13} + \lambda_C d_2 C_{33})\chi_{33}/M_R$ with M_R being the remanence magnetization. The above relationships show in a simple way the connections between magnetostriction and piezoelectric parameters via the strain! The following values for the parameters are being quoted: $\lambda_A = -15 \times 10^{-6}$; $\lambda_B \approx \lambda_C \approx \lambda_D \approx -\lambda_A$; $d_5 \approx -11 \times 10^{-12}$ m/v, $d_1 \approx 0.5d_5 \approx -0.5 d_2$; $M_R = 10$ a/m; $C_{11} \approx C_{33} = 1.5 \times 10^{11}$ Nt/m^2; $C_{13} = 0.6\ C_{11} = 1.2\ C_{44}$; and $\chi_\perp \approx \chi_\| \approx 30$. In Table 10.1, calculations of $\alpha_\|$ are shown for single crystal (s) and polycrystalline M- and Z-type hexaferrites and a composite material compared with measured values.

TABLE 10.1
ME Linear Coupling Parameter of Various Hexaferrites

	$\alpha_{\parallel}$(meas.) – s/m	$\alpha_{\parallel}$(calc.) – s/m
Z-type (s)	2.3×10^{-6}	2.4×10^{-6}
Z-type (p)	7.6×10^{-10}	10×10^{-9}
M-type (p)	2.4×10^{-10}	10×10^{-9}
M-type (p)	60.7×10^{-10}	10×10^{-9}
Composite	0.7×10^{-10}	2.2×10^{-7}

For the polycrystalline samples, the calculated values are consistently higher than the measured values. That is because the values of λ's and d's are considerably reduced.

DEPOSITION OF SINGLE-CRYSTAL ME HEXAFERRITE FILMS OF THE M-TYPE BY THE ATLAD

Clearly, there are two factors that affect the linear coupling in ME hexaferrites: (1) For barium hexaferrite, the Ba ion located at sites 2d (Wyckoff's designations) strongly distorts the octahedral site located at the 2b sites, giving rise to a bi-pyramidal fivefold coordination which induces a large uniaxial magnetic anisotropy parallel to the C-axis. In addition, Co^{2+} substitutions in ferrite structures are a well-known source of magnetoelastic coupling, λ, due to the large orbital moment of Co^{2+}. In order to demonstrate the validity of the above thermodynamic theory, it was helpful to observe the changes in α due to Co^{2+} substitutions (because of potential increase in λ) on hexaferrites that exhibited magnetoelectricity to begin with. The ideal test case was $SrCo_2Ti_2Fe_8O_{19}$, since it exhibited magnetoelectricity at room temperature. As barium hexaferrite was prepared by the ATLAD technique, the same technique should be successful in depositing single-crystal films of $SrCo_2Ti_2Fe_8O_{19}$ with the object of eventually placing Co^{2+} ions in the 2b trigonal or any other site. The key was to place Co^{2+} within the unit cell to affect α.

In bulk $SrCo_2Ti_2O_{19}$, the replacement of Ba with Sr ion altered the magnetic properties of the hexaferrite in the following sense. The smaller radius of Sr ions, compared to Ba ions, changed the superexchange bond angle in Fe-O-Fe sites (2a and 2b sites) near the 2d site (Sr site) from 116 degrees (Ba) to 123 degrees (Sr). In order to compensate the fact that the valence state of the Co ion is 2+, a transition metal ion of 4+ needed to complete the composition to electrically neutralize the full composition, hence Ti^{4+}. Published results indicated that the preferred atomic site for Ti ions is 12k. Placing Ti ions in between the S and R blocks decreased the exchange coupling between spins in the two blocks. Co^{2+} ions occupied octahedral sites $4f_2$ and 12k and the Fe^{3+} ions occupy the $4f_1$ (Tetrahedral) and 2a (Octahedral) sites in the S block, as well as all the other octahedral sites in the R block including the 2b (Trigonal) site. This implied that there were no Co^{2+} ions in the 2b site. Clearly, the purpose of the ATLAD technique was to place Co ions in the 2b site! Magnetoelectricity in $SrCo_2Ti_2O_{19}$ may then be explained in terms of reducing the antiferromagnetic exchange coupling between the two blocks (S and R) because Ti^{4+} is a non-magnetic (NM) ion. Thereby, spins spiral around the C-axis. As Dzyaloshinskii pointed out, it is a requirement for magnetoelectricity to exist in ferrite materials.

(2) The thermodynamic argument presented in the previous section demonstrated that α is proportional to the product of λ's and d's. For the moment, the focus was on enhancing λ's. Of all the transition metal magnetic ions, Cobalt cations are ideal for this purpose, because the magnetostriction constant of Cobalt compounds is notoriously relatively high. As shown in Chapter 4, the ATLAD technique is well suited to replace Ba with Sr and place Co cations in the trigonal site (2b). Since the trigonal site has lower symmetry than any other site in the unit cell, it is more likely to permeate strain upon the application of an electric field with the possibility of inducing a localized polarization at that site, implying that not only is λ affected but also d!

In order to simulate the chemical composition of $SrCo_2Ti_2Fe_8O_{19}$, the R block consisted of a target whose composition was $SrFe_{(4-\delta)}Co_{0.58}Ti_{0.58}O_7$. For the S block, the target composition was $Fe_{(1+0.25\delta)}Ti_{0.5(1-0.25\delta)}Co_{0.5(1-0.25\delta)}O_3$, where $\delta=0$, 0.2, 0.4, and 1.0. It is noticed that chemically $4Fe_{(1+0.25\delta)}Ti_{0.5(1-0.25\delta)}Co_{0.5(1-0.25\delta)}O_3+Fe_{(1+0.25\delta)}$ $Ti_{0.5(1-0.25\delta)}Co_{0.5(1-0.25\delta)}O_3=SrCo_2Ti_2Fe_8O_{19}$. Thus, the chemical composition does not change as the ATLAD distributes different amounts of Co and Ti cations at different sites in contrast to growing the same composition by the natural process! However, there is caveat to this procedure. For $\delta=0$, the Sr target contains **no** Cobalt ions. Hence, the ATLAD technique cannot possibly distribute those ions in the R block. However, for $\delta \neq 0$, the Sr target does contain Cobalt ions to distribute in the R block.

The targets were prepared by conventional ceramic techniques from high purity powders of CoO (99.995%), SrO (99.5%), TiO_2 (99.8%), and Fe_2O_3 (99.8%). Stoichiometric values of these powders were consistent with chemical formulas. The powders were ball-milled twice for 4 hours each followed by pressing in 2,000 psi to produce a disk. Sintering was set at 1,150°C for 16 hours in oxygen atmosphere. After the second ball-mill and before pressing, polyvinyl alcohol (PVA) was added to powders to increase their uniformity. The composition of all targets was checked utilizing energy dispersive X-ray spectroscopy (EDS).

Epitaxial single-crystal thin films of $SrCo_2Ti_2O_{19}$ ($\delta=0$) on (0001) were deposited on single-crystal sapphire (Al_2O_3) substrates by the ATLAD technique. The base pressure of vacuum chamber was in the order of 10^{-6} Torr and the partial pressure of the introduced high purity oxygen was controlled to be 200 mTorr during the deposition. Substrates were mounted on the resistive block heater with conductive silver paint and the temperature of substrates was maintained at 600°C during the deposition. This temperature is far below the conventional one for growth of M-type hexaferrites thin film from single targets, see Chapter 4. The distance between substrates and targets was approximately 5.5 cm. Laser beam from the KRF excimer laser with wavelength of 248 nm and energy of 400 mJ/pulse was optically focused on the targets.

Targets of $Fe_{(1+0.25\delta)}Ti_{0.5(1-0.25\delta)}Co_{0.5(1-0.25\delta)}O_3$ (S block) and $SrFe_{(4-\delta)}Co_{0.58}Ti_{0.58}O_7$ (R Block) were mounted on the carousel. To maximize the surface usage of targets which increased their lifetime and improve the film quality, all targets were rotated and rastered (the whole target is impinged upon by the pulsed laser) on the carousel during the deposition. In order to control the number of laser shots impinging on each target, the carousel was synchronized with the trigger of the laser via computer. After much iteration and calibration measurements of the deposition rate single target, the ATLAD procedure was initiated to deposit from the two targets. The number of pulses for each cycle of deposition was determined to be 17 pulses impinged upon a target of $Fe_{(1+0.25\delta)}Ti_{0.5(1-0.25\delta)}Co_{0.5(1-0.25\delta)}O_3$ (s block) and 8 on $SrFe_{(4-\delta)}Co_{0.5\delta}Ti_{0.5\delta}O_7$ (R block). The same numbers of pulse combinations were applied to the deposition of ME hexaferrites for $\delta = 0$, 0.2, 0.4 and 1.0.

It is noted that for the deposition of $BaFe_{12-x}Mn_xO_{19}$ by the ATLAD technique, see Chapter 4, the R block was simulated with a combination of targets, $BaFe_2O_4$ and Fe_2O_3, yielding $BaFe_4O_7$. Simple extension of the previous work dictated that the number of pulses impinging on the $Fe_{(1+0.25\delta)}Ti_{0.5(1-0.25\delta)}Co_{0.5(1-0.25\delta)}O_3$ target should have been about 22 instead of 17 pulses to simulate the S block. Thus, the essential steps were the calibration runs. However, the sticking coefficients are not necessarily the same as in Chapter 4 runs. Hence, this time it required many iterations to produce single-crystal films with correct compositions and X-ray diffraction patterns.

The films were post-annealed in a quartz tube furnace at 1,050°C with high purity oxygen flowing through it in order to compensate any oxygen loss and complete the oxidation of Fe ions which is important to maintain high resistivity in ME hexaferrites. Lack of oxygen may introduce ferrous ions (2+) which can decrease resistivity. The annealing of the films lasted for 40 minutes in order to prevent magnetic ions of the hexaferrite to diffuse into the substrate at the interface.

As in Chapter 4, the crystal structure and lattice constants were determined from X-ray diffraction θ–2θ from Cu K-alpha source. Strong peaks of (0 0 02n) were indexed, indicating an M-type hexaferrite. The crystals are oriented with the C-axis perpendicular to the film plane. The lattice parameters for $\delta = 0$, 0.2, and 0.4 are shown in Table 10.2.

The above lattice parameters are in good agreement to values published in the literature for $SrCo_2Ti_2Fe_8O_{19}$ grown by the natural process. However, as can be seen from Table 10.2, the trend in $a(A)$ is to decrease as δ increases. That explains why for $\delta = 1.0$ the quality of the crystal deteriorated, as its $a(A)$ value was not consistent with the literature value.

TABLE 10.2
Crystal Lattice Parameters of $SrCo_2Ti_2Fe_8O_{19}$

δ	$a(A)$	$c(A)$
0	5.87	23.05
0.2	5.84	23.05
0.4	5.84	23.05

MAGNETOMETRY AND MAGNETOELECTRIC MEASUREMENTS

Magnetic measurements were performed using a Vibrating Sample Magnetometer (VSM-SQUID). VSM magnetometry was measured for the applied field (H) in the plane of the film for temperatures between 5°K and 450°K.

Figure 10.1a–c shows the magnetization, $M(T)$, as a function of temperature in an applied field of 300 Oe after $H=0$ cooling (ZFC) and $H=30$ kOe cooling (FC). For $\delta=0.2$, 0.4, and 1.0 (FC) case, Figure 10.1b, increasing the temperature from 10°K causes $M(T)$ to steadily decrease as in a ferrimagnetic ferrite. At about 200°K, the slope of M versus T is nearly zero before decreasing with increasing temperature beyond 200°K. This temperature dependence of M was not observed for $\delta=0$, see Figure 10.1a. However, a peak is observed in all the samples for the (ZFC) case. The peak in the (ZFC) data, as seen in hexaferrites, is often interpreted as a change in the magnetic order from a spin spiral to a collinear spin arrangement. At room temperature, the coercivity for $\delta=0$, 0.2, and 0.4 is 720, 57 and < 10 Oe compared to 2,335, 1,085, and 700 Oe at 10°K. The large reduction in coercivity at 300°K is indicative of a large angle cone of magnetization.

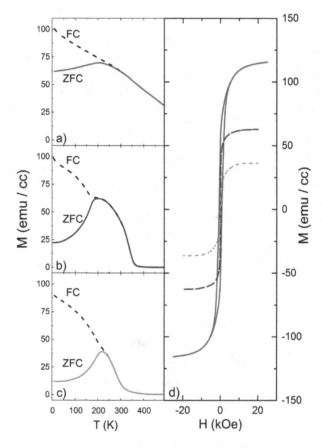

FIGURE 10.1   Magnetization versus temperature (a–c) and hysteresis (d) loop measurements.

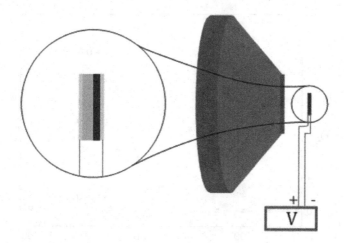

FIGURE 10.2 Setup for the magnetoelectric measurements.

The linear magnetoelectric coefficient, α, is defined as

$$\alpha = \mu_o \left(\partial M / \partial E \right) \approx \mu_o \left(\Delta M / \Delta E \right),$$

where $\Delta E = V/d$, with d being the combined thickness of substrate and the hexaferrite film assuming that the dielectric constants are the same and V being the voltage, free space permeability, and ΔM is the change in magnetization due to ΔE. Vibrating sample magnetometer (VSM) apparatus was deployed for the measurement of α, as shown in Figure 10.2.

An H field of 400 Oe was applied to remove any magnetic domains. H was applied normal to film plane which is parallel to the C-axis. The E-field is applied parallel to the C-axis and ΔM is measured along the same direction. As in the previous notations, $\alpha_\parallel$ is measured.

Typically, the film thickness was $\sim 0.5\,\mu m$ and the substrate thickness $\sim 250\,\mu m$. This implies that voltages less than 1 volt can produce the same results as in Figure 10.3, assuming that the film could be isolated from the substrate. The deduced magnetic parameters for $\delta = 0$, 0.2, and 0.4 are shown in Table 10.3.

The ME coupling $\alpha_\parallel$(s/m) was measured by VSM technique, see setup above. The $4\pi M_S$(G) was also measured by VSM. However, H_A(Oe) and the g factor (defined as the ratio of magnetic moment to the angular momentum) were deduced from FMR (ferromagnetic resonance) measurements. The reader should be mindful that for barium hexaferrite $g \approx 2$. For $\delta = 0.2$ and $\delta = 0.4$, their $\alpha_\parallel$ values are 36 and 50, respectively, times greater than the value for $\delta = 0$. However, for $\delta = 0$, the ATLAD technique purposely tried to avoid placing Co^{2+} in the R block in depositing a film of $SrCo_2Ti_2Fe_8O_{19}$. In contrast to $\delta = 0$ deposition, for $\delta = 0.2$ and 0.4, the aim of the ATLAD technique is to distribute Ti and Co cations in the R block while depositing films of $SrCo_2Ti_2Fe_8O_{19}$.

Sites $4f_2$ (octahedral) and 2b (trigonal) are located in the R block (no tetrahedral sites). Sites 2a (octahedral) and $4f_1$ (tetrahedral) are located in the S block. The 12k site straddles both the S and R blocks or the S^* and R^* blocks. X-ray absorption

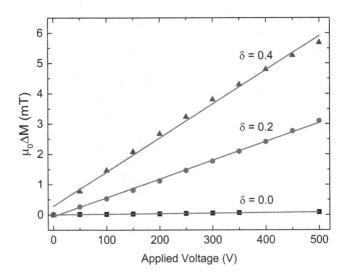

FIGURE 10.3   Induced magnetization versus applied voltage.

TABLE 10.3
Magnetic Parameters of ATLAD ME Hexaferrite Films of $SrCo_2Ti_2Fe_8O_{19}$

	$\delta = 0$	$\delta = 0.2$	$\delta = 0.4$
$\alpha_{\parallel}(s/m)$	5.2×10^{-9}	1.9×10^{-7}	2.6×10^{-7}
$4\pi M_S(G)$	1,280	700	1,360
$H_A(Oe)$	900	500	500
g(-)	3.0	3.2	1,000

spectroscopy (XAS) and X-ray magnetic dichroism are widely used techniques for determining the valence states of each transition metal cation, the site occupancy, and the local symmetry and/or electronic structure of matter. XMCD is a difference spectrum (resonance) of two XAS taken in a magnetic field (5–10 Tesla), one taken with left and the other with right circularly polarized light determining the spin and orbital magnetic moment. XAS data is obtained by tuning the photon energy to a range where core electrons are excited to the 3d empty states of the transition metal cations via electric dipole transitions.

The core electrons are identified by the quantum numbers $n = 1$, 2, and 3 corresponding to the K, L, and M edges, respectively. For example, the core electron labeled quantum mechanically $2p_{1/2}$ would be referred to as the L_2 edge core electron and $2p_{3/2}$ as the L_3 edge. Typically, electrons from $L_{2,3}$ are utilized to investigate the structure of the 3d cations. The energy states of the 3d cations are governed by the local or site symmetry (Octahedral, Tetrahedral, Trigonal, or some distorted coordination) of the 3d cations and oxygen anions. Chapter 3 illustrates simple example of crystal field calculations of energy levels of 3d states in various symmetries. It is the

TABLE 10.4

Site Occupancy of Transition Metal Cations in $SrCo_2Ti_2Fe_8O_{19}$

Block	R	S		S-R		R
Site	2d	$2a(O_h)$	$4f_1(T_d)$	$12k(O_h)$	$4f_2(O_h)$	2b(Trig.)
Bulk	Sr	Fe^{3+}	Fe^{3+} Co^{2+}	Fe^{3+} Ti^{4+}	Fe^{3+}	Fe^{3+}
$\delta = 0$	Sr	Fe^{3+} Co^{2+}	Fe^{3+} Co^{2+}	Fe^{3+} Ti^{4+}	Fe^{3+}	Fe^{3+}
$\delta = 0.2$	Sr	Fe^{3+} Co^{2+}	Fe^{3+} Co^{2+}	Fe^{3+} Ti^{4+}	Fe^{3+}	Fe^{3+} Co^{2+}
$\delta = 0.4$	Sr	Fe^{3+} Co^{2+}	Fe^{3+} Co^{2+}	Fe^{3+} Ti^{4+}	Fe^{3+}	Fe^{3+} Co^{2+}

particular selection of core electrons and the variation in symmetries that allow the detection of the site, valence, the symmetry, the magnetic moment, and local distortions of the cation in question. In Table 10.4, the results of these measurements are summarized for bulk grown $SrCo_2Ti_2Fe_8O_{19}$ and the same composition prepared by the ATLAD technique in which Co^{2+} was explicitly deposited in the R block ($\delta = 0$, 0.2, and 0.4). For $\delta = 0$ deposition, Co^{2+} was purposely not distributed in the R block, whereas for the other δ's, they were.

The 12k octahedral site is located in between S and R blocks. In comparing the data for the bulk sample (1) and $\delta = 0$ (2), they are similar except that the bulk sample showed no Co^{2+} substitutions into the 2a site! This may explain for the small discrepancy (<5%) in the measurement of the ME coupling (α). ATLAD is a technique that employs both omission and inclusion. For the $\delta = 0$ deposition, Co^{2+} was purposely omitted from the R block. For $\delta = 0.2$ and 0.4, Co^{2+} was purposely included in the R block. However, Table 10.4 may not be applicable if ferrous cations (Fe^{2+}) are present in the hexaferrite. In the above ATLAD films, the Fe^{2+} content was rather small. The occurrence of ferrous cations, probably due to incomplete oxidation during the deposition and annealing processes, is detrimental to ME ferrites, as it provides a conductive pathway between the Fe^{2+}/Fe^{3+} ions located at O_h sites. It suppresses electric polarization.

Site occupancy of 2b in hexaferrites has been a topic of interest in many other studies. In particular, magnetic relaxation of the uniform precession mode (FMR) to the lattice motion in barium hexaferrite has been of interest since the 1950s, see Chapter 6 and references therein. One model assumed relaxation to the lattice via the Fe^{3+} oscillations in a shallow double well at the 2b site. The oscillations are motions up and down along the C-axis near the 2d site (Ba^{2+}) and are coupled magnetostrictively to the lattice motion. The model was qualitative and speculative because it lacked quantitative measures of the double potential well and the relaxation times of the uniform mode to the Fe^{3+} oscillations at the 2b site and from there to the lattice. In the case we are considering now, Co^{2+} also occupies the 2b site. Quantitative estimations of the double well potentials for Fe^{3+} and Co^{2+} are shown.

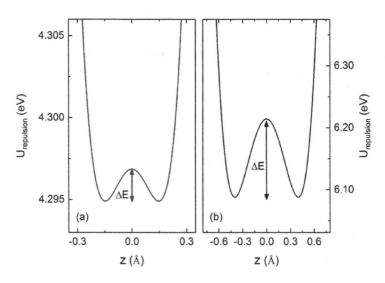

FIGURE 10.4   Potential wells for the occupation at (2b) crystal site for (a) Fe^{3+} and (b) Co^{2+}.

The z-axis is measured along the C-axis. The reference point $z = 0$ refers to the plane where it contains Sr and three oxygen ions (R block). In Figure 10.4a, the height of the double well potential is 2mev for Fe^{3+} at the 2b site. At room temperature, Fe^{3+} oscillates up and down the C-axis, 0.15 A above the reference plane and 0.15 A below. The equilibrium position is $z = 0$. This implies that the steady state induced electric dipole moment is zero, while there may be a small oscillating dipole moment. This is due to the fact that the center of mass for positive and negative ions has changed slightly, as Fe^{3+} oscillates. In Figure 10.4b, the height of the double well potential is 120 mev for Co^{2+}. This applies for $\delta = 0.2$ and 0.4, see Table 10.4. As such, there are no oscillations at room temperature, since the required temperature to excite the Cobalt ion over the potential barrier is about 1,117°C which is above the annealing temperature. At those temperatures, the material is paramagnetic.

As the annealing temperature is lowered below the Neel temperature, the Cobalt cation is "locked" either at $z \approx +0.4$ A or at $z \approx -0.4$ A. At either position, the induced electric dipole moment is static but not oscillatory, since the center of mass for positive and negative ions has changed permanently. This implies that a localized polarization, and therefore, the piezoelectric electric tensor parameters $[d]$ via the inverse modulus elastic tensor $[C]$, see the above section, have changed in addition to the matrix elements of the piezomagnetic tensor because of the inclusion of Co^{2+} in the 2b site. A priori, it is difficult to determine where exactly it is distributed ($+z$ or $-z$), since the internal field along the C-axis has uniaxial symmetry. However, it may be possible to select a z-position where Co^{2+} may reside by annealing the films in a magnetic field H, since its direction is unidirectional symmetry. According to reference, Ti^{4+} also "sits" in a double well potential similar to Co^{2+}. However, Ti cation is not affected by H during annealing, since the cation is not a magnetic ion.

Thus, the placement of Co^{2+} in the 2b site has two ramifications. Not only the magnetostriction parameters have been affected, but, surprisingly, also the d parameters! As pointed out from thermodynamic arguments above, the ME linear coupling tensor is the product of the piezoelectric and piezomagnetic tensors or simply put in this form symbolically as the product of parameters affected by the distribution of ions in the R block (see the previous section for definition of the parameters):

$$\alpha \propto \lambda d C d\chi / M. \tag{10.1}$$

FREE MAGNETIC ENERGY REPRESENTATION OF THE SPIN SPIRAL CONFIGURATION

The quantum representation of the spin spiral by Dzyaloshinski and Moriya included the following interaction form:

$$D \cdot \left(S_i \times S_j\right) D \cdot \left(S_i \times S_j\right), \tag{10.2}$$

where $\vec{D}$ is a vector proportional to exchange and local spin orbit and crystal fields interaction parameters at local sites i and j. In order for this interaction term to contribute to potential energy, spins at different sites could not be parallel to each other as in a spin spiral configuration, for example. Landau–Lifshitz introduced a semiclassical description of the above quantum equation form as follows (CGS):

$$F_{LL}\left(\text{ergs/cm}^3\right) = -\frac{K_{LL}}{M^2} M \cdot (\nabla \times M), \tag{10.3}$$

where F_{LL} is the free energy of a spin spiral configuration as derived by Landau–Lifshitz, K_{LL} is an anisotropy energy parameter in units of ergs/cm^2 and is proportional to the parameter D in Equation 10.2. M is the magnetization vector representing the spin variables over a relatively large volume compared to the crystal unit cell. Clearly, Equation 10.3 is microscopically applicable to single spins, whereas Equation 10.3 applies to macroscopic levels of interaction between magnetization vectors. Assuming that the transition from microscopic to macroscopic representations may be acceptable, the following predictions of Equation 10.3 may be construed. For simplicity, $\vec{M}$ may be expressed as follows:

$$M = M_{\parallel} a_z + M_{\perp}\left(a_x \cos \beta Z - a_y \sin \beta Z\right), \tag{10.4}$$

where a_x, a_y, and a_z are unit vectors in the x, y, and z directions, $M_{\parallel}$ is the component of the magnetization along the C-axis or the z-axis, see Figure 10.5, $M_{\perp}$ is the component perpendicular to the C-axis, $\beta = 2\pi/\lambda$, with λ being the wavelength of the helical or spin spiral configuration along the C-axis.

In order for this description to be valid, $\lambda \gg c$, where c is the lattice constant along the C-axis. Thus,

$$F_{LL} = -K_{LL}\beta\left(\frac{M_{\perp}}{M}\right)^2 = -K\left(\frac{M_{\perp}}{M}\right)^2. \tag{10.5}$$

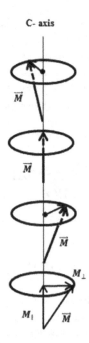

FIGURE 10.5 Spin spiral configuration along the C-axis. Note that $\mathbf{M}_{\parallel}$ is along the C-axis.

This is recognized as a classical uniaxial magnetic anisotropy energy term with a uniaxial axis along the C-axis. This is interesting to note that the macroscopic Landau–Lifshitz representation basically averages all local uniaxial anisotropy energies into one single expression, although the local uniaxial axis is at an oblique angle with respect to the C-axis, see Figure 10.5. This is not surprising in view of the fact that the sum of uniaxial anisotropy energies still results in one single uniaxial energy term as above. We will adopt the Landau–Lifshitz macroscopic representation and designate the uniaxial anisotropy energy parameter as K_{ss} or K_θ. Clearly, K_{ss} is related to the parameters K_{LL}, K, and D. The corresponding uniaxial magnetic anisotropy magnetic field, H_{ss}, is defined as $2K_{ss}/M$ and it applies to a spin spiral configuration in a ME hexaferrite. For the case $\alpha = 0$, as in normal hexaferrites, it is also meaningful to define a uniaxial anisotropy field, H_θ, equal to $2K_\theta/M$. Thus, K_{ss} plays a dual role depending on the value of α.

FREE ENERGY OF ME HEXAFERRITE

The total free energy, F(ergs/cm³), is comprised of magnetic, F_M, and electric, F_E, free energies or $F = F_M + F_E$, where (10.6)

$$F_M = -HM_z + 2\pi M_z{}^2 - K_{ss}\left(\frac{M_z}{M}\right)^2 - \left(\frac{A}{M^2}\right)M \cdot \nabla^2 M + \qquad (10.7)$$

The external magnetic field, H, is applied normal to the film plane and parallel to the C-axis, see Figure 10.6. The second term in F_M is the demagnetizing energy and

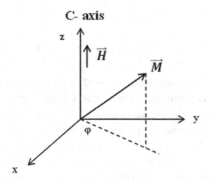

FIGURE 10.6 Magnetic field orientations relative to the C-axis.

the third term the macroscopic uniaxial anisotropy energy as described in the previous section. The energy parameter K_{ss} is analogous to the parameter $\vec{M}$, usually designated in typical hexaferrites. The subscript "ss" is to remind the reader that this parameter corresponds to a spin spiral configuration. The fourth term is the ordinary semiclassical exchange coupling term and is often referred to as the exchange stiffness constant, and the fifth term is the magnetic anisotropy energy corresponding to magnetization directions in the azimuth plane (normal to the C-axis), reflecting the sixfold symmetry. Typically, the magnetic anisotropy field associated with this energy term is in the order of 20–50 Oe. Finally, the last term is the magnetoelectric coupling energy term. The coupling parameter α is treated as a scalar. In general, it is anisotropic and may be represented as a tensor, and it may be represented as a dyadic vector. For now, it suffices for the purpose of demonstrating the formalism of the calculations. The polarization vector is a complicated function of the internal strain, as it must be for a ME medium like hexaferrites.

If anisotropic terms for the ME coupling were to be included, they would be of this form for the free energy. Extending Landau–Lifshitz formalism applicable for a material with cubic symmetry to hexagonal crystal structures, the free energy of the magnetoelastic coupling would take on this form

$$F_\lambda = (a_1\alpha_1{}^2 + \partial_3\alpha_3{}^2)\,\epsilon_{xx} + (\partial_1\alpha_2{}^2 + a_3\alpha_3{}^2)\,\epsilon_{yy} + [a_2\alpha_3{}^2 + a_3(\alpha_1{}^2 + \alpha_2{}^2)]\,\epsilon_{zz}$$

$$+\ a_4\alpha_1\alpha_2\,\epsilon_{xy} + a_5\alpha_1\alpha_3\,\epsilon_{xz} + a_5\alpha_1\alpha_2\,\epsilon_{yz}. \tag{10.8}$$

The a_i coefficients are related to the magnetostriction constants, see the previous section. The key point of the above equation is that the strain is coupled to the magnetic system, where α_i are the directional cosines of $\vec{M}$ relative to the x, y, and z coordinate system, with z being along the C-axis. The piezoelectric coupling between the strain and the electric field is introduced similarly as in Equation 10.8. Basically, the mediator that couples the magnetic system to the electric system is the strain field. Thus, for example, the excitation of an electric field is coupled via the piezoelectric strain coefficient to the strain field. The strain field in turn is coupled to the magnetic system via the magnetostriction coupling as in Equation 10.8. For hexagonal crystal symmetry, the piezoelectric coupling energy term takes on the following form $(-\mathbf{P} \cdot \mathbf{E})$:

$$F_d = -d_5 C_{44}(\epsilon_{zx} E_x + \epsilon_{zy} E_y) - d_1[(C_{11} + C_{12})(\epsilon_{xx} + \epsilon_{yy})$$

$$+ 2C_{13} \epsilon_{zz}]E_z - d_3[C_{13}(\epsilon_{xx} + \epsilon_{yy}) + C_{33} \epsilon_{zz}]E_z. \qquad (10.9)$$

Thus, the addition of Equations 10.8 and 10.9 to the free energy would replace the coupling energy term containing the scalar α. Clearly, more parameters are introduced, but it is needed to explain, for example, anisotropic coupling between the two systems. However, anisotropic coupling could also be analyzed using a tensor $[\alpha]$ rather than a scalar α. We will not dwell on the merit of the two methodologies, but simply choose the scalar form as a starting point, since the algebraic methodology is exactly the same for either case. The object is to introduce the formalism to calculate dynamic excitations in ME hexaferrites. However, some ME hexaferrites may also be characterized as being ferroelectric. In that case, terms like second, fourth order, and higher even powers of P may be added to the free energy. Furthermore, in polycrystalline ME hexaferrite of the Z-type, the material behaved electrostrictive implying that the relationship between strain and electric field was quadratic. In such cases, Equations 10.8 and 10.9 would have to be modified in order to allow for the quadratic dependence on strain.

ELECTROMAGNETIC WAVE DISPERSION OF MAGNETOELECTRIC HEXAFERRITE

The formalism introduced in Chapters 7 and 9 is sufficiently general to allow for nonlinear effects in ME ferrite materials. However, only linear response is considered. Basically, magnetic and electric fields and polarization vectors are decomposed into static and dynamic components. The product of two dynamic fields would be omitted in the formalism, since it would introduce nonlinear excitations. This approach is applicable for any form of the free energy, as the magnetic fields, elastic displacements, and electric polarizations are derived from the free energy and substituted into the magnetic and elastic equations of motion for the dynamic fields in terms of static fields. Requiring nontrivial solutions for them, a dispersion relation is obtained, see Chapters 7 and 9. Thus, the dispersion may be calculated as follows:

$$\left(\omega^2 - C_{33}k^2/\rho\right)\{(\omega^2 - C_{44}k^2/\rho)\left[\left(H_o + 2Ak^2/M^2\right)^2 - \omega^2/\gamma^2\right] +$$

$$2\left(\omega^2 - C_{44}k^2/\rho\right)(H_o + 2Ak^2/M^2)k^2 M \alpha d_5 C_{44}/\rho + k^4 M^2 (\alpha d_5 C_{44})^4/\rho^2)\} = 0. \quad (10.10)$$

The term $(\omega^2 - C_{33}k^2/\rho)$ in Equation 10.10 is identified as the longitudinal acoustic branch (labeled LA_1) and it is uncoupled to spin waves or the other two transverse acoustic branches. The dispersion relation may be expressed as

$\omega = k\sqrt{C_{33}/\rho}$, where

$v_L = \sqrt{C_{33}/\rho}$ = longitudinal wave velocity.

The two transverse acoustic branches or modes are degenerate for frequencies above and below the approximate FMR frequency or the crossing region between the spin wave branch and the two transverse acoustic modes, see Figure 10.7.

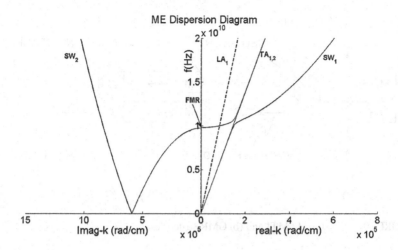

FIGURE 10.7 Plot of dispersion relations versus frequency and Re and Im(k) for $\alpha = 1$.

The unit of α (Gaussian) is dimensionless and it may be converted into MKS units by dividing by the velocity of light (3×10^8 m/s). The two transverse acoustic branches (labeled $TA_{1,2}$) are degenerate for frequencies above and below the FMR frequency, see the crossing region between the spin wave branch (labeled SW_1 and SW_2), Figure 10.7. The transverse acoustic velocity is $v_T = \sqrt{C_{44}/\rho}$. The following parameters were assumed for the plot of Figure 10.7: A (exchange stiffness constant) $= 0.4 \times 10^{-6}$ ergs/cm, ρ (density) $= 5.3$ g/cm^3, g $= 2.0$, $4\pi M_S = 1,400$ G, $d_5 = 1.1 \times 10^{11}$ m/v, $C_{44} = 1.0 \times 10^{12}$ ergs/cm^3, and $C_{33} = 3C_{44}$. The conversion of d_5 to CGS units is 3×10^4. The uniform mode ($k=0$) FMR excitations occur at $H_0 = 3,500$ Oe corresponding to a FMR frequency of 9.8 GHz. The spin wave branch (SW_1) intersects the two transverse acoustic branches, but only one transverse acoustic branch is dynamically coupled to SW_1 branch, splitting the two branches at the intersection. The other spin wave branch SW_2 consists only of imaginary propagation constants, see Figure 10.7. As such, this spin wave branch represents attenuating spin waves. It is often referred to in the literature as a surface spin wave mode, since the spin wave is mostly localized near the surface due to attenuation. It is usually referred to as the surface spin wave mode. Clearly, for an anisotropic α, Figure 10.7 would be considerably modified coupling the L_{A1} longitudinal acoustic branch to the spin wave branch.

ANALOG TO A SEMICONDUCTOR TRANSISTOR THREE-TERMINAL NETWORK

Three-terminal networks are the foundation of modern technologies in electrical engineering; yet, a three-port network is a rarity in ferrite technology. Three-port networks have existed as far back as the 1900s, when a control grid terminal was introduced by Lee De Forest in 1907 to control current flow in vacuum tube diodes (Audion). Similarly, a third terminal was introduced in the 1940s to control the current flow in a semiconductor diode. The common denominator in vacuum tube and semiconductor technologies is that the controlling element or terminal was

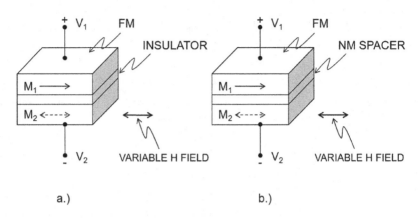

FIGURE 10.8   (a) Typical MTJ and (b) GMR structures.

compatible with the two-port diode, thus integrating all three terminals in one robust network.

The analog to the two-port diode network in ferrite technology is the conventional spin-valve MTJ (magnetic tunnel junction) and GMR (giant magnetoresistance) devices. An MTJ consists of an ultra-thin insulating layer that is used to separate two ferromagnetic (FM) metal electrodes, as shown in Figure 10.8a. Also, a GMR structure consists of two layers of FM materials that are separated by a NM conducting layer, as shown in Figure 10.8b. Although copper is an excellent conductor, electron scattering causes the resistance to increase significantly when it is only a few nm thick.

In this section, only the spin-valve (MTJ) device is considered, see Figure 10.8a. The direction of $\mathbf{M_1}$ is fixed in the plane of the layer; label it as the y-axis. In order to conduct electrical current through the two-terminal device, it needs to also be metallic. A metallic permanent magnetic may suffice or fabricate a composite in which the top layer consists of three layers. Two of the two layers are exchange coupled antiferromagnetically separated by a NM metal layer. The thickness of each layer is about 100 A. The reader should be mindful that more layers imply more scattering of conducting electrons. The insulating layer of choice between the top and bottom layers of the two-terminal device is MgO and the thickness is about 10 A. MgO is an excellent insulating material often used in MMIC circuits.

The bottom layer is a "soft" or "free" magnetic metal. By "soft," it means that the coercive field is very small, $\ll 1$ G. Permalloy and FeSi alloy are examples of many other amorphous metal magnetic alloys. Thus, for H as small as 1 G, $\mathbf{M_2}$ can be flipped back and forth between $+$ and $-y$, depending on the direction of H along the y-direction. $\mathbf{M_1}$ remains fixed in direction. Let's examine the electrical properties of these two-terminal MTJ devices.

For $\mathbf{M_1}$ in the opposite direction to $\mathbf{M_2}$ ($M_1 M_2 < 0$), the resistance across the two terminals is R_-. The current flow, I_-, is simply

$$I_- = V_1 / R_-, V_2 = 0.$$

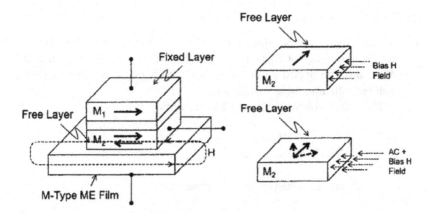

FIGURE 10.9   *H* field generated by a ME film.

Similarly, for $\mathbf{M_1}$ parallel to $\mathbf{M_2}$ $(M_1 M_2 > 0)$, current (I_+) flow would be

$$I_+ = V_1 / R_+, \text{ where}$$

$R_- \approx 1 k\Omega \approx 2 R_+$. The tunnel magnetoresistance ratio (TMR) is defined as

$$TMR = (R_- - R_+)/R_+, \text{ where } 0 < TMR < 2.$$

Thus, the analog of the diode is the two-terminal network shown in Figure 10.8a.

As in the diode case, a third terminal may be added to the network for the sole purpose of utilizing the fact that $R_- \neq R_+$. This means that the direction of $\mathbf{M_2}$ must be controlled by the third terminal. The direction of $\mathbf{M_2}$ may be "flipped" back and forth via an H field whose magnitude is slightly bigger than the coercive field of the bottom layer. There are a number of ways to generate such *H* field. The simplest way is to utilize a *ME* hexaferrite film to generate *H*. Conceptually, Figure 10.9 shows the physical arrangement for the *ME* film to be effective.

Figure 10.9 is only intended for visualization purpose to show the direction of *H* generated by a *ME* film to align in the bottom layer. For example, the *ME* film cannot be in contact with the bottom layer, since the *ME* film is an insulating layer, but it should be in close proximity. Thus, the third terminal drives a square wave voltage across the *ME* film. The application of an electric field induces the generation of magnetization either parallel or opposite to $\mathbf{M_1}$, since the voltage input is of either polarity to the *ME* film. The *H* field generated is approximately

$$H_{ME} \approx \pm 4\pi M_{ME}$$

PROBLEM

10.1 Calculate the gain of a potential three-terminal device assuming the scheme of Figure 10.9.

SOLUTION

10.1 Assume that a simple three-terminal device in which the top layer is attached to a resistor R_D and V_2 is attached to the ground terminal of the ME film. The opposite side of the ME film is driven by a square wave voltage with amplitude voltage $\pm V_{in}$.

The output voltage across the TMJ composite is

$$V_- \approx (V_1\, R_-/R_D) \text{ and}$$
$$V_+ \approx (V_1\, R_+/R_D), \text{ where } R_\pm < R_D.$$

The change in voltage is approximately

$$\Delta V = (V_- - V_+) \approx (V_1 TMR)(R_+/R_D), \text{ where}$$
$$TMR = (R_- - R_+)/R_+ = \text{tunnel magnetoresistance ratio.}$$
$$\text{Gain} = \Delta V/2V_{in} \approx 25, \text{ where}$$
$$TMR \approx 1, \; R+/R_D \approx 0.1, \; V_{in} \approx 10\,mV, \text{ and } V_1 \approx 5\,V.$$

BIBLIOGRAPHY

D.N. Astrov, *Soviet Phy. JETP*, **11**, 708, 1960.

J.E. Beevers, C.J. Love, V.K. Lazarov, S.A. Cavill, H. Izadkhah, and C. Vittoria, *Appl. Phys. Lett.*, **112**, 082401, 2018.

G.F. Dionne, *Magnetic Oxides*, Springer, Berlin, 2009.

I.J. Dzyaloshinskii. *Phys. Chem. Solids*, **4**, 241–255, 1958.

K. Ebnabbasi, Ph.D. Thesis, *Microwave Devices Utilizing Magnetoelectric Hexaferrite Materials*, Northeastern University, 2013.

V.J. Folen, G.T. Rado, and E.W. Stalder, *Phys. Rev. Lett.*, **607**, 102–105, 1961.

H. Izadkhah, S. Zare, and C. Vittoria, *Appl. Phys. Letters*, **106**, no. 14, 142905, 2015.

H. Izadkhah, S. Zare, S. Sivasubramanian, and C. Vittoria, *Appl. Phys. Lett.*, 106, 142905, 2015.

T. Kimura, *Annu. Rev. Condens. Matter Phys.*, **3**, 93, 2010.

L.D. Landau and E.M. Lifshitz, *Mechanics*, vol. 1, Pergamon Press, New York, 1976.

L.D. Landau and E.M. Lifshitz, *Electrodynamics of Continous Media*, Pergamon Press, Oxford Press, Oxford, 1984.

S.P. Marshall and J.B. Sokoloff, *Phys. Rev. B*, **44**, 619, 1991.

W. Mason, *Phys. Rev.*, **96**, 302, 1954.

M. Moriya, *Phys. Rev.*, **120**, 91–98, 1960.

G.T. Rado, C. Vittoria, J. Ferrari, and J.P. Remeika, *Phys. Rev. Lett.*, 41, 1253, 1978.

S.-P. Shen *et al*. *Phys. Rev. B*, **90**, 180404(R), 2014.

G. Srinivasan, *Annu. Rev. Mater.*, **40**, 153–178, 2010.

C. Vittoria and A. Widom, *Phys. Rev.*, B89, 134413, 2014.

11 Spin Surface Boundary Conditions

The number of allowable propagation modes in a magnetic medium is reduced if exchange fields are omitted in the equation of motion. As such, internal microwave field amplitudes may be determined uniquely from the application of electromagnetic boundary conditions only. Electromagnetic boundary conditions include the requirement that the tangential components of $\vec{e}$ and $\vec{h}$ be continuous across the surface boundaries. Inclusion of exchange fields in the equation of motion increases the number of allowable propagation constants by two. For this general case, electromagnetic boundary conditions alone are not sufficient to determine the internal microwave field amplitudes. Hence, we must introduce additional boundary conditions besides the electromagnetic boundary conditions at the surfaces of a magneto-dielectric medium. From a mathematical point of view, we must introduce as many boundary conditions as the number of unknown internal field parameters associated with each mode of propagation in a magnetic medium. In this chapter, we introduce spin boundary conditions at the surfaces of a magneto-dielectric medium, as the so-called additional boundary conditions, see Figure (11.1). For convenience, we choose the y-axis to be perpendicular to the pillbox. We start with the equation of motion for M (see Equations 5.19 and 6.15):

$$\frac{1}{\gamma}\frac{d\vec{M}}{dt} = \vec{M} \times \vec{H} - \vec{M} \times \left(\frac{2A}{M^2} \nabla^2 \vec{M} \right), \tag{11.1}$$

where $\vec{M}$ is the total magnetization, and $M = |\vec{M}|$. We rewrite the above equation and integrate over the volume of the magnetic medium:

$$\int_{\upsilon} \left(\frac{1}{\gamma}\frac{d\vec{M}}{dt} - \vec{M} \times \vec{H} \right) d\upsilon = -\int_{\upsilon} \left(\vec{M} \times \frac{2A}{M^2} \nabla^2 \vec{M} \right) d\upsilon. \tag{11.2}$$

We arbitrarily choose the volume of integration to be small, see Figure 11.1 over the pill box.

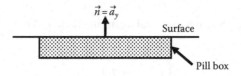

$$\vec{n} = \vec{a}_y$$

Surface

Pill box

FIGURE 11.1 Pill box configuration at the boundary surface of the magnetic film.

DOI: 10.1201/9781003431244-11

Thus, Equation 11.2 becomes

$$\int \vec{M} \times \frac{2A}{M^2} \nabla^2 \vec{M} d\upsilon = 0. \tag{11.3}$$

For simplicity, let's consider only one component of the above torque vector equation. The x-component of Equation 11.3 is then

$$\int_\upsilon \left(M_y \nabla^2 M_z - M_z \nabla^2 M_y \right) d\upsilon.$$

We apply Green's theorem, which states that

$$\int_\upsilon \left(\phi \nabla^2 \varphi - \varphi \nabla^2 \phi \right) d\upsilon = \int_S \left(\phi \vec{\nabla} \varphi - \varphi \vec{\nabla} \phi \right) \cdot \vec{n} da,$$

where
   n is the unit vector perpendicular to the surface
   da is the incremental area of the surface of the pillbox

We obtain that

$$\int_\upsilon \left(M_y \nabla^2 M_z - M_z \nabla^2 M_y \right) d\upsilon = \int_S \left(M_y \vec{\nabla} M_z - M_z \vec{\nabla} M_y \right) \cdot \vec{n} da.$$

We choose the y-axis to be normal to the pillbox. Thus,

$$\int_\upsilon \left(M_y \nabla^2 M_z - M_z \nabla^2 M_y \right) d\upsilon = \int_s \left(M_y \frac{\partial M_z}{\partial y} - M_z \frac{\partial M_y}{\partial y} \right) da.$$

By including the other two components of the torque in Equation 11.3, we may generalize the above result to the following:

$$\frac{2A}{M^2} \int_\upsilon \left(\vec{M} \times \nabla^2 \vec{M} \right) d\upsilon = \frac{2A}{M^2} \int_S \vec{M} \times \frac{\partial \vec{M}}{\partial y} da = 0, \tag{11.4}$$

where

$$da = dxdz.$$

The above is an indefinite integral, and assuming that $\vec{M}$ is not a function of x and z, the integral yields

$$\frac{2A}{M^2} \left(\vec{M} \times \frac{\partial \vec{M}}{\partial y} \right) + \vec{C} = 0. \tag{11.5}$$

The constant of integration $\vec{C}$ represents the essence of surface magnetism. Physically, it represents a surface torque $\vec{T}_s$, which arises solely from a surface magnetic anisotropy energy density, E_S. The units of E_s are ergs/cm^2. Thus,

$$\vec{C} = \vec{T}_s = -\vec{M} \times \left(\vec{\nabla}_M E_S\right). \tag{11.6}$$

Combining Equations 11.5 and 11.6, we obtain the general surface torque equation

$$\vec{M} \times \left[\frac{2A}{M^2}\frac{\partial \vec{M}}{\partial y} - \vec{\nabla}_M E_S\right] = 0. \tag{11.7}$$

We will show an example of E_s due to Neel.

A QUANTITATIVE ESTIMATE OF MAGNETIC SURFACE ENERGY

Based on Neel's model, the pair energy between two magnetic ions may be expressed as follows:

$$W(\phi) \cong g + l\left(\cos^2\phi - \frac{1}{3}\right) + q\left[\cos^4\phi - \frac{6}{7}\cos^2\phi + \frac{3}{35}\right] + \dots \tag{11.8}$$

The angle ϕ is measured between the direction of the magnetic dipole moment at the sites of the two magnetic ions and the bond axis connecting the two ions. The first term includes the exchange interaction between the two magnetic ions and it is an isotropic term. The second term describes the dipole–dipole interaction between the same two ions (see Chapter 5). The actual value of l can be evaluated from the values of magnetostriction constant for materials and from the uniaxial magnetic anisotropy energy in uniaxial crystal symmetry materials. In most cases of interest, it is about 100–1,000 times greater than the actual dipole–dipole interaction coefficient between two ions. The physical origin of l is still a topic of research to this day. Over the years, it has been related to anisotropic exchange and pseudo-dipolar interactions. The coefficient q can be related, for example, to the cubic magnetic anisotropy constant K_1 for cubic systems. Terms in l vanish because the nearest neighbors of a given atom are distributed about it with cubic symmetry (see Chapter 5). This is no longer the case when the atom considered is located at the surface of the crystal. In this case, the mean value of $l \cos^2 \phi$ generally is not zero.

We obtain the magnetic energy referred to an atom at the surface by summing over the values of $l \cos^2 \phi$ relating to all nearest neighboring atoms and multiplying the result by 1/2 in order not to count the same linkage energies twice. In this way, an energy E_s is calculated which depends both on orientation of the spontaneous magnetization relative to the film surface and bonding axes. By multiplying the number of atoms per unit area, the value of the surface energy density is obtained. Typically, E_s exhibits uniaxial symmetry and may be written as follows:

$$E_S = K_S \cos^2 \psi, \tag{11.9}$$

where ψ is the angle between the direction of the magnetization and the surface normal.

Assuming BCC structure, Neel showed that $E_s = 0$ for (111) and (100) surfaces. For (110) surfaces,

$$E_s = \frac{l\beta_2\beta_3}{a^2\sqrt{2}}, \tag{11.10}$$

where β_2 is the directional cosine between $\vec{M}$, magnetization, at the surface and the y-axis and β_3 between $\vec{M}$ and the z-axis. The z-axis was chosen to be in the film plane and parallel to the <001> cubic axis, and y is normal to the surface, and "a" is the lattice constant. We have related l to the magnetostriction constant of iron, for example, and we find

$$l = -\lambda_{100}\frac{9}{16}\frac{C_{11}-C_{12}}{N}, \tag{11.11}$$

where

$C_{11} - C_{12} = 0.95 \times 10^{12}$ ergs/cm³

$\lambda_{100} = 20.7 \times 10^{-6}$

$N = 4.3 \times 10^{22}$ site/cm³

C_{11} and C_{12} are elastic moduli, and λ_{100} is the magnetostriction constant in the [100] direction. Finally, K_s is deduced as follows:

$$K_s \cong 0.2\lambda_{100}(C_{11}-C_{12})a, \tag{11.12}$$

where $a = 2.86 \times 10^{-8}$ cm.

The estimate for iron is that $K_s = 0.11$ ergs/cm². The reader is reminded that $K_s = 0$ for (100) and (111) film surfaces assuming nearest neighbor sums. K_s may be nonzero if we account for next nearest neighbor interactions, for example. At this level of accuracy, it is appropriate to include the contribution from the third term in Equation 11.8 at the film surface. The main feature of Equation 11.12 is that K_s is expressed in terms of quantities that are measurable. There are no adjustable parameters in Equation 11.12, since λ_{100}, a, C_{11}, and C_{12} are well known in the literature.

ANOTHER SOURCE OF SURFACE MAGNETIC ENERGY

Thus, one mechanism for the source of E_s is the lack of crystal symmetry at the surface. Another potential mechanism is the anisotropic exchange coupling between two magnetic films separated by a small non-magnetic film (Vittoria). The separation may be in the order of one to five atomic lattice constants, as shown in Figure 11.2.

The exchange coupling between two spins located at the two surfaces of the films is given as

$$E = -J\vec{S}_1 \cdot \vec{S}_2. \tag{11.13}$$

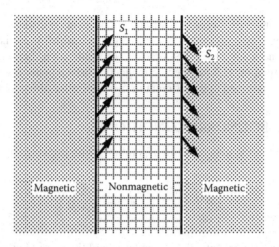

FIGURE 11.2 Exchange coupling model between spins at two magnetic film surfaces separated by a non-magnetic layer.

If J is positive, the coupling is ferromagnetic in nature. Equation 11.13 may be put in terms of local magnetization at the two surfaces:

$$E = \frac{-J\vec{M}_1 \cdot \vec{M}_2}{(N_1 g_1 \beta)(N_2 g_2 \beta)}, \tag{11.14}$$

where

β is the Bohr magneton
J is the exchange coupling
M_i is the magnetization at site i

Here, we have assumed that $\vec{S}_1$ and $\vec{S}_2$ are classic vectors and not quantum operators. We divide E by a^2 and obtain the surface energy as (a is the lattice parameter)

$$E_s = \frac{A_{12}}{M_1 M_2} \vec{M}_1 \vec{M}_2, \tag{11.15}$$

where

$$A_{12} = \frac{J}{a^2} S_1 S_2,$$

$$M_1 = N_1 g_1 \beta S_1,$$

and

$$M_2 = N_2 g_2 \beta S_2.$$

It is noted that for $J \sim 10^{-16}$ ergs (J is strongly dependent on distance between $\vec{S}_1$ and $\vec{S}_2$) and $a \sim 3 \times 10^{-16}$ cm, we obtain $A_{12} \sim 0.1$ ergs/cm^2, which is in the right order of magnitude for the value of E_s. The bracket is to indicate a thermal average. Let's now examine Equation 11.7 for specific magnetic surface boundary conditions. The g-factors may be assumed to be different in two dissimilar films, for example.

STATIC FIELD BOUNDARY CONDITIONS

The equation of motion may be written as follows with exchange field, see Equation 5.19 and 6.15:

$$\frac{1}{\gamma} \frac{d\vec{M}}{dt} = \vec{M} \times \left(-\frac{2A}{M^2} \nabla^2 \vec{M} + \vec{H} \right). \tag{11.16}$$

Since we are considering static field boundary conditions, we require that

$$\frac{d\vec{M}}{dt} = 0.$$

For simplicity, we assume again that the y-direction is normal to the film plane. Thus, Equation 11.16 reduces to

$$\vec{M} \times \left[-\frac{2A}{M^2} \frac{\partial^2}{\partial^2 y} \vec{M} + H \right] = 0. \tag{11.17}$$

This is a second-order differential equation. Two boundary conditions are needed to solve it uniquely. The boundary conditions at the two surfaces may be obtained from Equation 11.7 and they are

$$\vec{M} \times \left[-\frac{2A}{M^2} \frac{\partial \vec{M}}{\partial y} - \vec{\nabla}_M E_s \right] = 0. \tag{11.18}$$

Let's address a specific example in applying Equations 11.17 and 11.18. Assume $\vec{M}$ to be in the film plane and E_s is of the following form:

$$E_s = -K_s \left(\frac{M_x}{M} \right)^2. \tag{11.19}$$

Clearly, Equation 11.19 implies that the easy axis of magnetization is in the film plane and along the x-direction at the surface if $K_s > 0$. Also, assume that the applied static magnetic field is

$$\vec{H} = H\vec{a}_z,$$

implying that at the center of the film ($y = t/2$), the magnetization is along the z-direction. The nature of the problem is that as y changes from $t/2$ to 0 ($y = 0$ is the film surface), where t is the film thickness, $\vec{M}$ rotates in the film plane roughly through 90°. The question is: what is the rate of rotation? From Equation 11.17, we obtain

$$\frac{2A}{M^2}\left[-M_x\frac{\partial^2 M_z}{\partial y^2}+M_z\frac{\partial^2 M_x}{\partial y^2}\right]=M_x H, \tag{11.20}$$

where

$$M_x^2 + M_z^2 = M^2; \quad M_y = 0.$$

There results a differential equation of the form

$$\frac{2A}{M^2}\left[1+\left[\frac{M_z}{M_x}\right]^2\right]\left[\frac{d^2 M_z}{dy^2}+\frac{M_z}{M_x^2}\left[\frac{d^2 M_z}{dy^2}\right]^2\right]=H. \tag{11.21}$$

The two static field boundary conditions are

$$\frac{dM_z}{dy}=\left(\frac{K_s/A}{M}\right)M_x^2 M_z \quad \text{at} \quad y=\pm\frac{t}{2}. \tag{11.22}$$

Special care must be exercised in applying the above equation at the two surfaces, since d/dy changes sign. If the surface magnetic anisotropy field at the surface is dominant compared to H, then $M_z \simeq 0$ at the surface and $dM_z/dy \cong 0$.

DYNAMIC FIELD BOUNDARY CONDITIONS

The field configuration is shown in Figure 11.3.

In the static equilibrium condition, we assume that the magnetization makes an angle ψ with respect to the film normal at the film surface. The uniaxial axis direction is chosen to be normal to the film plane also. This is in contrast to the previous

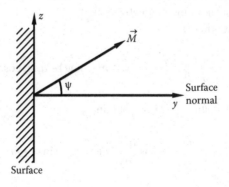

FIGURE 11.3 Magnetization direction relative to the film's normal direction.

example of the static case where the axis was along the x-axis. The free energy of the surface anisotropy energy density is

$$E_s = K_s \alpha_2^2,$$

where y is assumed to be perpendicular to the film plane. K_s are in units of ergs/cm² and $\alpha_2 = M_y/M$. The torque equation at the surface yields the following:

$$\frac{2A}{M^2} \vec{M} \times \frac{\partial \vec{m}}{\partial y} + \vec{T}_s = 0, \tag{11.23}$$

where

$$\vec{M} = \vec{M}_0 + \vec{m}$$

and

$$\vec{T}_s = \vec{M} \times \left(-\vec{\nabla} E_s\right).$$

Rewriting $\vec{T}_s$ in terms of K_s, we obtain

$$\vec{T}_s = -\left(\vec{M}_0 + \vec{m}\right) \times \left[\left(\frac{2K_s}{M^2}\right) M_y \vec{a}_y\right],$$

where

$$\vec{M} = M\left(\cos\psi \vec{a}_y + \sin\psi \vec{a}_z\right)$$

and

$$\vec{m} = m_x \vec{a}_x + m_y \vec{a}_y + m_z \vec{a}_z.$$

We have dropped the subscript "0" whenever it is self-evident that we are talking about a static component of the magnetization.

Thus, $\vec{T}_s$ may be written as

$$\vec{T}_s = \left(\frac{2K_s}{M}\right)\left[\vec{a}_x \left(m_y \sin\psi + m_z \cos\psi\right) - \vec{a}_z m_x \cos\psi\right]. \tag{11.24}$$

Substituting $\vec{T}_s$ into Equation 11.23, we obtain

$$\vec{a}_x \left[A\left(\cos\psi \frac{\partial m_z}{\partial y} - \sin\psi \frac{\partial m_y}{\partial y}\right) + K_s\left(m_y \sin\psi + m_z \cos\psi\right)\right]$$

$$+ \vec{a}_y A \sin\psi \frac{\partial m_x}{\partial y} - \vec{a}_z \cos\psi \left(A \frac{\partial m_x}{\partial y} + K_s m_x\right) = 0. \tag{11.25}$$

Let's examine Equation 11.25 for two limiting cases: $\psi = 0$ and $\psi = \pi/2$. The limit $\psi = 0$ implies that $\vec{M}$ is perpendicular to the film plane and Equation 11.25 reduces to the following:

$\psi = 0$:

$$A\frac{\partial m_z}{\partial y} + K_s m_z = 0 \tag{11.26}$$

and

$$A\frac{\partial m_x}{\partial y} + K_s m_x = 0. \tag{11.27}$$

In circularly polarized fields, we have

$$A\frac{\partial m_\pm}{\partial y} + K_s m_\pm = 0, \tag{11.28}$$

where

$$m_\pm = m_x \pm jm_z.$$

For $\psi = \pi/2$, $\vec{M}$ is parallel to the z-axis or in the film plane. The boundary conditions on the spin motion become as follows:

$\psi = \pi/2$:

$$A\frac{\partial m_y}{\partial y} - K_s m_y = 0 \tag{11.29a}$$

and

$$\frac{\partial m_x}{\partial y} = 0. \tag{11.29b}$$

For $K_s > 0$, the easy axis of magnetization is in the film plane. It is clear that with dynamic boundary conditions, the easy axis of magnetization has to be consistent with the static equilibrium conditions. For example, we cannot have the external field competing against the uniaxial field, i.e., H must point along the easy axis of magnetization.

APPLICATIONS OF BOUNDARY CONDITIONS

$\vec{H} \perp$ to the Film Plane

For microwave magnetic field excitation in a film, spin and electromagnetic boundary conditions at the film surfaces are important in determining the type of magnetic excitation induced by an external microwave magnetic field. As a specific example, we consider the case when an external static field, $\vec{H}$, is applied normal to the film plane, denoted as $\perp$ configuration. In this case, the normal modes of excitation are circularly polarized (see Chapter 8). One mode represents precessional magnetic resonance and may be mathematically written as

$$h = h_x - jh_z$$

and

$$m = m_x - jm_z.$$

The sense of rotation is CW, whereas in the anti-resonant mode, the sense of rotation is CCW (see Chapter 9). The internal electric field,

$$e = e_x - je_z,$$

is also circularly polarized, and it can be related to h via Maxwell's equation:

$$\vec{\nabla} \times \vec{h} = j\omega\varepsilon\vec{e}.$$

Assuming propagation in the $+y$-direction e^{-jky}, see Figure 11.3, we have components in the following form:

$$-jk\left[h_z\vec{a}_x - h_x\vec{a}_z\right] = j\omega\varepsilon\left(e_x\vec{a}_x + e_z\vec{a}_z\right).$$

It is simple to show that

$$e = -jZh,$$

where

$$Z = k/\omega\varepsilon.$$

For y–propagation,

$$e = +jZh.$$

We will not consider the anti-resonant mode of propagation in this example. However, a similar approach may be used. Let us now summarize the relevant boundary conditions. The boundary conditions may be stated as follows:

a. The tangential microwave magnetic field is continuous across the surface.
b. The tangential microwave electric field is continuous across the surface.
c. The spin boundary condition for $\perp$ field configuration is then

$$A\frac{\partial m}{\partial y} - K_s m = 0; \quad K_s > 0.$$

This form of boundary condition implies that the easy axis of magnetization is $\perp$ to the film plane, which is consistent with the direction of $\vec{H}$. Clearly, m is circularly polarized. We will express both e and h in terms of h. The relationship between m and h is obtained from Chapter 8:

$$m_i = \chi(k_i)h_i = -Q(k)h_i,$$

where the subscript i indicates which value of propagation constant, k_i, to be used in the relationship above.

$$\chi(k_i) = \frac{4\pi M}{H_0 + \dfrac{2A}{M}k_i^2 - \dfrac{\omega}{\gamma}},$$

$$Q(k_i) = \left(1 - \frac{k_i^2}{\omega^2 \varepsilon \mu_0}\right),$$

$$H_0 = H - 4\pi M, \quad \text{(note } H \text{ is the external field and } H_0 \text{ the internal field)}$$

and

$$\frac{\omega}{\gamma} \rightarrow \frac{\omega}{\gamma}(1 - j\alpha).$$

α is the Gilbert damping parameter. There are two values of k for which the resonant mode can propagate in the following format (see Chapter 8):

$$h_i^+ e^{-jk_i y} + h_i^- e^{jk_i y} = h_i(y). \tag{11.30}$$

The superscript ($\pm$) indicates propagation in the ($\pm$) y-direction, the i-index indicates which k-value one is considering. $h_i^\pm$ are the wave amplitudes for a given k, and $i = 1, 2$. The boundary conditions are as follows:

$$\sum_{i=1}^{2}\left(h_i^+ + h_i^-\right) = h_0, \quad y = 0, \tag{11.31}$$

$$\sum_{i=1}^{2}\left(h_i^+ e^{-jk_i t} + h_i^- e^{+jk_i t}\right) = h_t, \quad y = t, \tag{11.32}$$

$$\sum_{i=1}^{2}Z_i\left(h_i^+ - h_i^-\right) = e_0, \quad y = 0, \tag{11.33}$$

where

$$Z_i = \frac{k_i}{\omega \varepsilon}, \quad \varepsilon = \frac{\sigma}{j\omega}, \quad \text{metal}; \quad \varepsilon = \varepsilon' - j\varepsilon'', \quad \text{insulator} \tag{11.33a}$$

$$\sum_{i=1}^{2} Z_i \left(h_i^+ e^{-jk_i t} - h_i^- e^{+jk_i t} \right) = e_t, \quad y = t, \tag{11.34}$$

$$\sum_{i=1}^{2} \left(P_i^{(0)} h_i^+ + R_i^{(0)} h_i^- \right) = 0, \quad y = 0, \tag{11.35}$$

$$\sum_{i=1}^{2} \left(P_i^{(t)} h_i^+ e^{-jk_i t} + R_i^{(t)} h_i^- e^{+jk_i t} \right) = 0, \quad y = t. \tag{11.36}$$

The first four equations are standard electromagnetic boundary conditions. The sum is from $i = 1$ to 2, since there are two values of k for each frequency. The surface fields are h_0, h_t, e_0, and e_t, and they are circularly polarized. The last two equations are the spin boundary conditions at the two surfaces. K_s may be different at the two surfaces $K_s^{(0)}$ and $K_s^{(t)}$. The parameters $P_i^{(0,t)}$ and $R_i^{(0,t)}$ are defined as

$$P_i^{(0,t)} = \left(K_s^{(0,t)} + jk_i A \right) \chi(k_i),$$

$$R_i^{(0,t)} = \left(K_s^{(0,t)} - jk_i A \right) \chi(k_i), \quad \text{and}$$

$$\chi(k_i) = \frac{4\pi M}{(H - 4\pi M) + \dfrac{2A}{M} k_i^2 - \dfrac{\omega}{\gamma}(1 - j\alpha)},$$

and H is the external DC magnetic field.

The superscript $(0, t)$ implies either "0" or "t" for the corresponding value of K_S at surface $y = 0$ or $y = t$. $4\pi M$ is the saturation magnetization and α the Gilbert damping parameter.

The surface impedance, Z_S, is defined at $y = 0$ as

$$Z_S = \frac{e_0}{h_0}.$$

There is a factor of (j) floating around in the definition of impedances for linearly and circularly polarized fields. The above definition of surface impedance is consistent with its real part being real and positive giving rise to microwave absorption. The above definition must also be consistent with our definitions of the Z_i's (see Equation 11.33). The two surface impedances are equal to each other for symmetric microwave field excitations ($h_0 = h_t = h_S$). Specifically, Z_S may be calculated as follows:

$$Z_S = \frac{1}{h_0} \sum_{i=1}^{2} Z_i(k_i) \left(h_i^+ - h_i^- \right). \tag{11.37}$$

The internal fields $h_i^\pm$ are solved from Equations 11.31, 11.32, 11.35, and 11.36 in terms of h_0, and then substituted into Equation 11.37. This is the conventional way of calculating surface impedance. In Appendix 11.A, we plot the surface impedance of a permalloy film, as a function of frequency. Its parameters are given in Chapter 9.

It is instructive to calculate Z_S in an "unconventional" way using the Poynting integral, as introduced in Chapter 1. According to Equation 1.27, the Poynting integral may be expressed as follows (slightly modified from Chapter 1):

$$Z_S = \frac{1}{2} j\omega t \left(\mu_V - j\varepsilon^* |Z_S|^2 \right), \tag{11.38}$$

where

$$\mu_V = \frac{\mu_0}{t|h_0|^2} \int_0^t dy \left[\left| \sum_{i=1}^{2} h_i(y) \right|^2 + \sum_{i=1}^{2} \left(\chi(k_i) |h_i(y)|^2 \right) \right] \tag{11.39}$$

and

$$\varepsilon_V = \frac{\varepsilon}{t|e_0|^2} \int_0^t dy \left| \sum_{i=1}^{2} e_i(y) \right|^2. \tag{11.40}$$

As in Chapter 1, $\varepsilon = \sigma/j\omega$ for metals. Equation 11.38 is slightly different from Equation 1.27 in the sense that in this case we have symmetrical excitation and a factor of 1/2 is introduced since there are two surfaces contributing to the surface impedance. In Equation 1.27, H_S is used rather than h_0. In Equation 11.39, the factor $h_i(y)$ is defined in Equation 11.30 and it is very similar to Equation 1.27. While in Equation 1.27 only one propagation constant is assumed, here there are two. In Equation 11.40, the factor inside the parenthesis is, by definition, $e_i(y) = Z_i \left(h_i^+ e^{-jk_i y} - h_i^- e^{+jk_i y} \right)$. The reader should be mindful that e is solved in terms of h_0, not e_0. As in Chapter 1, it is meaningful to introduce the quantity $|Z_V|$, which connects the internal electric field magnitude with the surface magnetic field.

Indeed, Equation 11.38 does yield exactly the same result as Equation 11.37. The reader should examine the MATLAB® code for both methods of calculations. Let's now define the quantities in Equation 11.38. The quantity μ_V is interpreted as the average permeability measured in a microwave scattering experiment. It is considerably smaller than the expression

$$\mu = + \frac{4\pi M}{H_0 - \dfrac{\omega}{\gamma}}$$

which does not include conductivity, exchange, etc. For example, at FMR, the above expression yields about 300 compared to ~ 50 predicted by Equation 11.39. Parameters for permalloy were assumed. Again, this comparison may be explored in the MATLAB® code in Appendix 11.A. In Appendix 11.A, μ_V (labeled as MuS) is plotted as a function of frequency. The quantity ε_V is interpreted as the average value of permittivity. In Appendix 11.A, we plot ε_V (labeled as epsS in Appendix 11.A) as a function of frequency. The average conductivity, σ_V, may be defined as follows:

$$\sigma_V = j\omega\varepsilon_V.$$

In Appendix 11.A, we plot σ_V (labeled as σ_S in Appendix 11.A) as a function of frequency. The implications of the above arguments are that the magnetic film may be characterized by ε_V and μ_V and be able to do a one-dimensional (one k-value) calculation, as outlined in Chapter 1 to determine the surface impedance.

$\vec{H}$ // TO THE FILM PLANE

For the case of $\vec{H}$ in the film plane and along the z-direction, the magnetic field excitation is linearly polarized along the x-direction, h_x. The electric field, e_z, is also linearly polarized and is related to h_x as

$$e_z = Zh_x; \quad Z = \frac{k}{\omega\varepsilon}.$$

The normal modes are elliptically polarized as a result of the microwave demagnetizing field being "nonuniform." For example, for this case (see Chapter 9),

$$h_y = -m_y, \quad \text{MKS}$$

and

$$h_x = -\frac{1}{Q(k)}m_x,$$

where

$$Q(k) = 1 - \frac{k^2}{\omega^2\varepsilon\mu_0}.$$

However, the film is excited by a linearly polarized wave in the x-direction or $\vec{h}_0 = h_0\vec{a}_x$. For simplicity, we write the excitation as h_0. Along the x-direction, there are three internal magnetic fields that are also polarized in the x-direction and we designate them as $h_i^{\pm}$. The subscript i is to indicate the specific wave number k_i, $i = 1$, 2, 3. For each $h_i(y)$, there are three waves propagating in the $+y$-direction and three others in the $-y$-direction.

ELECTROMAGNETIC SPIN BOUNDARY CONDITIONS

There are four boundary conditions for spin motion at the two surfaces, since there are two dynamic components at each surface. For simplicity, first consider the boundary condition

$$\frac{\partial m_x}{\partial y} = 0.$$

Since

$$m_x = -Q(k)h_x,$$

we may relate m_{ix} to h_i, the internal electromagnetic field along the x-direction, simply as

$$m_{ix} = -Q(k_i)h_i.$$

The internal field, $h_i(y)$, includes three components for wave propagation in the (+) y-direction and three for wave propagation in the (–)y-direction, see definition in Equation 11.30.

Finally, the spin boundary condition

$$A\frac{\partial m_y}{\partial y} - K_s m_y = 0$$

requires that we relate m_y to h_x or h_i. Let us do that.

For // field configuration of $\vec{H}$, we have from Chapter 8 (Equations 8.8a and 8.8b),

$$\theta = \frac{\pi}{2}, \quad \theta = \frac{\pi}{2},$$

resulting in

$$\left(k^2 - \omega^2\varepsilon\mu_0\mu_{xx}\right)h_x - \omega^2\varepsilon\mu_0 j\chi h_y = 0$$

and

$$\omega^2\varepsilon\mu_0 j\chi h_x - \omega^2\varepsilon\mu_0\mu_{yy}h_y = 0.$$

This leads to the relation that

$$h_y = j\frac{\chi(k_i)}{\mu_{yy}}h_x,$$

where

$$\mu_{xx} = \mu_{yy} = \mu = 1 + \frac{4\pi MH}{\Omega^2},$$

$$\underline{H} = H_0 + \frac{2A}{M}k^2 = H + \frac{2A}{M}k^2,$$

$$\Omega^2 = \underline{H}^2 - \frac{\omega^2}{\gamma^2},$$

and

$$\chi = \frac{\omega}{\gamma}\left(\frac{4\pi M}{\Omega^2}\right).$$

In this chapter, we have used the symbol H to designate the external magnetic field. Let's designate the latter as $\underline{H}$ and H as the external magnetic field in this chapter.

Thus,

$$m_y = -\frac{j\chi}{\mu}h_x.$$

The internal fields may be written as

$$m_{iy} = -\frac{j\chi(i)}{\mu(i)}h_i,$$

where

$$\frac{\chi(i)}{\mu(i)} = \frac{\dfrac{\omega}{\gamma}\dfrac{4\pi M}{\gamma\Omega^2(i)}}{1+\dfrac{4\pi M\underline{H}(i)}{\Omega^2(i)}} = \frac{\omega}{\gamma}\frac{4\pi M}{\left(\Omega^2(i)+4\pi M\underline{H}(i)\right)}.$$

Summarizing the spin and Maxwell boundary conditions,

$$h_{0x} = \sum_{i=1}^{3}\left(h_i^+ + h_i^-\right), \quad y=0 \tag{11.41}$$

$$h_{tx} = \sum_{i=1}^{3}\left(h_i^+ e^{-jk_it} + h_i^- e^{+jk_it}\right), \quad y=t \tag{11.42}$$

$$e_{0z} = \sum_{i=1}^{3}Z_1\left(h_i^+ - h_i^-\right), \quad y=0; \quad Z_i = \frac{k_i}{\omega\varepsilon} \tag{11.43}$$

$$e_{tz} = \sum_{i=1}^{3}Z_i\left(h_i^+ e^{-jk_it} - h_i^- e^{+jk_it}\right), \quad y=t \tag{11.44}$$

$$\sum_{i+1}^{3}Q(k_i)jk_i\left(h_i^+ - h_i^-\right)=0, \quad y=0 \tag{11.45}$$

$$\sum_{i+1}^{3}Q(k_i)jk_i\left(h_i^+ e^{-jk_it} - h_i^- e^{+jk_it}\right)=0, \quad y=t \tag{11.46}$$

$$-\sum_{i=1}^{3}\frac{j\chi(k_i)}{\mu(k_i)}\Big[\big(K_s^{(0)}+jk_iA\big)h_i^+ +\big(K_s^{(0)}-jk_iA\big)h_i^-\Big]=0, \quad y=0 \qquad (11.47)$$

$$-\sum_{i=1}^{3}\frac{j\chi(k_i)}{\mu(k_i)}\Big[\big(K_s^{(t)}+jk_iA\big)h_i^+e^{-jk_it} +\big(K_s^{(t)}-jk_iA\big)h_i^-e^{-jk_it}\Big]=0, \quad y=t. \qquad (11.48)$$

Note that

$$-\frac{j\chi(k_i)}{\mu(k_i)}\rightarrow \frac{1}{H^2-\dfrac{\omega^2}{\gamma^2}+4\pi M\underline{H}(i)}.$$

The above convergence is appropriate, since the right-hand side of Equation 11.44 is equal to 0. Dividing 0 by constant factors does not alter the equation. In Appendix 11.A, a MATLAB® code is provided to calculate FMR absorption curves for H applied $\perp$ and $//$ to the film plane.

Equations 11.37–11.40 are still applicable for the case of H in the film plane except that (1) the sum over the index I is from 1 to 3 since there are three normal modes of propagation in this case and (2) Z_S is now defined in terms of linearly polarized fields at the surface:

$$Z_S=\frac{e_{0z}}{h_{0x}}.$$

In solving e_{0z} in terms of h_{0x} requires one to solve for $h_i^{\pm}$ ($i=1,2,3$) from Equations 11.41, 11.42, and 11.45–11.48 in terms of h_{0x}, where $h_{0x}=h_{tx}$ (symmetrical excitation).

In Appendix 11.A, we plot Z_S (Z_S in Appendix 11.A), μ_V (MuS), ε_V (epsS), and σ_V (sigmaS) as a function of frequency. The same interpretations apply here as in the previous case.

Most often in magnetic recording applications, a quantity of interest is the so-called magneto-impedance, Z. We now wish to relate Z to Z_S. In Figure 11.4, a sketch is shown of possible recording sensors.

By Ohm's law, the magneto-impedance is

$$Z=\frac{V}{I}=\frac{E_S\ell}{H_S 2W}=Z_S\left(\frac{\ell}{2W}\right)=j\omega t\left(\frac{\ell}{2W}\right)\big(\mu_V-j\varepsilon^*|Z_S|^2\big).$$

For a magnetic nanowire, we find that

$$Z=\frac{V}{I}\approx\frac{E_S\ell}{H_S 2\pi R}=Z_S\left(\frac{\ell}{2\pi R}\right)=j\omega\rho\left(\frac{\ell}{2\pi R}\right)\big(\mu_V-j\varepsilon^*|Z_S|^2\big),$$

where δ is the skin depth of a magnetic metallic nanowire.

$$\delta=\sqrt{\frac{2}{\omega\sigma_V\mu_V}}$$

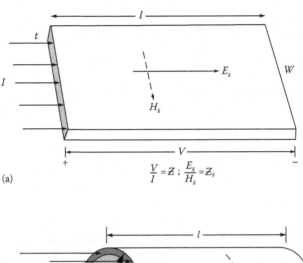

(a)

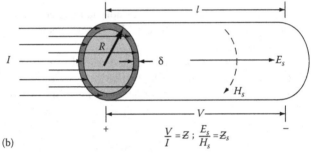

(b)

FIGURE 11.4   (a) Thin film configuration and (b) nanowire configuration.

PROBLEMS

11.1   (a) Re-derive Equations 11.26 and 11.27 for $E_s = K_s \alpha_1^2$. (b) Re-derive (a) for $E_s = K_s \alpha_3^2$.

11.2   (a) Re-derive Equations 11.29 and 11.30 for $E_s = K_s \alpha_1^2$. (b) Re-derive (a) for $E_s = K_s \alpha_3^2$.

11.3   Derive Equation 11.21 and solve it subject to the boundary conditions (Equation 11.22).

11.4   Within a length of $l = \sqrt{A/K}$, spins rotate by $180°$ in a domain wall, the order of the so-called exchange coherence length. Would you expect roughly the same in solving Equation 11.21? Explain.

11.5   Assume a simple cubic crystal structure in which the surface is a [100] plane. Calculate the dipole–dipole pair interaction energy at the surface assuming (a) nearest neighbor sum only and (b) next nearest neighbor sum only.

11.6   Repeat Problem 11.5 but assume BCC structure.

11.7   Repeat Problem 11.5 but assume FCC structure.

11.8   Repeat Problem 11.5 but assume hexagonal structure in which the base of the hexagon is the plane of the surface.

APPENDIX 11.A

PERPENDICULAR CASE (FIGURES 11.A.1–11.A.4)

```
% Permalloy, Perpendicular FMR.
% Calculating the surface impedance of a permalloy film
with magnetization
% Perpendicular to the plane. Theta_k = 0.
% CGS units.
clc;
clear all;
close all;
global d V
```

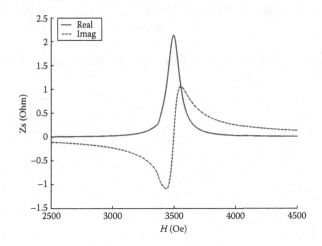

FIGURE 11.A.1 Surface impedance of 0.3 μm permalloy film. *H* applied normal to film plane.

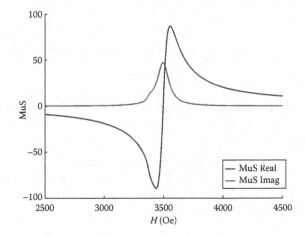

FIGURE 11.A.2 Average permeability of permalloy film. H is applied perpendicular to film plane.

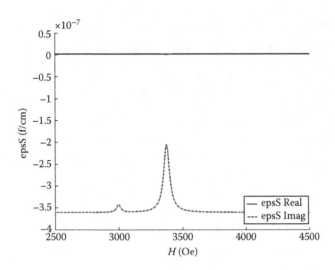

FIGURE 11.A.3   Average permittivity of permalloy film. H is applied perpendicular to film plane.

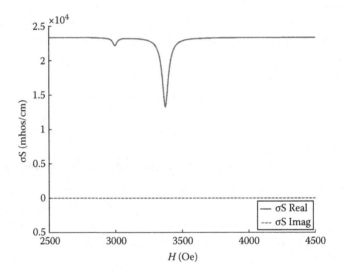

FIGURE 11.A.4   Average conductivity of permalloy film. H is applied perpendicular to film plane.

```
% Calculate the dispersion relation first.
global w mu0 eps Xi1s Xi2s k1s k2s Z1s Z2s h1r h1l h2r h2l
mu0 = 4*pi*10^(-9); % henry/cm
eps0 = 1/(36*pi)*10^(-11); % F/cm
Z0 = sqrt(mu0/eps0);
sigma = 0.7*10^5; % mho/cm
A = 1.14e-6; % erg/cm
```

```
g = 2.10;
M = 10000/(4*pi); % Gauss
Ho = 2500:1:4500; % Oe
% Ho = 2500:1:9000; % Oe
L = length(Ho);
gamma = 2*pi*1.4*10^6*g; % Hz/Oe
w = gamma*3500; % GHz
f = w/(2*pi);
dH = 2*4.33e-3*w/gamma; % Oe
omega = w*(1-j*4.33e-3);
w = omega;
eps = -j*sigma/omega; % F/cm
k0 _ sq = -j*omega*sigma*mu0;
% Resonant mode.
c1 = ones(1,L);
c2 = -(k0 _ sq + (w/gamma-Ho)*M/(2*A));
c3 = M/(2*A)*k0 _ sq*(w/gamma-Ho-4*pi*M);
k1 _ sq = -c2./(2*c1) + sqrt(c2.^2 - 4*c1.*c3)./(2*c1);
k2 _ sq = -c2./(2*c1) - sqrt(c2.^2 - 4*c1.*c3)./(2*c1);
k1 = sqrt(k1 _ sq);
k1 = sign(-imag(k1)).*k1; %   Pick   up   the   physical   mode
(non-growing).
k2 = sqrt(k2 _ sq);
k2 = sign(-imag(k2)).*k2;
% % Non-resonant mode.
% c1 = ones(1,L);
% c2 = -(k0 _ sq+(-w/gamma-Ho).*M/(2*A));
% c3 = M/(2*A)*k0 _ sq.*(-w/gamma-Ho-4*pi*M);
% k3 _ sq = -c2./(2*c1) +sqrt(c2.^2 - 4*c1.*c3)./(2*c1);
% k4 _ sq = -c2./(2*c1) - sqrt(c2.^2 - 4*c1.*c3)./(2*c1);
%
% k3 = sqrt(k3 _ sq);
% k3 = sign(-imag(k3)).*k3;
% k4 = sqrt(k4 _ sq);
% k4 = sign(-imag(k4)).*k4;
% Mu _ r = 1+4*pi*M./(Ho-w/gamma);
% ZV = j.*sqrt(mu0.*Mu _ r./eps);%volume char. imp.
figure(1);%k vs H
hold on;
plot(real(k1),Ho,'b','Linewidth',2);
plot(imag(k1),Ho,'b--','Linewidth',2);
plot(real(k2),Ho,'r','Linewidth',2);
plot(imag(k2),Ho,'r--','Linewidth',2);
% plot(Ho,real(k3),'g','Linewidth',2);
% plot(Ho,imag(k3),'g--','Linewidth',2);
% plot(Ho,abs(real(k4)),'cyan','Linewidth',2);
% plot(Ho,imag(k4),'cyan--','Linewidth',2);
```

```
ylabel('Ho(Oe)','Fontsize',24);
% xlabel('Im(k) & Re(k) (cm^-^1)','Fontsize', 24);
xlabel('k(rad/cm)','Fontsize', 24);
h=gca;
set(h,'Fontsize',22);
title('Permalloy Dispersion Diagram','FontSize',24);
hold off;
% Calculate the surface impedance, only consider resonant
modes
d=0.3E-4; % cm
Xi1 = 4*pi*M./(Ho+2*A/M.*k1 _ sq-w/gamma); %
Xi2 = 4*pi*M./(Ho+2*A/M.*k2 _ sq-w/gamma); %
Ks _ 0 = 0; Ks _ d = 0;
n _ minus _ 2pi = 0;
n _ plus _ 2pi = 0;
indicator = 0;
for i = 1:L
S = [1 1 1 1; exp(-j*k1(i)*d) exp(j*k1(i)*d) exp(-j*k2(i)*d)
exp(j*k2(i)*d);...
Xi1(i)*(Ks _ 0+j*k1(i)*A) Xi1(i)*(Ks _ 0-j*k1(i)*A)
Xi2(i)*(Ks _ 0+j*k2(i)*A) Xi2(i)*(Ks _ 0-j*k2(i)*A);...
Xi1(i)*(Ks _ d+j*k1(i)*A)*exp(-j*k1(i)*d)
Xi1(i)*(Ks _ d-j*k1(i)*A)*exp(j*k1(i)*d)
Xi2(i)*(Ks _ d+j*k2(i)*A)*exp(-j*k2(i)*d)
Xi2(i)*(Ks _ d-j*k2(i)*A)*exp(j*k2(i)*d)];
b = [1 1 0 0]'; % boundary condition
h _ vec(:,i) = inv(S)*b;  %  h _ vec(:,i) = [h1+(i)   h1-(i)   h2+(i)
h2-(i)]'
Z1(i) =k1(i)/(omega*eps);
Z2(i) =k2(i)/(omega*eps);
% Zs _ 0(i) =Z1(i)*(h _ vec(1,i)-h _ vec(2,i)) +Z2(i)*(h _ vec(3,i)-
h _ vec(4,i));
% Zs _ d(i) =
Z1(i)*(h _ vec(1,i)*exp(-j*k1(i)*d)-h _ vec(2,i)*exp(j*k1(i)*d))...
+Z2(i)*(h _ vec(3,i)*exp(-j*k2(i)*d)-h _ vec(4,i)*exp(j*k2(i)*d));
SP = [j*Z1(i)*exp(-j*k1(i)*d) -j*Z1(i)*exp(j*k1(i)*d)
j*Z2(i)*exp(-j*k2(i)*d) -j*Z2(i)*exp(j*k2(i)*d); ...
exp(-j*k1(i)*d) exp(j*k1(i)*d) exp(-j*k2(i)*d) exp(j*k2(i)*d);
...
Xi1(i)*(Ks _ 0+j*k1(i)*A) Xi1(i)*(Ks _ 0-j*k1(i)*A)
Xi2(i)*(Ks _ 0+j*k2(i)*A) Xi2(i)*(Ks _ 0-j*k2(i)*A);...
Xi1(i)*(Ks _ d+j*k1(i)*A)*exp(-j*k1(i)*d)
Xi1(i)*(Ks _ d-j*k1(i)*A)*exp(j*k1(i)*d) ...
Xi2(i)*(Ks _ d+j*k2(i)*A)*exp(-j*k2(i)*d)
Xi2(i)*(Ks _ d-j*k2(i)*A)*exp(j*k2(i)*d)];
A _ mat =A _ matrix(SP, Z1(i), Z2(i));
a11(i) =A _ mat(1, 1);
```

```
D = A _ mat(1, 1)*A _ mat(2, 2)-A _ mat(1, 2)*A _ mat(2, 1);
Z _ square = A _ mat(1, 2)/A _ mat(2, 1);
Z(i) = sign(imag(Z _ square))*sqrt(Z _ square);
% calculation of alpha and beta: keff = beta+jalpha
ZS(i) = -j*(A _ mat(1,1)-1)/A _ mat(2,1);
emcoupl(i) = (j*ZS(i))/Z(i);
C(i) = A _ mat(1,1) + (A _ mat(1,2))/Z(i);%1/S21,matched imp.
% C = exp(jwt) = 1/S21(S11 = 0)
thet(i) = angle(C(i));
beta(i) = thet(i)/d;
alpha(i) = log(abs(C(i)))/d;
% if beta(i)>0
% beta(i) = -2*pi/d+beta(i);
% end
% if real(C(i))<0
% R(i) = imag(C(i))/real(C(i));
% beta(i) = -atan(R(i))/d;
% end
% if beta(i)<-pi/(2*d)
% beta(i) = -beta(i)-pi/d;
% elseif beta(i)> pi/(2*d)
% beta(i) = -beta(i) +pi/d;
% end
keff(i) = beta(i)-j*alpha(i);
% ZS(i) = Z(i)*tan(keff(i)*(d/2));
% emcoupl(i) = j*tan(keff(i)*(d/2));
% ZS(i) = A _ mat(1,1);
% ZS(i) = cos(keff(i)*d);
% ZS(i) = (j*Z(i))*sin(keff(i)*d);
% ZS(i) = (j/Z(i))*sin(keff(i)*d);
Mueff(i) = ((keff(i))*Z(i))/(w.*mu0);
Mueff(i) = -j*Mueff(i);
epseff(i) = (keff(i))/(Z(i)*w);
epseff(i) = j*epseff(i);
sigma _ eff = epseff.*j.*w;
Zeff(i) = j*sqrt((Mueff(i)*mu0)./epseff(i));
Zs _ 0(i) = Z1(i)*(h _ vec(1,i)-h _ vec(2,i)) + Z2(i)*(h _ vec(3,i)-
h _ vec(4,i));
Zs _ d(i) =
Z1(i)*(h _ vec(1,i)*exp(-j*k1(i)*d)-h _ vec(2,i)*exp(j*k1(i)*d))
...
+ Z 2 ( i ) * ( h _ v e c ( 3 , i ) * e x p ( - j * k 2 ( i ) * d )
-h _ vec(4,i)*exp(j*k2(i)*d));
% Check the result by integrating the poynting vector.
Xi1s = Xi1(i); Xi2s = Xi2(i); k1s = k1(i); k2s = k2(i); Z1s = Z1(i); Z2s
= Z2(i);
```

```
h1r = h _ vec(1,i);    h1l = h _ vec(2,i);    h2r = h _ vec(3,i);h2l = h _
vec(4,i);
Integ(i) = quadgk(@myfun,0,d);  % 2Zs(i)
Integ1(i) = quadgk(@myfun1,0,d)/d;  % Effective permeability
by volume
integration
% Integ2(i) = quadgk(@myfun2,0,d)/(Integ(i)/2)^2/d; % Effective
permittivity by volume integration
%Integ3(i) = quadgk(@myfun3,0,d);
Integ2(i) = quadgk(@myfun2,0,d)/d;
end
figure(2);%char. impedance,Z, vs H (circ. pol.)
hold on;
plot(Ho,real(Z),'r','Linewidth',2);
plot(Ho,imag(Z),'r--','Linewidth',2);
% plot(Ho,real(ZV),'b','Linewidth',2);
% plot(Ho,imag(ZV),'b--','Linewidth',2);
% plot(Ho, -real(Integ/2),'b','LineWidth',1);
% plot(Ho, -imag(Integ/2),'b--','LineWidth',1);
xlabel('H(Oe)','Fontsize',24);
ylabel('Z(Ohm)','Fontsize', 24);
% ylabel('Im(Z) & Re(Z) (ohms)','Fontsize', 24);
% h = gca;
% set(h,'Fontsize',22);
% hold off;
% axis([2500 4500 -100 100]);
h = gca;
set(h,'Fontsize',22);
title('characteristic impedance of 0.3 \mum Permalloy
Film','FontSize',24);
legend('Z Real','Z Imag',2);
% legend('y = 0 Real','y = 0 Imag','y = d Real','y = d Imag','Integ
Real', 'Integ
Imag',4);
hold off;
figure(3);%effective k vs H (calc. from 2-2 matrix)
hold on;
plot(real(keff),Ho,'r','Linewidth',2);
plot(imag(keff),Ho,'r--','Linewidth',2);
ylabel('H(Oe)','Fontsize',24);
xlabel('keff(rad/cm)','Fontsize', 24);
h = gca;
set(h,'Fontsize',22);
title('Propogation constant of 0.3 \mum Permalloy Film',
'FontSize',24);
legend('real(keff)','imag(keff)',2);
```

```
hold off;
figure(4);% surface impedance, Zs, vs H (from 4x4 matrix)
hold on;
plot(Ho, real(Zs _ 0),'r','Linewidth',2);
plot(Ho, imag(Zs _ 0),'r--','Linewidth',2);
% plot(Ho,phi,'r','Linewidth',2);
xlabel('H(Oe)','Fontsize',24);
ylabel('Zs(Ohm)','Fontsize', 24);
h = gca;
set(h,'Fontsize',22);
title('Zs _ 0 of 0.3 \mum Permalloy Film', 'FontSize',24);
legend('real','Imag',2);
hold off;
figure(5);%surface impedance,ZS,vs H (from 2x2 matrix)
hold on;
plot(Ho,real(ZS),'b','Linewidth',2);
plot(Ho,imag(ZS),'b--','Linewidth',2);
xlabel('H(Oe)','Fontsize',24);
ylabel('ZS(ohm)','Fontsize', 24);
h = gca;
set(h,'Fontsize',22);
title('ZS of 0.3 \mum Permalloy Film', 'FontSize',24);
legend('real','Imag',2);
hold off;
figure(6);%mueff vs H (from poynting vector;mu vs H(from
tensor mu)
hold on;
plot(Ho, real(Mueff),'r','LineWidth',1);
plot(Ho, imag(Mueff),'r--', 'LineWidth',1);
plot(Ho, real(Integ1),'b','LineWidth',1);
plot(Ho, imag(Integ1),'b--', 'LineWidth',1);
xlabel('H(Oe)', 'FontSize', 24);
ylabel('Mueff', 'FontSize',24);
%legend('Mueff Real','Mueff Imag','MuS Real','MuS Imag',1);
h = gca;
set(h,'Fontsize',22);
title('Permeability Comparison','FontSize',24);
hold off;
figure(7);%epseff vs H
hold on;
plot(Ho, real(epseff),'b','LineWidth',2);
plot(Ho, imag(epseff),'b--', 'LineWidth',2);
xlabel('H(Oe)', 'FontSize', 24);
ylabel('\epsilon _ e _ f _ f(f/cm)', 'FontSize',32);
legend('\epsilon _ e _ f _ f Real','\epsilon _ e _ f _ f Imag',4);
h = gca;
```

```
set(h,'Fontsize',30);
title('Effective Average Permittivity','FontSize',24);
hold off;
figure(8);%sigmaeff vs H (see fig. 7)
hold on;
plot(Ho, real(sigma _ eff),'b','LineWidth',2);
plot(Ho, imag(sigma _ eff),'r--','LineWidth',2);
plot(Ho, sigma,'r','LineWidth',2);
h = gca;
set(h,'Fontsize',30);
hold off;
xlabel('H(Oe)', 'FontSize', 30);
ylabel('sigma _ e _ f _ f(mhos/cm)', 'FontSize',32);
title('Effective conductivity','FontSize',30);
legend('sigma _ e _ f _ f Real','sigma _ e _ f _ f Imag',4);
figure(9);%Zeff vs H (calculated from sqrt(mu0*mueff/epseff)
hold on;
plot(Ho, real(Zeff),'LineWidth',2);
plot(Ho, imag(Zeff),'r','LineWidth',2);
h = gca;
set(h,'Fontsize',30);
hold off;
xlabel('H(Oe)', 'FontSize', 30);
ylabel('Z _ e _ f _ f(Ohm)', 'FontSize',32);
title('Z _ e _ f _ f','FontSize',30);
legend('Z _ e _ f _ f Real','Z _ e _ f _ f Imag',4);
ZSM2 = (abs(Zs _ 0)).^2;
MuS = mu0*Integ1;
%epsS = -2.*Zs _ 0./(j.*w.*ZSM2.*d) +MuS./ZSM2;
epsS = Integ2./ZSM2;
%epsS = conj(epsS);
sigmaS = epsS.*j.*w;
Zsv = 0.5.*(j.*w.*d.*(MuS-(conj(epsS).*ZSM2)));% also, Zs
figure(10);%epsS vs H (from poynting vector)
hold on;
plot(Ho, real(epsS),'b','LineWidth',2);
plot(Ho, imag(epsS),'b--', 'LineWidth',2);
xlabel('H(Oe)', 'FontSize', 24);
ylabel('epsS(f/cm)', 'FontSize',32);
legend('epsS Real','epsS Imag',4);
h = gca;
set(h,'Fontsize',30);
title('Surface Average Permittivity','FontSize',24);
hold off;
figure(11);%sigmaS vs H (see fig. 7)
hold on;
```

```
plot(Ho, real(sigmaS),'b','LineWidth',2);
plot(Ho, imag(sigmaS),'r--','LineWidth',2);
plot(Ho, sigmaS,'r','LineWidth',2);
h=gca;
set(h,'Fontsize',30);
hold off;
xlabel('H(Oe)', 'FontSize', 30);
ylabel('\sigmaS(mhos/cm)', 'FontSize',32);
title('Surface conductivity','FontSize',30);
legend('\sigmaS Real','\sigmaS Imag',4);
figure(12);%Zsv vs H (calculated from poynting integral)
hold on;
plot(Ho, real(Zsv),'LineWidth',2);
plot(Ho, imag(Zsv),'r','LineWidth',2);
h=gca;
set(h,'Fontsize',30);
hold off;
xlabel('H(Oe)', 'FontSize', 30);
ylabel('Zsv(Ohm)', 'FontSize',32);
title('Zsv','FontSize',30);
legend('Zsv Real','Zsv Imag',4);
figure(13);%emcoupl(i)
hold on;
plot(Ho, real(emcoupl),'b','LineWidth',2);
plot(Ho, imag(emcoupl),'b--', 'LineWidth',2);
xlabel('H(Oe)', 'FontSize', 24);
ylabel('emcoupl)', 'FontSize',32);
legend('emcoupl real','emcoupl Imag',4);
h=gca;
set(h,'Fontsize',30);
title('em coupling','FontSize',24);
hold off;
% Plot out the h(y) at Ho=2500 Oe.
N=501;
y=0:d/200:d;
hy=h_vec(1,N)*exp(-j*k1(N)*y) +h_vec(2,N)*exp(j*k1(N)*y) +
h_vec(3,N).*exp(-j*k2(N)*y) +h_vec(4,N).*exp(j*k2(N)*y);
% figure(3);
% hold on;
% plot(y,real(hy),'LineWidth',2);
% % plot(y,imag(hy),'g','LineWidth',2);
% % plot(y,abs(hy),'r','LineWidth',2);
% h=gca;
% set(h,'Fontsize',30);
% % axis([0d 0.8 1]);
% xlabel('y(cm)','Fontsize',30);
```

```
% ylabel('Re(h(y))','Fontsize',30);
% title(['Ho = ', num2str(Ho(N)), 'Oe'],'Fontsize',30);
% hold off;
%X. Save raw data file for figure
1%%%%%%%%%%%%%%%%%%%%%%%%%%%%%%%%%%%%%%%%%%%%%%%%
% Plot out the h(y) at Ho = 2500 Oe.
N = 501;
y = 0:d/200:d;
hy = h _ vec(1,N)*exp(-j*k1(N)*y) +h _ vec(2,N)*exp(j*k1(N)*y) +
h _ vec(3,N).*exp(-j*k2(N)*y) +h _ vec(4,N).*exp(j*k2(N)*y);
% figure(3);
% hold on;
% plot(y,real(hy),'LineWidth',2);
% % plot(y,imag(hy),'g','LineWidth',2);
% % plot(y,abs(hy),'r','LineWidth',2);
% h = gca;
% set(h,'Fontsize',30);
% % axis([0 d 0.8 1]);
% xlabel('y(cm)','Fontsize',30);
% ylabel('Re(h(y))','Fontsize',30);
% title(['Ho = ', num2str(Ho(N)), 'Oe'],'Fontsize',30);
% hold off;
%
```

IN-PLANE CASE (FIGURES 11.A.5–11.A.8)

```
% Permalloy in plane FMR
clc;
clear all;
```

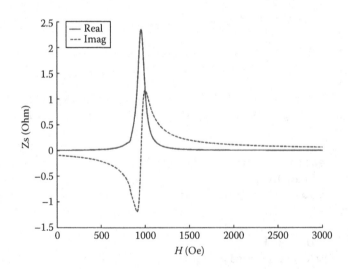

FIGURE 11.A.5 Surface impedance of a permalloy film. *H* is applied in the film plane.

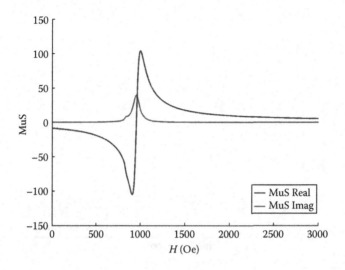

FIGURE 11.A.6 Average permeability of a permalloy film. H is applied in the film plane.

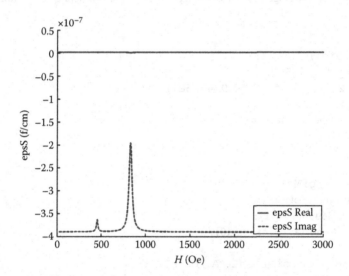

FIGURE 11.A.7 Average permittivity of a permalloy film. H is applied in the film plane.

```
close all;
global w mu0 Qi1s Qi2s Qi3s k1s k2s k3s h1r h1l h2r h2l h3r
h3l Z1s Z2s Z3s eps
g = 2.1;
fourpiMs = 10000;
% Hext = 1103.3;
dH = 30;
sigma = 0.7e5;
A = 1.14e-6;
mu0 = 4*pi*10^-9;
```

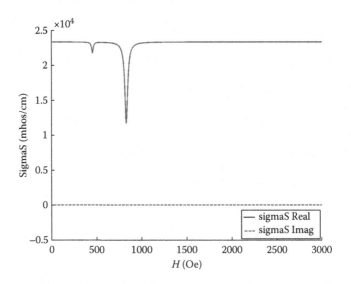

FIGURE 11.A.8   Average conductivity of a permalloy film. H is applied in the film plane.

```
f = 9.53e9;
e0 = 1/(36*pi)*10^-11;
gamma = g.*0.8805e7;
omega = 2.*pi*f-j.*(dH.*0.924525e7);
w = omega;
topn = j.*fourpiMs.*w/gamma;
H0 = 10:1:3000;
L = length(H0);
H1 = H0;
H2 = H0+fourpiMs;
e = -j.*sigma./omega;
M = fourpiMs./(4.*pi);
k0 = sqrt(omega.^2.*e.*mu0);
a1 = 1;
a2 = (2.*H0+fourpiMs)./(2.*A./M)-k0.^2;
a3 = (H0.*(H0+fourpiMs)-(omega./gamma).^2)./(4.*A.^2./M.^2)-2.*
(H0+fourpiMs)./(2.*A./M).*k0.^2;
a4 = ((omega./gamma).^2-(H0+fourpiMs).^2)./(4.*A.^2./M.^2).*k0.^2;
a = a2;
b = a3;
c = a4;
[k1 _ sq,k2 _ sq,k3 _ sq] = solve3order(a,b,c);
k0 _ sq = -j.*omega.*sigma.*mu0;%k0squared
k1 = -sign(imag(k1 _ sq)).*sqrt(k1 _ sq);
k2 = -sign(imag(k2 _ sq)).*sqrt(k2 _ sq);
k3 = -sign(imag(k3 _ sq)).*sqrt(k3 _ sq);
figure(1);%k vs H
```

```
hold on;
plot(real(k1),H0,'b','Linewidth',2);
plot(imag(k1),H0,'b--','Linewidth',2);
plot(real(k2),H0,'r','Linewidth',2);
plot(imag(k2),H0,'r--','Linewidth',2);
plot(real(k3),H0,'g','Linewidth',2);
plot(imag(k3),H0,'g--','Linewidth',2);
% plot(H0,real(k3),'g','Linewidth',2);
% plot(H0,imag(k3),'g--','Linewidth',2);
% plot(H0,abs(real(k4)),'cyan','Linewidth',2);
% plot(H0,imag(k4),'cyan--','Linewidth',2);
ylabel('H0(Oe)','Fontsize',24);
% xlabel('Im(k) & Re(k)  (cm^-^1)','Fontsize', 24);
xlabel('k(rad/cm)','Fontsize', 24);
% legend('Re(k)>0','Im(k)<0',2);
h=gca;
set(h,'Fontsize',22);
title('Permalloy Dispersion Diagram','FontSize',24);
hold off;
% Calculating the surface impedance of a permalloy film
with
magnetization
% Parallel to the plane. Theta_ k=pi/2.
% CGS units.
% clc;
% clear all;
% close all;
% global d V
% global w mu0 eps Xi1s Xi2s Xi3s k1s k2s k3s Z1s Z2s Z3s
h1r h1l h2r h2l h3r
h3l
eps0 = 1/(36*pi)*10^-11; % F/cm
Z0 = sqrt(mu0/eps0);
eps = -j*sigma./w; % F/cm
Mu_ r = 1+fourpiMs.*H2./(H1.*H2-(w./gamma)^2);
% Calculate the surface impedance, only consider resonant
modes
d = 0.3E-4; % cm
Di1 = (H1+2.*(A./M).*k1_ sq).*(H2+2.*(A./M).*k1_ sq)-(w./
gamma).^2;
Di2 = (H1+2.*(A./M).*k2_ sq).*(H2+2.*(A./M).*k2_ sq)-(w./
gamma).^2;
Di3 = (H1+2.*(A./M).*k3_ sq).*(H2+2.*(A./M).*k3_ sq)-(w./
gamma).^2;
Xi1 = topn./Di1;
Xi2 = topn./Di2;
Xi3 = topn./Di3;
```

```
Qi1 = (1-k1 _ sq./k0 _ sq);
Qi2 = (1-k2 _ sq./k0 _ sq);
Qi3 = (1-k3 _ sq./k0 _ sq);
Ks _ 0 = 0; Ks _ d = 0;
n _ minus _ 2pi = 0;
n _ plus _ 2pi = 0;
indicator = 1;
for i = 1:L
S = [1 1 1 1 1 1;…
exp(-j*k1(i)*d) exp(j*k1(i)*d) …
exp(-j*k2(i)*d) exp(j*k2(i)*d) …
exp(-j*k3(i)*d) exp(j*k3(i)*d); …
Xi1(i)*(Ks _ 0+j*k1(i)*A) Xi1(i)*(Ks _ 0-j*k1(i)*A) …
Xi2(i)*(Ks _ 0+j*k2(i)*A) Xi2(i)*(Ks _ 0-j*k2(i)*A) …
Xi3(i)*(Ks _ 0+j*k3(i)*A) Xi3(i)*(Ks _ 0-j*k3(i)*A); …
Xi1(i)*(Ks _ d+j*k1(i)*A)*exp(-j*k1(i)*d)
Xi1(i)*(Ks _ d-j*k1(i)*A)*exp(j*k1(i)*d) …
Xi2(i)*(Ks _ d+j*k2(i)*A)*exp(-j*k2(i)*d)
Xi2(i)*(Ks _ d-j*k2(i)*A)*exp(j*k2(i)*d) …
Xi3(i)*(Ks _ d+j*k3(i)*A)*exp(-j*k3(i)*d)
Xi3(i)*(Ks _ d-j*k3(i)*A)*exp(j*k3(i)*d); …
j.*k1(i)*Qi1(i) -j.*k1(i)*Qi1(i) …
j.*k2(i)*Qi2(i) -j.*k2(i)*Qi2(i) …
j.*k3(i)*Qi3(i) -j.*k3(i)*Qi3(i); ….
j.*k1(i)*Qi1(i)*(exp(-j*k1(i)*d))
-j.*k1(i)*Qi1(i)*(exp(j*k1(i)*d)) …
j.*k2(i)*Qi2(i)*(exp(-j*k2(i)*d))
-j.*k2(i)*Qi2(i)*(exp(j*k2(i)*d)) …
j.*k3(i)*Qi3(i)*(exp(-j*k3(i)*d))
-j.*k3(i)*Qi3(i)*(exp(j*k3(i)*d))];
%N = [1e-6 0 0 0 0 0; 0 1e-6 0 0 0 0; 0 0 1e-6 0 0 0; 0 0 0
1e-6 0 0; 0 0 0 0
1e-6 0; 0 0 0 0 0 1e-6];
b = [1 1 0 0 0 0]'; % boundary condition
%b = N*b;
%S = N*S;
h _ vec(:,i) = inv(S)*b;
Z1(i) = k1(i)/(omega*eps);
Z2(i) = k2(i)/(omega*eps);
Z3(i) = k3(i)/(omega*eps);
SP = [Z1(i)*exp(-j*k1(i)*d) -Z1(i)*exp(j*k1(i)*d) …
Z2(i)*exp(-j*k2(i)*d) -Z2(i)*exp(j*k2(i)*d) …
Z3(i)*exp(-j*k3(i)*d) -Z3(i)*exp(j*k3(i)*d); ….
exp(-j*k1(i)*d) exp(j*k1(i)*d) …
exp(-j*k2(i)*d) exp(j*k2(i)*d) …
exp(-j*k3(i)*d) exp(j*k3(i)*d); …
Xi1(i)*(Ks _ 0+j*k1(i)*A) Xi1(i)*(Ks _ 0-j*k1(i)*A) …
```

```
Xi2(i)*(Ks _ 0+j*k2(i)*A)  Xi2(i)*(Ks _ 0-j*k2(i)*A) ...
Xi3(i)*(Ks _ 0+j*k3(i)*A)  Xi3(i)*(Ks _ 0-j*k3(i)*A); ...
Xi1(i)*(Ks _ d+j*k1(i)*A)*exp(-j*k1(i)*d)
Xi1(i)*(Ks _ d-j*k1(i)*A)*exp(j*k1(i)*d) ...
Xi2(i)*(Ks _ d+j*k2(i)*A)*exp(-j*k2(i)*d)
Xi2(i)*(Ks _ d-j*k2(i)*A)*exp(j*k2(i)*d) ...
Xi3(i)*(Ks _ d+j*k3(i)*A)*exp(-j*k3(i)*d)
Xi3(i)*(Ks _ d-j*k3(i)*A)*exp(j*k3(i)*d); ...
j.*k1(i)*Qi1(i)  -j.*k1(i)*Qi1(i) ...
j.*k2(i)*Qi2(i)  -j.*k2(i)*Qi2(i) ...
j.*k3(i)*Qi3(i)  -j.*k3(i)*Qi3(i); ....
j.*k1(i)*Qi1(i)*(exp(-j*k1(i)*d))
-j.*k1(i)*Qi1(i)*(exp(j*k1(i)*d)) ...
j.*k2(i)*Qi2(i)*(exp(-j*k2(i)*d))
-j.*k2(i)*Qi2(i)*(exp(j*k2(i)*d)) ...
j.*k3(i)*Qi3(i)*(exp(-j*k3(i)*d))
-j.*k3(i)*Qi3(i)*(exp(j*k3(i)*d))];
% SP = N*SP;
B _ mat = B _ matrix(SP, Z1(i), Z2(i), Z3(i));
a11(i) = B _ mat(1, 1);
D = B _ mat(1, 1)*B _ mat(2, 2)-B _ mat(1, 2)*B _ mat(2, 1);
Z _ square = B _ mat(1, 2)/B _ mat(2, 1);
Z(i) = sqrt(Z _ square);
% calculation of alpha and beta: keff = beta+jalpha
ZS(i) = (B _ mat(1,1)-1)/B _ mat(2,1);
emcoupl(i) = (ZS(i))/Z(i);
C(i) = B _ mat(1,1)+(B _ mat(1,2))/Z(i);%1/S21,matched imp.
%C = exp(jwt) = 1/S21(S11 = 0)
thet(i) = angle(C(i));
beta(i) = thet(i)/d;
alpha(i) = -log(abs(C(i)))/d;
% if beta(i)>0
% beta(i) = -2*pi/d+beta(i);
% end
if real(C(i))<0
R(i) = imag(C(i))/real(C(i));
beta(i) = -atan(R(i))/d;
end
keff(i) = beta(i)+j*alpha(i);
emcoupl(i) = j*tan(keff(i)*(d/2));
% emcoupl(i) = ZS(i)/Z(i);
% ZS(i) = j.*Z(i)*tan(keff(i)*(d/2));
% ZS(i) = B _ mat(1,1);
Mueff(i) = (keff(i)*Z(i))/(w.*mu0);
epseff(i) = (keff(i))/(Z(i)*w);
sigma _ eff = epseff.*j.*w;
Zeff(i) = sqrt((Mueff(i)*mu0)./epseff(i));
```

```
Zs _ 0(i) = Z1(i)*(h _ vec(1,i)-h _ vec(2,i))
+Z2(i)*(h _ vec(3,i)-h _ vec
(4,i))+ Z3(i)*(h _ vec(5,i)-h _ vec(6,i));
Zs _ d(i) = Z1(i)*(h _ vec(1,i)*exp(-j*k1(i)*d)-h _ vec(2,i)*
exp(j*k1(i)*d))          +Z2(i)*(h _ vec(3,i)*exp(-j*k2(i)*d)-h _
vec(4,i)*exp
(j*k2(i)*d))+…
Z3(i)*(h _ vec(5,i)*exp(-j*k3(i)*d)-h _ vec(6,i)*exp(j*k3(i)*d));
% Check the result by integrating the poynting vector.
Qi1s = Qi1(i); Qi2s = Qi2(i); Qi3s = Qi3(i); k1s = k1(i); k2s = k2(i);
k3s = k3(i); Z1s = Z1(i); Z2s = Z2(i);Z3s = Z3(i);
h1r = h _ vec(1,i);       h1l = h _ vec(2,i);       h2r = h _ vec(3,i);
h2l = h _ vec(4,i);
h3r = h _ vec(5,i);h3l = h _ vec(6,i);
Integ(i) = quadgk(@myfun4,0,d); % 2Zs(i)
Integ1(i) = quadgk(@myfun5,0,d)/d;  % Effective  permeability
by volume
integration
% Integ2(i) = quadgk(@myfun6,0,d)/(Integ(i)/2)^2/d; % Effective
permittivity by volume integration
% Integ3(i) = quadgk(@myfun3,0,d);
Integ2(i) = quadgk(@myfun6,0,d)/d;
end
figure(2);%char. impedance,Z, vs H (circ. pol.)
hold on;
plot(H0,real(Z),'r','Linewidth',2);
plot(H0,imag(Z),'r--','Linewidth',2);
% plot(H0, -real(Integ/2),'b','LineWidth',1);
% plot(H0, -imag(Integ/2),'b--','LineWidth',1);
xlabel('H(Oe)','Fontsize',24);
ylabel('Zeff(Ohm)','Fontsize', 24);
% axis([2500 4500 -100 100]);
h = gca;
set(h,'Fontsize',22);
title('characteristic  impedance  of  0.3\mum  Permalloy
Film','FontSize',24);
legend('Zeff Real','Zeff Imag',2);
% legend('y = 0 Real','y = 0 Imag','y = d Real','y = d Imag','Integ
Real', 'Integ
Imag',4);
hold off;
figure(3);%effective k vs H (calc. from 2x2 matrix)
hold on;
plot(real(keff),H0,'r','Linewidth',2);
plot(imag(keff),H0,'r--','Linewidth',2);
ylabel('H(Oe)','Fontsize',24);
xlabel('keff(rad/cm)','Fontsize', 24);
```

```
h = gca;
set(h,'Fontsize',22);
title('Propogation    constant    of    0.3    \mum    Permalloy
Film','FontSize',24);
legend('keff real','keff imag',2);
hold off;
%figure(3);%effective k vs H (calc. from 2×2 matrix)
%hold on;
%plot(H0,beta,'r','Linewidth',2);
%plot(H0,alpha,'r--','Linewidth',2);
%xlabel('H(Oe)','Fontsize',24);
%ylabel('k(rad/cm)','Fontsize', 24);
%h = gca;
%set(h,'Fontsize',22);hold off;
%title('Propogation    constant    of    0.3    \mum    Permalloy
Film','FontSize',24);
% legend('beta','alpha',2);
figure(4);% surface impedance, Zs, vs H (from 4×4 matrix)
hold on;
plot(H0, real(Zs _ 0),'r','Linewidth',2);
plot(H0, imag(Zs _ 0),'r--','Linewidth',2);
% plot(Ho,phi,'r','Linewidth',2);
xlabel('H(Oe)','Fontsize',24);
ylabel('Zs(Ohm)','Fontsize', 24);
h = gca;
set(h,'Fontsize',22);
title('Zs _ 0 of 0.3 \mum Permalloy Film','FontSize',24);
legend('real','Imag',2);
hold off;
figure (5);%surface impedance,ZS,vs H (from 2×2 matrix)
hold on;
plot(H0,real(ZS),'b','Linewidth',2);
plot(H0,imag(ZS),'b--','Linewidth',2);
xlabel('H(Oe)','Fontsize',24);
ylabel('ZS(Ohm)','Fontsize', 24);
h = gca;
set(h,'Fontsize',22);
title('ZS of 0.3 \mum Permalloy Film','FontSize',24);
legend('real','Imag',2);
hold off;
figure(6);%mueff vs H (from poynting vector;mu vs H(from
tensor mu)
hold on;
plot(H0, real(Mueff),'b','LineWidth',1);
plot(H0, -imag(Mueff),'b--', 'LineWidth',1);
%plot(H0, real(Integ1),'b','LineWidth',1);
%plot(H0, imag(Integ1),'b--', 'LineWidth',1);
```

```
xlabel('H(Oe)', 'FontSize', 24);
ylabel('mueff', 'FontSize',24);
%legend('Mueff Real','Mueff Imag','MuS Real','MuS Imag',1);
h = gca;
set(h,'Fontsize',22);
title('Permeability','FontSize',24);
hold off;
ZSM2 = (abs(Zs _ 0)).^2;
MuS = mu0*Integ1;
epsS = Integ2./ZSM2;
sigmaS = epsS.*(j.*w);
figure(7);%epseff vs H (from poynting vector)
hold on;
plot(H0, real(epseff),'b','LineWidth',2);
plot(H0, imag(epseff),'b--', 'LineWidth',2);
xlabel('H(Oe)', 'FontSize', 24);
ylabel('\epsilon _ e _ f _ f(f/cm)', 'FontSize',32);
legend('\epsilon _ e _ f _ f Real','\epsilon _ e _ f _ f Imag',4);
h = gca;
set(h,'Fontsize',30);
title('Effective Average Permittivity','FontSize',24);
hold off;
figure(8);
hold on;
plot(H0, real(sigma _ eff),'LineWidth',2);
plot(H0, imag(sigma _ eff),'r--','LineWidth',2);
plot(H0, sigma,'r','LineWidth',2);
h = gca;
set(h,'Fontsize',30);
hold off;
xlabel('H(Oe)', 'FontSize', 30);
ylabel('sigma _ e _ f _ f(mhos/cm)', 'FontSize',32);
title('Effective conductvity','FontSize',30);
legend('sigma _ e _ f _ f Real','sigma _ e _ f _ f Imag',4);
figure(9);%Zeff vs H (calculated from sqrt(mu0*mueff/epseff)
hold on;
plot(H0, real(Zeff),'LineWidth',2);
plot(H0, imag(Zeff),'r','LineWidth',2);
h = gca;
set(h,'Fontsize',30);
hold off;
xlabel('H(Oe)', 'FontSize', 30);
ylabel('Z _ e _ f _ f(Ohm)', 'FontSize',32);
title('Z _ e _ f _ f','FontSize',30);
legend('Z _ e _ f _ f Real','Z _ e _ f _ f Imag',4);
ZSM2 = (abs(Zs _ 0)).^2;
MuS = mu0*Integ1;
```

```
%epsS = -2.*Zs _ 0./(j.*w.*ZSM2.*d) +MuS./ZSM2;
epsS = Integ2./ZSM2;
%epsS = conj(epsS);
sigmaS = epsS.*j.*w;
Zsv = 0.5.*(j.*w.*d.*(MuS-(conj(epsS).*ZSM2)));% also, Zs
figure(10);%epsS vs H (from poynting vector)
hold on;
plot(H0, real(epsS),'b','LineWidth',2);
plot(H0, imag(epsS),'b--', 'LineWidth',2);
xlabel('H(Oe)', 'FontSize', 24);
ylabel('epsS(f/cm)', 'FontSize',32);
legend('epsS Real','epsS Imag',4);
h = gca;
set(h,'Fontsize',30);
title('Surface Average Permittivity','FontSize',24);
hold off;
figure(11);%sigmaS vs H (see fig. 7)
hold on;
plot(H0, real(sigmaS),'b','LineWidth',2);
plot(H0, imag(sigmaS),'r--','LineWidth',2);
plot(H0, sigmaS,'r','LineWidth',2);
h = gca;
set(h,'Fontsize',30);
hold off;
xlabel('H(Oe)', 'FontSize', 30);
ylabel('sigmaS(mhos/cm)', 'FontSize',32);
title('Surface conductivity','FontSize',30);
legend('sigmaS Real','sigmaS Imag',4);
figure(12);%Zsv vs H (calculated from poynting integral)
hold on;
plot(H0, real(Zsv),'LineWidth',2);
plot(H0, imag(Zsv),'r','LineWidth',2);
h = gca;
set(h,'Fontsize',30);
hold off;
xlabel('H(Oe)', 'FontSize', 30);
ylabel('Zsv(Ohm)', 'FontSize',32);
title('Zsv','FontSize',30);
legend('Zsv Real','Zsv Imag',4);
figure(13);%emcoupl(i)
hold on;
plot(H0, real(emcoupl),'b','LineWidth',2);
plot(H0, imag(emcoupl),'b--', 'LineWidth',2);
xlabel('H(Oe)', 'FontSize', 24);
ylabel('emcoupl)', 'FontSize',32);
legend('emcoupl real','emcoupl Imag',4);
h = gca;
```

```
set(h,'Fontsize',30);
title('em coupling','FontSize',24);
hold off;
% % Plot out the h(y) at Ho = 2500 Oe.
%
% N = 501;
%
% y = 0:d/200:d;
%  hy = h _ vec(1,N)*exp(-j*k1(N)*y)  +h _ vec(2,N)*exp(j*k1(N)*y)
+
h _ vec(3,N).*exp(-j*k2(N)*y)  +h _ vec(4,N).*exp(j*k2(N)y);
%
% figure(3);
% hold on;
% plot(y,real(hy),'LineWidth',2);
% % plot(y,imag(hy),'g','LineWidth',2);
% % plot(y,abs(hy),'r','LineWidth',2);
% h = gca;
% set(h,'Fontsize',30);
% % axis([0d 0.8 1]);
% xlabel('y(cm)','Fontsize',30);
% ylabel('Re(h(y))','Fontsize',30);
% title(['Ho = ', num2str(Ho(N)),  'Oe'],'Fontsize',30);
% hold off;
% X. Save raw data file for figure
1%%%%%%%%%%%%%%%%%%%%%%%%%%%%%%%%%%%%%%%%%
% % Plot out the h(y) at Ho = 2500 Oe.
%
% N = 501;
%
% y = 0:d/200:d;
%  hy = h _ vec(1,N)*exp(-j*k1(N)*y)  +h _ vec(2,N)*exp(j*k1(N)*y)
+
h _ vec(3,N).*exp(-j*k2(N)*y)  +h _ vec(4,N).*exp(j*k2(N)*y);
%
% figure(3);
% hold on;
% plot(y,real(hy),'LineWidth',2);
% % plot(y,imag(hy),'g','LineWidth',2);
% % plot(y,abs(hy),'r','LineWidth',2);

% h = gca;
% set(h,'Fontsize',30);
% % axis([0d 0.8 1]);
% xlabel('y(cm)','Fontsize',30);
% ylabel('Re(h(y))','Fontsize',30);
% title(['Ho = ', num2str(Ho(N)),  'Oe'],'Fontsize',30);
% hold off;
```

BIBLIOGRAPHY

W.S. Ament and G.T. Rado, *Phys. Rev.*, 97, 1558, 1955.
P. Ciureanu, L.G.C. Melo, D. Seldaoui, D. Menard, and A. Yelon, *J. Appl. Phys.*, 102, 073908, 2007.
L. Neel, *J. Phys. Radium*, 15, 225, 1954.
C. Vittoria, *Phys. Rev. B*, 37, 2387, 1987.
J.R. Weertman and G.T. Rado, *Phys. Chem. Solids*, 11, 315, 1959.

SOLUTIONS

11.1

a. For $E_s = K_s\alpha_1^2$, where $\alpha_1 = M_x/M_0$ (Figure S11.1):
Torque equation at the surface is given by

$$\frac{2A}{M_0^2}\vec{M}\times\frac{\partial\vec{m}}{\partial y}+\vec{T}_s = 0,$$

where $\vec{T}_s = -\vec{M}\times\vec{\nabla}_M E_s$ and $\vec{M} = \vec{M}_0 + \vec{m}$.

$\vec{M} = \left(M_{0x} + m_x\right)\vec{a}_x + \left(M_{0y} + m_y\right)\vec{a}_y + \left(M_{0z} + m_z\right)\vec{a}_z$, where

$M_{0x} = M_0\sin\theta\cos\varphi, M_{0y} = M_0\sin\theta\sin\varphi,$ and $M_{0z} = M_0\cos\theta.$

Therefore, the torque equation on the film surface is

$$\begin{aligned}
&\left(AM_{0y}\frac{\partial m_z}{\partial y} - AM_{0z}\frac{\partial m_y}{\partial y}\right)\vec{a}_x \\
&+\left(AM_{0z}\frac{\partial m_x}{\partial y} - K_s M_{0z}m_x - AM_{0x}\frac{\partial m_z}{\partial y} - K_s M_{0x}m_z\right)\vec{a}_y \\
&+\left(AM_{0x}\frac{\partial m_y}{\partial y} - K_s M_{0x}m_y - AM_{0y}\frac{\partial m_x}{\partial y} + K_s M_{0y}m_x\right)\vec{a}_z = 0
\end{aligned}$$

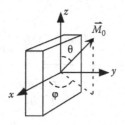

FIGURE S.11.1 Geometry of the problem with magnetization distribution.

At the limit of

$$
\theta = \frac{\pi}{2} \text{ and } \varphi = 0 \left| \begin{array}{ll} A\dfrac{\partial m_y}{\partial y} + K_s m_y = 0 & \text{and} \quad A\dfrac{\partial m_z}{\partial y} + K_s m_z = 0 \\[2ex] A\dfrac{\partial m^\pm}{\partial y} + K_s m^\pm = 0 & \text{where} \quad m^\pm = m_y \pm jm_z \end{array} \right.
$$

b. For $E_s = K_s \alpha_3^2$, where $\alpha_3 = \dfrac{M_z}{M_0}$:

$$
\begin{aligned}
&\left(AM_{0y}\frac{\partial m_z}{\partial y} - K_s M_{0y} m_z - AM_{0z}\frac{\partial m_y}{\partial y} - K_s M_{0z} m_y \right)\vec{a}_x \\
&+ \left(AM_{0z}\frac{\partial m_x}{\partial y} + K_s M_{0z} m_x - AM_{0x}\frac{\partial m_z}{\partial y} + K_s M_{0x} m_z \right)\vec{a}_y \\
&+ \left(AM_{0x}\frac{\partial m_y}{\partial y} - AM_{0y}\frac{\partial m_x}{\partial y} \right)\vec{a}_z = 0
\end{aligned}
$$

At the limit of

$$
\theta = 0 \text{ and } \varphi = \frac{\pi}{2}, \left| \begin{array}{ll} A\dfrac{\partial m_x}{\partial y} + K_s m_x = 0 & \text{and} \quad A\dfrac{\partial m_y}{\partial y} + K_s m_y = 0 \\[2ex] A\dfrac{\partial m^\pm}{\partial y} + K_s m^\pm = 0, & \text{where} \quad m^\pm = m_x \pm jm_y \end{array} \right.
$$

11.2

a. For $E_s = K_s \alpha_1^2$:

$$
\text{From} \quad \begin{aligned}
&\left(AM_{0y}\frac{\partial m_z}{\partial y} - AM_{0z}\frac{\partial m_y}{\partial y} \right)\vec{a}_x \\
&+ \left(AM_{0z}\frac{\partial m_x}{\partial y} - K_s M_{0z} m_x - AM_{0x}\frac{\partial m_z}{\partial y} - K_s M_{0x} m_z \right)\vec{a}_y \\
&+ \left(AM_{0x}\frac{\partial m_y}{\partial y} - K_s M_{0x} m_y - AM_{0y}\frac{\partial m_x}{\partial y} + K_s M_{0y} m_x \right)\vec{a}_z = 0
\end{aligned}
$$

At the limit of $\theta = 0$ and $\varphi = \dfrac{\pi}{2}$, $\dfrac{\partial m_y}{\partial y} = 0$ and $A\dfrac{\partial m_x}{\partial y} - K_s m_x = 0$

b. For $E_s = K_s \alpha_3^2$:

$$\left|\begin{array}{l}\left(AM_{0y}\dfrac{\partial m_z}{\partial y}-K_sM_{0y}m_z-AM_{0z}\dfrac{\partial m_y}{\partial y}-K_sM_{0z}m_y\right)\vec{a}_x\\[3mm]+\left(AM_{0z}\dfrac{\partial m_x}{\partial y}+K_sM_{0z}m_x-AM_{0x}\dfrac{\partial m_z}{\partial y}+K_sM_{0x}m_z\right)\vec{a}_y\\[3mm]+\left(AM_{0x}\dfrac{\partial m_y}{\partial y}-AM_{0y}\dfrac{\partial m_x}{\partial y}\right)\vec{a}_z=0\end{array}\right.$$

From

At the limit of $\theta=\dfrac{\pi}{2}$ and $\varphi=0$, $\dfrac{\partial m_y}{\partial y}=0$ and $A\dfrac{\partial m_z}{\partial y}-K_sm_z=0$

11.3

1. Equation of the motion: $\dfrac{2A}{M_0^2}\left(-M_x\dfrac{\partial^2 M_z}{\partial y^2}+M_z\dfrac{\partial^2 M_x}{\partial y^2}\right)=M_xH_0$,
where $M_0^2=M_x^2+M_z^2$.

$$\dfrac{\partial^2 M_x}{\partial y^2}=\left[\left(\dfrac{1}{M_x}-\dfrac{M_z^2}{M_x^3}\right)\left(\dfrac{\partial M_z}{\partial y}\right)^2+\dfrac{M_z}{M_x}\dfrac{\partial^2 M_z}{\partial y^2}\right]$$

Thus, the equation of the motion is given by

$$-\dfrac{2A}{M_0^2}\left[1-\left(\dfrac{M_z}{M_x}\right)^2\right]\left[\dfrac{d^2M_z}{dy^2}-\left(\dfrac{M_z}{M_x^2}\right)\left(\dfrac{dM_z}{dy}\right)^2\right]=H_0$$

2. Surface boundary condition: $\dfrac{2A}{M_0^2}\left(-M_x\dfrac{\partial M_z}{\partial y}+-M_z\dfrac{\partial M_x}{\partial y}\right)+$

$\dfrac{2K_s}{M_0^2}M_xM_z=0$, where $\dfrac{\partial M_x}{\partial y}=\dfrac{M_z}{M_x}\dfrac{\partial M_z}{\partial y}$.

Therefore, the surface boundary condition is given by

$$\dfrac{dM_z}{dy}=M_z\left[1-\left(\dfrac{M_z}{M_x}\right)^2\right]^{-1}\dfrac{K_s}{A}$$

11.4

From the two static boundary conditions:

$$\dfrac{dM_z}{dy}=M_z\left[1-\left(\dfrac{M_z}{M_x}\right)^2\right]^{-1}\dfrac{K_s}{A}\int\left(\dfrac{1}{M_z}-\dfrac{M_z}{M_0^2-M_z^2}\right)dM_z=\dfrac{K_s}{A}\sqrt{\dfrac{A}{K_s}}\int_0^\theta\dfrac{1}{\cos\theta}\,d\theta,$$

where $dy=\sqrt{A}\dfrac{1}{\sqrt{E_s(\theta)}}\,d\theta$ and $E_s(\theta)=K_s\cos^2\theta$ from $180°$ domain wall.

$$y = \sqrt{\frac{A}{K_s}} \int_0^\theta \frac{d\theta}{\cos\theta} = \sqrt{\frac{A}{K_s}} \ln \tan\left(\frac{\theta}{2} + \frac{\pi}{4}\right),$$

where $l = \sqrt{\dfrac{A}{K_s}}$.

Magnetization rotates by 90° from surface to center of the film (Figure S11.4).

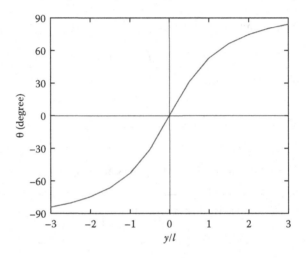

FIGURE S.11.4   Angular rotation of the magnetization.

11.5

a. SC [100] plane surface (Figure S11.5a)

Dipole–dipole pair interaction energy in the plane is given by

$$U = \frac{1}{4\pi\mu_0 r^3}\left[\vec{M}_1 \cdot \vec{M}_2 - \frac{3}{r^2}\left(\vec{M}_1 \cdot \vec{r}\right)\left(\vec{M}_2 \cdot \vec{r}\right)\right]$$

where $\vec{M}_1 = M\sin\theta\left(\cos\phi\,\vec{a}_x + \sin\phi\,\vec{a}_y\right)$ and $\vec{M}_1 = \vec{M}_2$.

Dipole–dipole interaction energy for the nearest neighbors on the x-axis and on the y-axis:

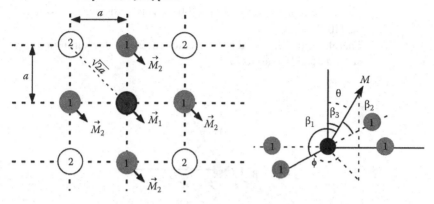

FIGURE S.11.5a Ions distribution at the surface.

$$U_x = \frac{1}{4\pi\mu_0 r^3}\left(M^2\sin\theta\right)\left[1-3\cos^2\phi\right] \quad \text{and}$$

$$U_y = \frac{1}{4\pi\mu_0 r^3}\left(M^2\sin\theta\right)\left[1-3\sin^2\phi\right].$$

Therefore, the total sum of dipole–dipole interaction energy on [100] plane with a distance a is

$$\boxed{U_T = U_x + U_y = \frac{-M^2}{4\pi\mu_0 r^3}\sin^2\theta = \frac{-M^2}{4\pi\mu_0 a^3}\left(1-\cos^2\theta\right)}$$

Based on Neel's model, the surface energy is

$$W(\theta) \cong g + l\left(\cos^2\theta - \frac{1}{3}\right) + q\left[\cos^4\theta - \cdots\right] + \cdots,$$

where the second term describes the dipole–dipole interaction between two ions.

Since there are four nearest neighbor atoms on [100] plane of SC symmetry, we have the surface energy as

$$\boxed{\begin{aligned} W(\theta) &\cong \frac{1}{2}\sin^2\theta\left[\frac{1}{2}\cos^2\phi + \frac{1}{2}\cos^2\left(\frac{\pi}{2}+\phi\right) + \frac{1}{2}\cos^2\left(\frac{\pi}{2}-\phi\right) + \frac{1}{2}\cos^2\left(\pi-\phi\right)\right] \\ &= \frac{l}{2}\left(1-\cos^2\theta\right) \cong -\frac{l}{2}\cos^2\theta \end{aligned}}$$

where the constant term is ignored.

  b. For next nearest neighbor sum only (see Figure S11.5b):
      There are four next nearest neighbors with a distance $r = \sqrt{2}a$ on the [100] plane.
      Therefore, the dipole–dipole interaction energy for summing next nearest neighbor atoms is

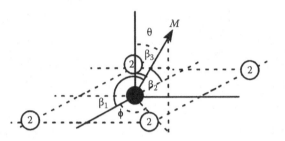

FIGURE S.11.5b Ions distribution at the surface.

$$U_T = U_x + U_y = \frac{-M^2}{4\pi\mu_0 r^3}\sin^2\theta = \frac{-M^2}{8\sqrt{2}\pi\mu_0 a^3}\left(1 - \cos^2\theta\right)$$

  Also, the surface energy of the plane is

$$W(\theta) \cong \frac{1}{4}\sin^2\theta\left[\cos^2\left(\phi - \frac{\pi}{4}\right) + \cos^2\left(\phi + \frac{\pi}{4}\right) + \cos^2\left(\phi + \frac{3\pi}{4}\right) + \cos^2\left(\frac{3\pi}{4} - \phi\right)\right]$$

$$= \frac{l}{2}\left(1 - \cos^2\theta\right) \cong -\frac{l}{2}\cos^2\theta$$

11.6

  a. For BCC [100] plane surface (see Figure S11.6):

Body center cubic [100] plane

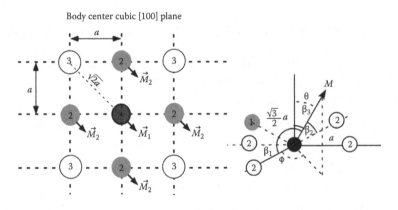

FIGURE S.11.6 Ions distribution at the surface.

There are four next nearest neighbor atoms with a distance $r=a$, but zero nearest neighbor atoms (with a distance $r=\left(\sqrt{3}/2\right)a$) on [100] plane of BCC symmetry. Therefore, the surface dipole–dipole interaction energy for summing nearest neighbor atoms is $\boxed{U_T=0}$. Also $\boxed{W(\theta)=0}$.

b. Since, there are four next nearest neighbor atoms with $r=a$ on [100] plane of BCC symmetry, the dipole–dipole interaction and surface energy are $\boxed{U_T = \dfrac{-M^2}{4\pi\mu_0 a^3}\left(1-\cos^2\theta\right)}$ and also $\boxed{W(\theta)\cong -\dfrac{l}{2}\cos^2\theta}$.

11.7

a. For FCC [100] plane surface (see Figure S11.7):

There are four nearest neighbors with a distance $r=\left(\sqrt{2}/2\right)a$ on [100] plane of FCC symmetry. Therefore, the dipole–dipole interaction and surface energy are $\boxed{U_T = \dfrac{-M^2}{\sqrt{2}\pi\mu_0 a^3}\left(1-\cos^2\theta\right)}$ and $\boxed{W(\theta)\cong -\dfrac{l}{2}\cos^2\theta}$.

b. There are four next nearest neighbor atoms shown in [100] plane of FCC with a distance $r=a$. Therefore, the energy is

$$\boxed{U_T = \dfrac{-M^2}{4\pi\mu_0 a^3}\left(1-\cos^2\theta\right)} \text{ and } \boxed{W(\theta)\cong -\dfrac{l}{2}\cos^2\theta}$$

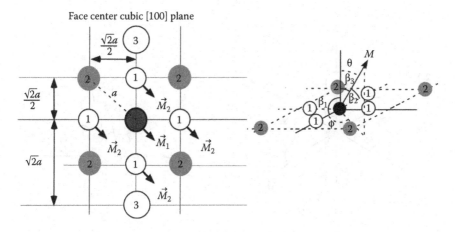

Face center cubic [100] plane

FIGURE S.11.7 Ions distribution at the surface.

11.8

a. For the basal plane of the hexagon surface (see Figure S11.8):
There are six nearest neighbor atoms with a distance of $r=a$ on basal plane of hexagonal symmetry.
Dipole–dipole interaction energy:

1.
$$\vec{r} = r\vec{a}_x$$
$$U_x = \frac{1}{4\pi\mu_0 r^3}\left(M^2 \sin\theta\right)\left[1 - 3\cos^2\phi\right]$$

2.
$$\vec{r} = r\left[\cos\left(\frac{\pi}{3}\right)\vec{a}_x + \sin\left(\frac{\pi}{3}\right)\vec{a}_y\right] = \frac{r}{2}\left(\vec{a}_x + \sqrt{3}\vec{a}_y\right)$$
$$U_{\frac{\pi}{3}} = \frac{M^2 \sin\theta}{4\pi\mu_0 r^3}\left[1 - \frac{3}{4}\left(\cos^2\phi + 2\sqrt{3}\cos\phi\sin\phi + 3\sin^2\phi\right)\right]$$

3.
$$\vec{r} = r\left[\cos\left(\frac{2\pi}{3}\right)\vec{a}_x + \sin\left(\frac{2\pi}{3}\right)\vec{a}_y\right] = \frac{r}{2}\left(-\vec{a}_x + \sqrt{3}\vec{a}_y\right)$$
$$U_{\frac{2\pi}{3}} = \frac{M^2 \sin\theta}{4\pi\mu_0 r^3}\left[1 - \frac{3}{4}\left(\cos^2\phi - 2\sqrt{3}\cos\phi\sin\phi + 3\sin^2\phi\right)\right]$$

Therefore, the total sum of dipole–dipole interaction energy on hexagonal basal plane with a distance a is

$$\boxed{U_T = U_x + U_{\frac{\pi}{3}} + U_{\frac{2\pi}{3}} = \frac{-3}{2}\frac{M^2}{4\pi\mu_0 r^3}\sin^2\theta = \frac{-3M^2}{8\pi\mu_0 a^3}\left(1 - \cos^2\theta\right)}$$

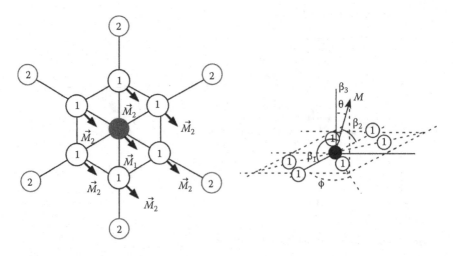

FIGURE S.11.8 Ions distribution at the surface.

The surface energy for sum of nearest neighbor atoms in basal hexagonal plane is

$$W(\theta) \cong \frac{1}{6}\sin^2\theta\left[1 + \cos^2\left(\frac{\pi}{3}\right) + \cos^2\left(\frac{2\pi}{3}\right) + \cos^2(\pi) + \cos^2(3\pi) + \cos^2\left(\frac{5\pi}{3}\right)\right].$$

$$= \frac{l}{2}\sin^2\theta \cong -\frac{l}{2}\cos^2\theta$$

b. There are six next nearest neighbor atoms with a distance $r = 2a$ on hexagonal symmetry. Therefore, the sum of dipole–dipole pair interaction energy is $U_T = \dfrac{-3M^2}{64\pi\mu_0 a^3}(1 - \cos^2\theta)$.

12 Matrix Representation of Wave Propagation

MATRIX REPRESENTATION OF WAVE PROPAGATION IN SINGLE LAYERS

Let's consider the case of an electromagnetic wave incident upon a single layer of magnetic material. The field excitations are the following: the external magnetic field, $\vec{H}$, is applied either in the plane of the layer or film (//) or perpendicular to the plane of the film ($\perp$). The microwave excitation field is assumed linearly polarized and in the plane of the film. We consider the electromagnetic response in terms of scattering S-parameters for either case (//) or ($\perp$).

(//) CASE

The microwave magnetic field excitation is assumed to be linearly polarized, but perpendicular to $\vec{M}_0$, the internal static magnetization. The response is also linearly polarized, as shown in Figure 12.1.

The incident, reflected, and transmitted RF fields are linearly polarized in the x-direction, although the precessional motion of the magnetization is elliptical, see Figure 12.2.

As in Chapter 11, there are eight basic boundary condition equations and they are

$$\sum_{i}^{0}\left(h_i^+ + h_i^-\right) = h_1, \tag{12.1}$$

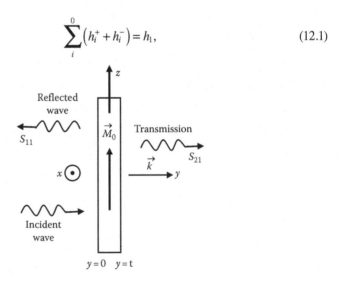

FIGURE 12.1 Electromagnetic wave field configurations about a magnetic film.

DOI: 10.1201/9781003431244-12

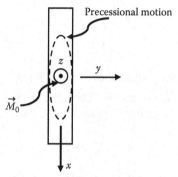

Figure 12.2   Sketch of normal mode motion as viewed from the z-*axis*.

$$\sum_i \left(h_i^+ e^{-jk_it} + h_i^- e^{+jk_it} \right) = h_2, \tag{12.2}$$

$$\sum_i Z_i \left(h_i^+ - h_i^- \right) = e_1, \tag{12.3}$$

$$\sum_i Z_i \left(h_i^+ e^{-jk_it} - h_i^- e^{+jk_it} \right) = e_2, \tag{12.4}$$

$$\sum_i Q(k_i) ik_i \left(h_i^+ + h_i^- \right) = 0, \tag{12.5}$$

$$\sum_i Q(k_i) jk_i \left(h_i^+ e^{-jk_it} - h_i^- e^{+jk_it} \right) = 0, \tag{12.6}$$

$$\sum_i \frac{i\chi(k_i)}{\mu(k_i)} \left(\left(K_S^{(0)} + jk_i A \right) h_i^+ - \left(K_S^{(0)} - jk_i A \right) h_i^- \right) = 0, \tag{12.7}$$

$$\sum_i \frac{i\chi(k_i)}{\mu(k_i)} \left(\left(K_S^{(0)} + jk_i A \right) h_i^+ e^{-jk_it} - \left(K_S^{(0)} - jk_i A \right) h_i^- e^{+jk_it} \right) = 0. \tag{12.8}$$

All of the above parameters have been defined in Chapter 11. It suffices to say that there are three normal modes of propagation for the (//) case, and hence, $i = 1, 2, 3$. The electromagnetic surface fields at $y = 0$ are e_1 (electric) and h_1 (magnetic) and e_2 and h_2 at $y = t$. The surface fields are polarized along the x-direction and are continuous across the two surfaces (first four equations). The last four equations are statements about the constraints on the spin motion at the surfaces (see Chapter 11). The object is to relate (e_1, h_1) to (e_2, h_2). The relationship between the two sets of surface fields is governed by a 2×2 matrix, since there are only two surface fields to consider:

$$\begin{bmatrix} e_1 \\ h_1 \end{bmatrix} = [A] \begin{bmatrix} e_2 \\ h_2 \end{bmatrix} = \begin{bmatrix} a_{11} & a_{12} \\ a_{21} & a_{22} \end{bmatrix} \begin{bmatrix} e_2 \\ h_2 \end{bmatrix}. \tag{12.9}$$

The mathematical procedure is as follows: Solve for $h_i^{\pm}$ from Equations 12.2 and 12.4 through 12.8 in terms of e_2 and h_2. Then, the $h_i^{\pm}$ solutions are substituted into Equations 12.1 and 12.3. Assuming solutions of the form

$$h_i^{\pm} = \alpha_i^{\pm} e_2 + \beta_i^{\pm} h_2, \tag{12.10}$$

we may express the matrix elements of $[A]$ in terms of $\alpha's$ and $\beta's$. Thus, we may write

$$a_{11} = \sum_i^3 Z_i \left(\alpha_i^+ - \alpha_i^- \right),$$

$$a_{12} = \sum_i^3 Z_i \left(\beta_i^+ - \beta_i^- \right),$$

$$a_{21} = \sum_i^3 \left(\alpha_i^+ + \alpha_i^- \right),$$

and

$$a_{22} = \sum_i^3 \left(\beta_i^+ - \beta_i^- \right).$$

Consider now the calculation of reflection and transmission scattering S-parameters, S_{11} and S_{21}, through the layer. S_{11} and S_{21} are readily calculated once $[A]$ is defined (Vittoria) (Figure 12.3).

We may write the following:

$$S_{11} = \frac{a_{12} + a_{11} Z_{0R} - a_{22} Z_{0L} - a_{21} Z_{0R} Z_{0L}}{a_{12} + a_{11} Z_{0R} + a_{22} Z_{0L} + a_{21} Z_{0R} Z_{0L}} \tag{12.11}$$

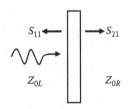

FIGURE 12.3 Definition of S_{11} and S_{21}.

and

$$S_{21} = \frac{2Z_{0R}}{a_{12} + a_{11}Z_{0R} + a_{22}Z_{0L} + a_{21}Z_{0R}Z_{0L}}. \qquad (12.12)$$

Z_{0R} and Z_{0L} are the characteristic impedances for the media to the right and left of the layer, respectively. Let's assume $Z_{0R} = Z_{0L} = Z_0$. Then,

$$S_{11} = \frac{a_{12} - a_{21}Z_0^2}{a_{12} + 2a_{11}Z_0 + a_{21}Z_0^2}$$

and

$$S_{21} = \frac{2Z_0}{a_{12} + 2a_{11}Z_0 + a_{21}Z_0^2}.$$

For a single layer, $a_{11} = a_{22}$ and the "feeder" and "output" lines have characteristic impedance of Z_0. In the expressions for S_{11} and S_{21}, the a's are a function of k_i's.

Let's now express the a's in terms of the S scattering parameters, which are usually measurable quantities. It is simple to show from the above set of equations that the relationships between the a's and the S's take the following form:

$$\begin{bmatrix} 1 + S_{11} \\ 1 - S_{11} \\ 1 - \left(S_{11}^2 + S_{21}^2 \right) \end{bmatrix} = S_{21} \begin{bmatrix} 1 & 1 & 0 \\ 1 & 0 & 1 \\ 0 & 1 - S_{11} & 1 + S_{11} \end{bmatrix} \begin{bmatrix} a_{11} \\ a_{12} \\ a_{21} \end{bmatrix}. \qquad (12.13)$$

In deriving the above relationship, we have utilized

$$a_{11}a_{22} - a_{12}a_{21} = 1$$

due to conservation law. Also, we have assumed a single layer so that $a_{11} = a_{22}$. Applying Kramer's rule for solving algebraic equations, we obtain

$$2S_{11}a_{11} = 1 - S_{11}^2 + S_{21}^2,$$
$$2S_{21}a_{12} = Z_0 \left[(1 + S_{11})^2 - S_{21}^2 \right], \text{ and}$$
$$2S_{21}a_{21} = \frac{\left[(1 - S_{11})^2 - S_{21}^2 \right]}{Z_0}.$$

The $[A]$ matrix may be related to an effective characteristic impedance, Z, and propagation constant, k, of the layer or

$$[A] = \begin{bmatrix} a_{11} & a_{12} \\ a_{21} & a_{22} \end{bmatrix} = \begin{bmatrix} \cos kt & jZ \sin kt \\ \dfrac{j \sin kt}{Z} & \cos kt \end{bmatrix}. \qquad (12.14)$$

Finally, k and Z may be solved in terms of a's by utilizing the above definition of $[A]$ and obtain

$$Z = \sqrt{\frac{a_{12}}{a_{21}}} \qquad (12.15)$$

and

$$k = \frac{1}{t}\cos^{-1}(a_{11}).$$

where t is the thickness of the layer. Thus, we have reduced the complexity of various modes of propagations into an equivalent k and Z values.

By representing wave propagation in a film as a transfer function matrix, it may now be possible to consider a film enclosed in a microwave cavity as a whole system. For example, in Chapter 8, we represented a microwave cavity with an iris coupling hole by two transfer function matrices and simulated electrically the reflection, S_{11}, from a microwave cavity. Now, with the insertion of a magnetic film in the microwave cavity, the resultant 2×2 matrix is the product of four matrices: the iris hole, the spacing between the iris hole (length, $(l-t)/2$), the magnetic film (length, t), and the spacing between the film and the shorted end of the microwave cavity (length, $(l-t)/2$). The overall length of the cavity is l, as shown in Figure 12.4. The matrix representation of the microwave cavity and magnetic film is summarized in Figure 12.5.

The matrix $[C]$ represents the iris hole (see Chapter 11), and $[B]$ the spacing or transmission line between film and the shorting ends of waveguide and $[A]$ the film.

The whole system may be represented by a 2×2 matrix of the form:

$$[D] = [C][B][A][B],$$

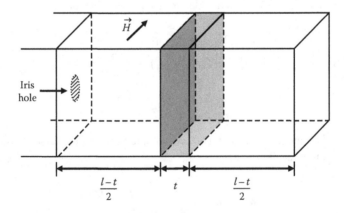

FIGURE 12.4 Microwave cavity system together with magnetic film at the center.

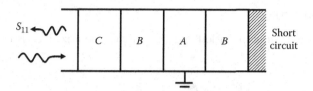

FIGURE 12.5 Transfer function matrices representation of microwave cavity in Figure 12.4.

from which S_{11} may be calculated as before. In the limit $t \to o$, the matrix $[D]$ is diagonal so that $[D]$ would represent the empty microwave cavity, as analyzed in Chapter 11. Experimentally, usually the film only covers part of the cross area of a microwave cavity. Thus, our analysis maximizes the calculation of S_{11}. For small sizes of the film, one needs to apply boundary conditions over the entire cross section of the waveguide and film in order to determine the effective propagation constant, k, within the film as well as its characteristic impedance. This is rather mathematically tedious but doable. The point is that even for partial coverage of the film across the waveguide cross section, it is meaningful to introduce the $[A]$ transfer function matrix.

($\perp$) CASE

We assume linear polarization of the RF field excitation. However, now the precessional motion of the magnetization is circular for the resonance or anti-resonance mode of propagation (Figure 12.6).

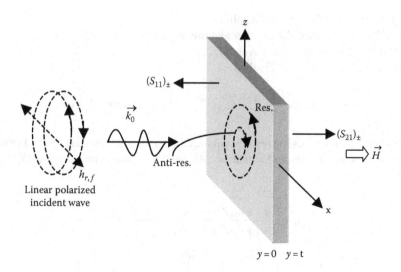

FIGURE 12.6 Display of various modes of excitations.

THE INCIDENT FIELD

The incident field in terms of left and right circular polarized fields is given by (in complex notation)

$$\vec{E} = E_0 e^{-jk_0 y}\vec{a}_z = \frac{1}{2}\left(\vec{E}_+ + \vec{E}_-\right)e^{-jk_0 y},$$

where

$$\vec{E}_{\pm} = E_0\left(\vec{a}_z \pm j\vec{a}_x\right).$$

The subscript $\pm$ indicates the sense of rotation of the circular polarization.

The incident voltage, V, may be expressed in time domain simply as

$$V(y,t) = V_0 \cos(\omega t - k_0 y)$$

V_0 may be related to E_0 as follows:

$$E_0 = \frac{V_0}{h}, \text{microstripline,}$$

$$E_0 = \frac{V_0}{(b-a)\ln(b/a)}, \text{coaxial line,}$$

and

$$E_0 = V_0\sqrt{\frac{2}{ab}}, \text{waveguide,}$$

where

     h is the height of the micro-stripline

     b and a are the coaxial line outer and inner radii

     ab is the cross area of the waveguide

     The incident electric field is then

$$\vec{E}(y,t) = \text{Re}\left[\frac{1}{2}\left(\vec{E}_+ + \vec{E}_-\right)e^{j(\omega t - k_0 y)}\right] = E_0 \cos(\omega t - k_0 y)\vec{a}_z.$$

The (−) notation denotes the resonant mode and (+) the anti-resonant mode. For each mode, there corresponds an [A] matrix, [A]$_+$ and [A]$_-$, with corresponding $(S_{11})_-$ and $(S_{21})_-$:

$$(S_{11})_{\pm} = \left(\frac{a_{12} + a_{11}Z_{0R} - a_{22}Z_{0L} - a_{21}Z_{0R}Z_{0L}}{a_{12} + a_{11}Z_{0R} + a_{22}Z_{0L} + a_{21}Z_{0R}Z_{0L}}\right)_{\pm},$$

$$(S_{21})_{\pm} = \left(\frac{2Z_{0R}}{a_{12} + a_{11}Z_{0R} + a_{22}Z_{0L} + a_{21}Z_{0R}Z_{0L}}\right)_{\pm}.$$

The reflected signal is the sum of two waves $(S_{11})_-$ since the input is two circularly polarized waves:

$$\vec{E}_R = \frac{1}{2}\left[\left(\vec{E}_R\right)_+ + \left(\vec{E}_R\right)_-\right],$$

or

$$\vec{E}_R = \frac{E_0}{2}\left[(S_{11})_+\left(\vec{a}_z + j\vec{a}_x\right) + (S_{11})_-\left(\vec{a}_z - j\vec{a}_x\right)\right]e^{jk_0 y},$$

or

$$\vec{E}_R = \frac{E_0}{2}\left[\left((S_{11})_+ + (S_{11})_-\right)\vec{a}_z + j\left((S_{11})_+ - (S_{11})_-\right)\vec{a}_x\right]e^{jk_0 y}.$$

Clearly, $\vec{E}_R$ is elliptically polarized, since

$$(S_{11})_+ + (S_{11})_- \neq (S_{11})_+ - (S_{11})_-.$$

Similarly, the transmitted signal is

$$\vec{E}_T = \frac{1}{2}\left[\left(\vec{E}_T\right)_+ + \left(\vec{E}_T\right)_-\right],$$

or

$$\vec{E}_T = \frac{E_0}{2}\left[(S_{11})_+\left(\vec{a}_z + j\vec{a}_x\right) + (S_{11})_-\left(\vec{a}_z - j\vec{a}_x\right)\right]\bar{e}^{jk_0 y},$$

or

$$\vec{E}_R = \frac{E_0}{2}\left[\left((S_{11})_+ + (S_{11})_-\right)\vec{a}_z + j\left((S_{11})_+ - (S_{11})_-\right)\vec{a}_x\right]\bar{e}^{jk_0 y}.$$

The propagation constant in the media outside of the magnetic layer is k_0, free space medium or feeder lines. Again, the transmitted signal is elliptically polarized, and therefore, polarization is not conserved. The incident wave is linearly polarized, whereas the reflected and transmitted waves are elliptically polarized. This is a result of not exciting the magnetic layer in its normal mode of excitation. Let's trace E_R in the time domain in order to show the ellipticity pattern. Considering only E_R at $y=0$ (at the surface), we have

$$E_x = j\frac{E_0}{2}\left[(S_{11})_+ + (S_{11})_-\right] = Ae^{j\phi_A}$$

and

$$21\ E_z = \frac{E_0}{2}\left[(S_{11})_+ + (S_{11})_-\right] = Be^{j\phi_B}.$$

In the time domain, we have simply

$$E_x(0,t) = A\cos\left(\omega t + \phi_A\right)$$

and

$$E_z(0,t) = B\cos\left(\omega t + \phi_B\right).$$

A time trace pattern of an ellipse may be traced much like as we did in Chapter 1.

Let's now calculate the matrix elements of the $[A]_+$ and $[A]_-$ matrices. In Chapter 11, we wrote the following boundary conditions for the ($\perp$) case:

$$\sum_{i}^{2}\left(h_i^+ + h_i^-\right) = h_1,\ y = 0 \tag{12.16}$$

$$\sum_{i}^{2}\left(h_i^+ e^{-jk_i t} + h_i^- e^{+jk_i t}\right) = h_2,\ y = t \tag{12.17}$$

$$\sum_{i}^{2} Z_i\left(h_i^+ - h_i^-\right) = e_1,\ y = 0 \tag{12.18}$$

$$\sum_{i}^{2} Z_i\left(h_i^+ e^{-jk_i t} - h_i^- e^{+jk_i t}\right) = e_2,\ y = t \tag{12.19}$$

$$\sum_{i}^{2}\left(P_i h_i^+ + R_i h_i^-\right) = 0,\ y = 0 \tag{12.20}$$

$$\sum_{i}^{2}\left(P_i h_i^+ e^{-jk_i t} + R_i h_i^- e^{+jk_i t}\right) = 0,\ y = t. \tag{12.21}$$

Variables are defined in Chapter 11. We have assumed in the above set of equations that the surface magnetic anisotropies are the same at the two surfaces. Using Equations 12.17 and 12.19 through 12.21, $h_i^\pm$ are solved in terms of e_2 and h_2.

Substituting the above relationships into Equations 12.16 and 12.18, we obtain the $[A]_\pm$ matrices, where

$$[A]_\pm = \begin{bmatrix} a_{11} & a_{12} \\ a_{21} & a_{22} \end{bmatrix}_\pm .$$

The subscript (−) denotes the resonant mode and (+) the anti-resonant mode. The anti-resonant case is simply attained by putting $\omega = -\omega$ into the dispersion relation for the propagation constants k_i (see Chapter 11). Assume solutions of $h_i^\pm$ in terms of e_2 and h_2 of the form

$$h_i^\pm = \alpha_i^\pm e_2 + \beta_i^\pm h_2,$$

For either mode (resonant or anti-resonant), we obtain

$$a_{11} = \sum_i^2 Z_i \left(\alpha_i^+ - \alpha_i^- \right), \qquad (12.22)$$

$$a_{12} = \sum_i^2 Z_i \left(\beta_i^+ - \beta_i^- \right), \qquad (12.23)$$

$$a_{21} = \sum_i^2 \left(\alpha_i^+ + \alpha_i^- \right), \qquad (12.24)$$

and

$$a_{22} = \sum_i^2 \left(\beta_i^+ + \beta_i^- \right). \qquad (12.25)$$

The superscripts here indicate wave propagation directions.

FERROMAGNETIC RESONANCE IN COMPOSITE STRUCTURES: NO EXCHANGE COUPLING

Specifically, we consider the following problem: we assume normal incidence of an electromagnetic wave upon an asymmetric periodic magnetic layered structure composed of layers that are magnetic, dielectric, magnetic, dielectric, magnetic, etc. The external static magnetic field, $\vec{H}$, is applied normal to the plane of the layered structure, as shown in Figure 12.7. It is assumed that the dielectric layer is characterized by a scalar dielectric constant rather than a tensor quantity. For these separation distances, the exchange coupling between magnetic layers is very small in comparison to surface magnetic anisotropy energies. The magnetostatic energy coupling between

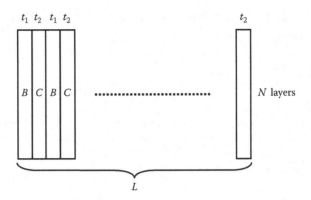

t_1 t_2 t_1 t_2 t_2

B | C | B | C ···························· N layers

L

FIGURE 12.7 Superlattice structure.

layers is also ignored, since there is no microwave magnetization component normal to the plane of the layered structure. The magnetization $\vec{M}$, exchange-stiffness constant A, g value, and the conductivity σ of the magnetic film may be well known in the literature. Spin-pinning boundary conditions at the surfaces of each magnetic layer are also included in this theoretical formulation.

We require a transformation which expresses the internal fields of each layer in terms of the microwave electric and magnetic fields at the surface, which are both circularly polarized, at the two surfaces. It is then mathematically convenient to define a matrix transfer function that relates the microwave surface fields at the two surfaces. The relationship between the microwave surface and the internal fields for $\vec{H}$ normal to a layered structure plane is given below. The total $[A_T]$ matrix may readily be derived for this case for the entire multilayer structure.

Define

$$[A] = [B][C].$$

Then, the cumulative matrix becomes

$$[A_T] = [A]^n,$$

where

$$[AT] = a_0[I] + a_1[A] \tag{12.26}$$

and

$$[A] = \begin{bmatrix} a_{11} & a_{12} \\ a_{21} & a_{22} \end{bmatrix},$$

where n is the number of pair layers. $[I]$ is the unit matrix, and a_0 and a_1 are arbitrary coefficients defined as

$$\lambda_1^n = a_0 + a_1\lambda_1 \tag{12.27a}$$

and

$$\lambda_2^n = a_0 + a_1\lambda_2. \tag{12.27b}$$

λ_1 and λ_2 are the eigenvalues of the $[A]$ matrix or

$$\det \begin{bmatrix} a_{11} - \lambda & a_{12} \\ a_{21} & a_{22} - \lambda \end{bmatrix} = \lambda^2 - \lambda(a_{11} + a_{22}) + 1 = 0.$$

Thus,

$$\lambda_{1,2} = \frac{b}{2} \pm \sqrt{\frac{b^2}{4} - 1}; \; b = a_{11} + a_{22}. \tag{12.28}$$

From Equations 12.27 and 12.28, we solve for a_0 and a_1 and obtain

$$a_0 = \frac{\lambda_1\lambda_2^n - \lambda_2\lambda_1^n}{\lambda_1 - \lambda_2}$$

and

$$a_0 = \frac{\lambda_1^n - \lambda_2^n}{\lambda_1 - \lambda_2}.$$

For example, consider the case when each layer is l thick such that $Nl = L$ and each layer is simply a dielectric layer. If all the layers are identical, then $[A_T]$ is trivial and equal to

$$[A_T] = \begin{bmatrix} \cos kL & jZ\sin kL \\ \dfrac{j\sin kL}{Z} & \cos kL \end{bmatrix}.$$

However, let's prove that

$$\begin{bmatrix} \cos kl & jZ\sin kl \\ \dfrac{j\sin kl}{Z} & \cos kl \end{bmatrix}^n \overset{?}{=} \begin{bmatrix} \cos k(nl) & jZ\sin k(nl) \\ \dfrac{j\sin k(nl)}{Z} & \cos k(nl) \end{bmatrix}.$$

The eigenvalues of the single layer $[A]$, where $[A]$ is the matrix on the LHS of the above equation, are

$$\lambda_{1,2} = e^{\pm jkl}.$$

The coefficients a_0 and a_1 now are defined as

$$a_0 = \frac{-\sin k(n-1)l}{\sin kl}$$

and

$$a_1 = \frac{\sin knl}{\sin kl}.$$

Substituting the above relationships into Equation 12.26, we have indeed that

$$[A_T] = a_0 \begin{bmatrix} 1 & 0 \\ 0 & 1 \end{bmatrix} + a_1 \begin{bmatrix} \cos kL & jZ \sin kL \\ \dfrac{j \sin kL}{Z} & \cos kL \end{bmatrix}$$

$$= \begin{bmatrix} \cos k(nl) & jZ \sin k(nl) \\ \dfrac{j \sin k(nl)}{Z} & \cos k(nl) \end{bmatrix}, \text{QED.}$$

We have calculated the absorption of a single layer versus that of a multilayer, N pairs. The metal layer was characterized as a magnetic metal permalloy and the dielectric as a lossless dielectric. A plot of the comparison between the two absorptions is shown in Figure 12.8. In each case, the surface impedance was calculated.

Let's calculate the surface impedance, Z_S, given that an $[A]$ matrix is known. Usually, one may write in general

$$\begin{pmatrix} e_1 \\ h_1 \end{pmatrix} = \begin{bmatrix} a_{11} & a_{12} \\ a_{21} & a_{22} \end{bmatrix} \begin{pmatrix} e_2 \\ h_2 \end{pmatrix}.$$

The LHS of the equation contains surface electromagnetic electric and magnetic fields (e and h) with the subscript 1 indicating the input side of the layer. The subscript 2 on the electromagnetic fields indicates the output side. A simple expression for the surface impedance is obtained for symmetrical excitations of the incident electromagnetic fields, where

$$h_2 = h_2 = h_S.$$

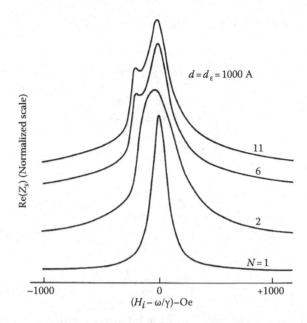

FIGURE 12.8 Microwave absorption as a function of an effective field. The number of layers is varied in the superlattice.

Expanding the matrix, one obtains the equation

$$e_1 = a_{11}e_2 + a_{12}h_2. \tag{12.29}$$

Since the layer is symmetrically excited, one may assume that the surface impedances are also symmetrical. Ignore for now non-reciprocal effects as the surface fields are linearly polarized. Thus, assume

$$e_1 = Z_s h_s \text{ and } e_2 = -Z_s h_s.$$

The (–) sign is important as it directs energy into the layer. Equation 12.29 now becomes

$$Z_s h_s = a_{11}(-Z_s h_s) - a_{12}h_s.$$

Solving for the surface impedance, one obtains

$$Z_s = \frac{a_{12}}{1 + a_{11}}.$$

The reader is cautioned here that for multilayers, the surface impedances may be different even for symmetrical excitations. It is simple to show that for a multilayer, the following equation is obtained for symmetrical excitation (after expanding the matrix):

$$Z_{S1} + Z_{S2}a_{11} = a_{12},$$

$$0 + Z_{S2}a_{21} = a_{22} - 1.$$

The solutions are

$$Z_{S2} = \frac{a_{11} - 1}{a_{21}}$$

and

$$Z_{S2} = \frac{a_{22} - 1}{a_{21}}.$$

For multilayers, $a_{11} \neq a_{22}$ in contrast to a single layer (see Vittoria). Z_{S1} and Z_{S2} are the surface impedances at surfaces 1 and 2, respectively.

In Figure 12.8, we see that a composite of layers yields indeed a different FMR response from a single layer ($N = 1$). Details of the analysis are provided in the reference list.

FERROMAGNETIC RESONANCE IN COMPOSITE STRUCTURES: EXCHANGE COUPLING

We consider two cases: $\vec{H}$ in the plane (//) and perpendicular ($\perp$) to the film plane.

($\perp$) CASE

We begin by considering the layered structure (shown in Figure 12.9) composed of $2N + 1$ alternate layers of ferromagnetic material A and dielectric layer B, both characterized by $[A]$ and $[B]$ matrices.

N is the number of pair layers. The layers extend to infinity in the x–z plane and the interfaces are assumed to be abrupt. There is a static external magnetic field in the y-direction and both materials have their static magnetization along the external field. Identical microwave excitations are normally incident on the structure from the $+y$ and $-y$ directions. The internal microwave electric and magnetic fields in each layer are denoted by e_i and h_i. Because of the interaction with the spin wave modes inside the magnetic layers, there are four allowable circularly polarized propagating electromagnetic modes (each with its own complex propagation vector k). The k vectors for these modes can be found at each value of the internal static magnetic field by numerically solving the relevant dispersion relation (see Chapter 11).

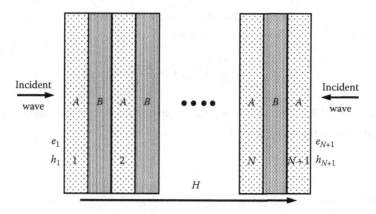

FIGURE 12.9   Exchange-coupled superlattice structure.

Two of the four values of k correspond to magnetic resonance modes and the other two to anti-resonant modes. Only the two resonant modes are of interest, since we are calculating FMR line shapes of layered structures. We therefore take only the resonant active polarization modes for each direction of propagation. This means that the microwave properties are reciprocal with respect to the direction of the incident electromagnetic wave. We are assuming that the incident circularly polarized wave impinging from the left or right of the multilayers only couples to the resonant mode. However, for linearly polarized incident waves, the anti-resonant modes must be included in the analysis and that can be quite complicated. At the interfaces, Maxwell's equations give us the condition that the tangential components of the microwave e and h fields must be continuous. We express the fields at each surface in terms of internal fields. This allows us to relate the fields at two surfaces.

Let's obtain the $[A]$ matrix for the above case. The two magnetic layers are bounded between $y=0$ and $y=t_1$; $y=t_2$ and $y=t_3$. In between t_1 and t_2, we have a dielectric layer, B. Consider first the surface of magnetic layer 1 coupled to magnetic layer 2. As explained in Chapter 11, the spins at the surfaces couple to each other by the exchange interactions. This manifests itself in the form of an extraneous torque at the surface:

$$\frac{2A_1}{M_1^2}\vec{M}_1 \times \frac{\partial \vec{M}_1}{\partial y} + \vec{T}_S = 0, \tag{12.30a}$$

where $\vec{T}_S = \vec{M}_1 \times \left(-\vec{\nabla}_M E_S\right)$. E_S contains two terms:

$$E_S = K_S \left(\frac{M_{1y}}{M_1}\right)^2 - \frac{2A_{12}}{M_1 M_2}\vec{M}_1 \cdot \vec{M}_2.$$

The subscripts "1" and "2" refer to magnetic layers 1 and 2. The exchange coupling is between $\vec{M}_1$ and $\vec{M}_2$ at the respective surfaces. The first term in E_S is the uniaxial magnetic anisotropy energy at the surface of layer 1. The second term is the exchange coupling between the layers. Rewrite Equation 12.30a as

$$\frac{2A_1}{M_1^2}\vec{M}_1 \times \frac{\partial \vec{m}_1}{\partial y} + \frac{2K_S}{M_1^2}\vec{M}_1 \times \vec{m}_1 + \frac{2A_{12}}{M_1 M_2}\vec{M}_1 \times \vec{M}_2 = 0. \qquad (12.30b)$$

In Equations 12.30a and 12.30b, we have assumed that the static magnetization is perpendicular to the film plane. Defining $\vec{M}_1$ and $\vec{M}_2$ as

$$\vec{M}_1 = \vec{M}_{01} + \vec{m}_1 \text{ and } \vec{M}_2 = \vec{M}_{02} + m_2,$$

we obtain

$$\vec{M}_1 \times \left[\frac{A_1}{M_1^2}\frac{\partial \vec{m}_1}{\partial y} + \frac{K_S}{M_1^2}\vec{m}_1 - \frac{A_{12}}{M_1^2}\left(\vec{m}_1 - \frac{M_1}{M_2}\vec{m}_2\right) \right] = 0.$$

Finally, the boundary condition may be written as

$$A_1 \frac{\partial \vec{m}_1}{\partial y} + K_S \vec{m}_1 - A_{12}\left(\vec{m}_1 - \frac{M_1}{M_2}\vec{m}_2\right) = 0. \qquad (12.31a)$$

It is understood that K_S in the above equation applies to the first layer. The effect of exchange coupling to the other magnetic layer is twofold. First, the coupling gives rise to a surface static uniaxial anisotropy field with easy axis along the static magnetization direction at surface 1. At the other surface, we would expect the same. The total magnitude of this uniaxial field at the surface of layer 1 is

$$H_A = \frac{2(A_{12} + |K_S|)}{M_1}, \quad K_S < 0.$$

$A_{12} > 0$ implies ferromagnetic coupling, and therefore, also implies an easy axis of magnetization normal to the film in this case. Second, the other effect due to the coupling is dynamic whereby the motion of $\vec{m}_1$ at the surface of layer 1 is restrained by the surface dynamic moment at layer 2, $\vec{m}_2$.

At layer 2 surface, we have from symmetry

$$A_2 \frac{\partial \vec{m}_2}{\partial y} + K_S \vec{m}_2 - A_{12}\left(\vec{m}_2 - \frac{M_2}{M_1}\vec{m}_1\right) = 0. \qquad (12.31b)$$

Our discussion about Equation 12.31a applies equally well to Equation 12.31b. Now we must be careful about the origin of K_S or at which surface it applies, 1 or 2. Rewrite Equations 12.31a and 12.31b as

$$A_1 \frac{\partial \vec{m}_1}{\partial y} - \left(K_S^{(1)} + A_{12}\right)\vec{m}_1 = -A_{12}\frac{M_1}{M_2}\vec{m}_2, \qquad (12.31c)$$

$$A_2 \frac{\partial \vec{m}_2}{\partial y} - \left(K_S^{(2)} + A_{12} \right) \vec{m}_2 = -A_{12} \frac{M_2}{M_1} \vec{m}_1. \tag{12.31d}$$

Both $K_S^{(1)}$ and A_{12} are positive. For simplicity, let $K_S^{'(1)} = K_S^{(1)} + A_{12}$ and $K_S^{'(II)} = K_S^{(2)} + A_{12}$, and furthermore, we assume that $K_S^{'(I,II)} > 0$. Then, we can simply rewrite Equations 12.31c and 12.31d as

$$A_1 \frac{\partial \vec{m}_1}{\partial y} - K_S^{'(I)} \vec{m}_1 = -A_{12} \frac{M_1}{M_2} \vec{m}_2, \tag{12.31e}$$

$$A_2 \frac{\partial \vec{m}_2}{\partial y} - K_S^{'(II)} \vec{m}_2 = -A_{12} \frac{M_2}{M_1} \vec{m}_1. \tag{12.31f}$$

The exchange constant for layer 1 is A_1 and layer 2 is A_2.

Let's now obtain the [A] matrix for the layered structure, where the dielectric layer is characterized by

$$\mu = \mu_0 \text{ and } \varepsilon = \varepsilon' - j\varepsilon''.$$

BOUNDARY CONDITIONS

$y = 0$:

$$\sum_i^2 \left(h_i^+ + h_i^- \right) = h_1, \tag{12.32}$$

$$\sum_i^2 Z_i \left(h_i^+ - h_i^- \right) = e_1, \tag{12.33}$$

$$\sum_i^2 \left(P_i h_i^+ + R_i h_i^- \right) = 0. \tag{12.34}$$

The first two equations are the usual Maxwell surface boundary conditions and the last equation is the spin boundary condition at $y = 0$ (layer 1). As such, the last equation is not coupled to any other layer via exchange.

$y = t_1$:

$$\sum_i^2 \left(h_i^+ e^{-jk_i t_1} + h_i^- e^{+jk_i t_1} \right) = h_0^+ e^{-jk_0 t_1} + h_0^- e^{+jk_0 t_1}, \tag{12.35}$$

$$\sum_i^2 Z_i \left(h_i^+ e^{-jk_it_1} - h_i^- e^{+jk_it_1} \right) = Z_0 \left(h_0^+ e^{-jk_0t_1} - h_0^- e^{+jk_0t_1} \right), \tag{12.36}$$

$$\sum_i^2 \left(P_i' h_i^+ e^{-jk_it_1} + R_i' h_i^- e^{+jk_it_1} \right) = -A_{12} \frac{M_1}{M_2} \left[\sum_i^2 Q(k_i) \left(h_{i+2}^+ e^{-jk_it_2} + h_{i+2}^- e^{+jk_2} \right) \right]. \tag{12.37}$$

The first two equations are electromagnetic boundary conditions between magnetic layer 1 and the dielectric layer. Notice that $h_0^\pm$ are internal fields associated with the dielectric layer, and Z_0 and k_0 are the characteristic impedance and propagation constant, respectively. The last term shows the exchange coupling between the surface moments in layers 1 and 2. Notice, for example, on the RHS, the term in the square bracket represents m_2 evaluated at $y=t_2$.

$y=t_2$:

$$\sum_i^2 \left(h_{i+2}^+ e^{-jk_it_2} + h_{i+2}^- e^{+jk_it_2} \right) = h_0^+ e^{-jk_0t_2} + h_0^- e^{+jk_0t_2}, \tag{12.38}$$

$$\sum_i^2 Z_i \left(h_{i+2}^+ e^{-jk_it_2} - h_{i+2}^- e^{+jk_it_2} \right) = Z_0 \left(h_0^+ e^{-jk_0t_2} - h_0^- e^{+jk_0t_2} \right), \tag{12.39}$$

$$\sum_i^2 \left(P_i' h_{i+2}^+ e^{-jk_it_2} + R_i' h_{i+2}^- e^{+jk_it_2} \right) = -A_{12} \frac{M_2}{M_1} \left[\sum_i^2 Q(k_i) \left(h_i^+ e^{-jk_1t_1} + h_i^- e^{+jk_1t_1} \right) \right]. \tag{12.40}$$

The first two equations are very similar to the boundary conditions at $y=t_1$, but now it is applied at $y=t_2$. Again, the last equation is the reverse of the similar equation as above, with the exception that the term in the square bracket is m_1 evaluated at $y=t_1$.

$y=t_3$:

$$\sum_i^2 \left(h_{i+2}^+ e^{-jk_it_3} + h_{i+2}^- e^{+jk_it_3} \right) = h_2, \tag{12.41}$$

$$\sum_i^2 Z_i \left(h_{i+2}^+ e^{-jk_it_3} - h_{i+2}^- e^{+jk_it_3} \right) = e_2, \tag{12.42}$$

$$\sum_i^2 \left(P_i h_{i+2}^+ e^{-jk_it_3} + R_i h_{i+2}^- e^{+jk_it_3} \right) = 0. \tag{12.43}$$

The boundary conditions here are very similar to the ones at $y=0$ except now $y=t_3$.

In above definitions of P_i' and R_i', one needs to replace K_S by $(K_S + A_{12})$ relative to the definitions of P_i and R_i in Chapter 11. The above set of equations only considers three layers, two magnetic and one dielectric. The reader is reminded that there are in total $2N + 1$ similar equations.

PROCEDURE FOR SOLUTION

By collecting all of the terms in the internal fields $\left(h_i^\pm\right)_1$ on the right-hand side, we may summarize the boundary condition equations simply as

$$[f]_n = [a]_n [h]_n,$$

where n indicates the particular surface or interface in question. Except for $n = 1$ and $n = 2N + 1$ (the first and last surfaces), $[f]_n$ is a (4×1) column vector matrix in which all the elements are equal to zero; $[a]$ is a (4×8) matrix whose matrix elements are the coefficients multiplying the internal field variables $\left(h_i^\pm\right)_1$ and $\left(h_i^\pm\right)_2$.

At the first and last layer, there is only one spin boundary condition involving only surface fields. Thus,

$$[f]_1 = \begin{bmatrix} h_1 \\ e_1 \\ 0 \\ 0 \end{bmatrix}$$

and

$$[f]_{2N+1} = \begin{bmatrix} 0 \\ 0 \\ h_2 \\ e_2 \end{bmatrix},$$

where h and e are the surface magnetic and electric fields at the first and last surfaces, respectively. However, $[h]_1$ and $[h]_{2N+1}$ are still (8×1) column matrices. There are four internal fields in layer 1 and four internal fields in layer 2. We can now put all of the boundary equations into one compact matrix and that is

$$[F] = [A][h],$$

where

$$[F] = \begin{bmatrix} [f]_1 \\ [f]_2 \\ \vdots \\ [f]_{2N+1} \end{bmatrix} ; [h] = \begin{bmatrix} [h]_1 \\ [h]_2 \\ \vdots \\ [h]_{2N+1} \end{bmatrix}$$

and

$$
[A] = \begin{bmatrix}
[A]_1 & 0 & 0 & 0 & 0 \\
0 & [A]_2 & 0 & 0 & 0 \\
0 & 0 & [A]_2 & 0 & 0 \\
0 & 0 & 0 & . & 0 \\
0 & 0 & 0 & 0 & [A]_{2N+1}
\end{bmatrix}.
$$

The $[A]$ matrix is non-diagonal of dimensionality $[4(2N)+6] \times [8(2N+2)]$. The $[F]$ and $[h]$ matrices are column matrices with dimensionalities $[4(2N+6)] \times 1$ and $[8(2N+2)] \times 1$, respectively. $[F]$ may be referred to as the "excitation" matrix, since it contains the incident field amplitudes. $[h]$ contains all internal field amplitudes. The matrix $[A]$ may be viewed as the "grand" transfer function matrix of the system.

The object of this calculation is to express h_1 and e_1 in terms of h_2 and e_2. For the case of no exchange coupling between layers, the relationship between the two sets of surface fields is straightforward, as shown in the previous section. This relationship is simply represented by a 2×2 matrix. From this 2×2 matrix, it may be possible to calculate the surface impedance, Z_S, at the two surfaces, see the previous section. For the case of the exchange coupling between layers, the problem is that $[A]_n$ are non-square matrices as a result of the coupling. However, the two sets of surface fields are still represented by a 2×2 matrix. Clearly, once this relationship is established, we can simply take over all of the algebraic steps used to obtain Z_S in the previous section. One can always use standard matrix inversion of the whole set of equations to solve for the internal fields and construct the response for surface impedance, for example. However, if there are too many layers, this standard procedure can be demanding on a computer memory. We have adopted a form of Gaussian elimination procedure to eliminate the variables of each layer sequentially.

Using the four equations at the interface between the first two layers 1, for example, we can eliminate the variables $\left(h_i^\pm\right)_1$ from the three equations at surface 1. The subscript 1 indicates internal variables in layer 1. We then have three equations for h_1 and e_1 in terms of $\left(h_i^\pm\right)_2$. Using the equations for the next interface, we can then eliminate $\left(h_i^\pm\right)_2$ and relate $\left(h_i^\pm\right)_1$ to e_1 to $\left(h_i^\pm\right)_3$ and on and on. This allows us to construct a "transmission matrix" so that h_1 and e_1 can be expressed in terms of h_2 and e_2.

We may express the electromagnetic fields at the first surface in terms of fields at the last surface by writing

$$
\begin{bmatrix} e_2 \\ h_1 \end{bmatrix} = \begin{bmatrix} a_{11} & a_{12} \\ a_{21} & a_{22} \end{bmatrix} \begin{bmatrix} e_2 \\ h_2 \end{bmatrix}.
$$

Since we have more than one layer, $a_{11} \neq a_{22}$, but $\det[A] = 1$.

The matrix elements a_{11}, a_{12}, a_{21}, and a_{22} are complicated functions of A, A_{12}, g, and H. The essential point here is that, even for exchange coupling between layers, it may be possible to express the microwave fields at the two surfaces by a simple 2×2 matrix. The ratio of the electric to magnetic fields gives the surface impedance of the sample. The real part of the surface impedance is related to the sample FMR absorption. Absorption curves are plotted in Figure 12.10 for a typical magnetic metal. Refer to the previous section for the calculation of surface impedance.

In Figure 12.11, only the exchange coupling between layers is varied in plotting the shift from the uniform FMR mode in units of Oe. The major advantage of our approach over standard calculational procedure is that since we sequentially

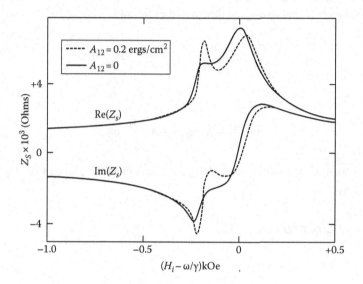

FIGURE 12.10 FMR spectra of exchange-coupled magnetic layers.

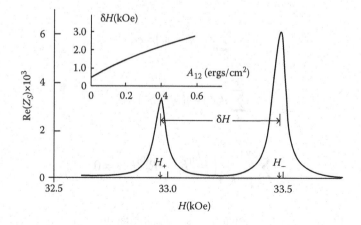

FIGURE 12.11 FMR spectra of exchange-coupled superlattice.

eliminate variables using the boundary conditions at each interface, we never have to operate with a system of equations with more than seven linear equations. Adding additional layers increases the computation time only linearly, rather than by $N!$ as in standard matrix inversion approaches.

(//) CASE

There are now 16 equations instead of 12 for each set of coupling layers. There is one extra spin precessing boundary condition at each surface. However, the same procedure used in the previous section in solving for internal fields is applicable. The spin boundary conditions at the first layer are of the same form. The coupling between magnetic layers (1 and 2) may be designated as follows:

$$A\frac{\partial m_{1y}}{\partial x} - \left(K_S^{(1)} + A_{12}\right)m_{1y} = -A_{12}\frac{M_1}{M_2}m_{2y} \tag{12.44}$$

and

$$A\frac{\partial m_{1x}}{\partial y} - A_{12}m_{1x} = -A_{12}\frac{M_1}{M_2}m_{2x}. \tag{12.45}$$

The easy axis of magnetization is along the z-axis or in the plane of the layer. Both $K_S^{(1)}$ and $A_{12} > 0$.

Boundary Conditions (// FMR)

$y = 0$:

$$\sum_i^3 \left(h_i^+ + h_i^-\right) = h_1, \tag{12.46}$$

$$\sum_i^3 Z_i\left(h_i^+ - h_i^-\right) = e_1, \tag{12.47}$$

$$\sum_i^3 jk_iQ(k_i)\left(h_i^+ - h_i^-\right) = 0, \tag{12.48}$$

$$\sum_i^3 \frac{\chi(k_i)}{\mu(k_i)}\left(P_i^{(0)}h_i^+ + R_i^{(0)}h_i^-\right) = 0. \tag{12.49}$$

Although the above set of equations is very similar to Equations 12.32 through 12.34, there appears to be an extra equation. The extra equation is a result of an extra spin condition for H in the film plane (compare Equations 11.30a and 11.30b with Equation 11.29).

$y = t_1$:

$$\sum_i^3 \left(h_i^+ e^{-jk_i t_1} + h_i^- e^{+jk_i t_1} \right) = h_0^+ e^{-jk_0 t_1} + h_0^- e^{+jk_0 t_1}, \tag{12.50}$$

$$\sum_i^3 Z_i \left(h_i^+ e^{-jk_i t_1} - h_i^- e^{+jk_i t_1} \right) = Z_0 \left(h_0^+ e^{-jk_0 t_1} - h_0^- e^{+jk_0 t_1} \right), \tag{12.51}$$

$$\sum_i^3 \left(P_i' h_i^+ e^{-jk_i t_1} + R_i' h_i^- e^{+jk_i t_1} \right) = -A_{12} \frac{M_1}{M_2} \left[\sum_i^3 Q(k_i) \left(h_{i+3}^+ e^{-jk_i t_2} + h_{i+3}^- e^{+jk_i t_2} \right) \right], \tag{12.52}$$

$$\sum_i^3 \frac{\chi(k_i)}{\mu(k_i)} \left(P_i'' h_i^+ e^{-jk_i t_1} + R_i'' h_i^- e^{+jk_i t_1} \right)$$

$$= -A_{12} \frac{M_1}{M_2} \left[\sum_i^3 \frac{\chi(k_i)}{\mu(k_i)} Q(k_i) \left(h_{i+3}^+ e^{-jk_i t_2} + h_{i+3}^- e^{+jk_i t_2} \right) \right]. \tag{12.53}$$

The quantity inside the square bracket in Equation 12.52 is identified as m_{2x} evaluated at $y = t_2$. The quantity inside the square bracket in Equation 12.53 is m_{2y} evaluated at the same surface.

$y = t_2$:

$$\sum_i^3 \left(h_{i+3}^+ e^{-jk_i t_2} + h_{i+3}^- e^{+jk_i t_2} \right) = h_0^+ e^{-jk_0 t_2} + h_0^- e^{+jk_0 t_2}, \tag{12.54}$$

$$\sum_i^3 Z_i \left(h_{i+3}^+ e^{-jk_i t_2} - h_{i+3}^- e^{+jk_i t_2} \right) = Z_0 \left(h_0^+ e^{-jk_0 t_2} - h_0^- e^{+jk_0 t_2} \right), \tag{12.55}$$

$$\sum_i^3 \left(P_i' h_{i+3}^+ e^{-jk_i t_2} + R_i' h_{i+3}^- e^{+jk_i t_2} \right) = -A_{12} \frac{M_2}{M_1} \left[\sum_i^3 Q(k_i) \left(h_i^+ e^{-jk_i t_1} + h_i^- e^{+jk_i t_1} \right) \right], \tag{12.56}$$

$$\sum_i^3 \frac{\chi(k_i)}{\mu(k_i)} \left(P_i'' h_{i+3}^+ e^{-jk_i t_2} + R_i'' h_{i+3}^- e^{+jk_i t_2} \right)$$

$$= -A_{12} \frac{M_2}{M_1} \left[\sum_i^3 \frac{\chi(k_i)}{\mu(k_i)} Q(k_i) \left(h_i^+ e^{-jk_i t_1} + h_i^- e^{+jk_i t_1} \right) \right]. \tag{12.57}$$

The exchange coupling ensures that the interaction is reciprocated and the quantity inside the square bracket in Equation 12.56 is identified as m_{1x} evaluated at $y=t_1$. The quantity inside the square bracket in Equation 12.57 is m_{1y} evaluated at the same surface.

$y=t_3$:

$$\sum_{i}^{3}\left(h_{i+3}^{+}e^{-jk_it_3} + h_{i+3}^{-}e^{+jk_it_3}\right) = h_2, \tag{12.58}$$

$$\sum_{i}^{3} Z_i\left(h_{i+3}^{+}e^{-jk_it_3} - h_{i+3}^{-}e^{+jk_it_3}\right) = e_2, \tag{12.59}$$

$$\sum_{i}^{3} jk_i Q(k_i)\left(h_{i+3}^{+}e^{-jk_it_3} - h_{i+3}^{-}e^{+jk_it_3}\right) = 0, \tag{12.60}$$

$$\sum_{i}^{3} \frac{\chi(k_i)}{\mu(k_i)}\left(P_i^{(t_3)}h_{i+3}^{+}e^{-jk_it_3} + R_i^{(t_3)}h_{i+3}^{-}e^{+jk_it_3}\right) = 0. \tag{12.61}$$

In the above definitions of P_i', R_i', P_i'', and R_i'', one needs to replace K_S by (K_S+A_{12}) relative to the definitions of P_i and R_i (see Chapter 11).

There is no exchange coupling terms between layers here and the four equations are very similar to Equations 12.46 through 12.49 (the opposite surface). There are six internal fields for each magnetic layer and two internal fields for the dielectric layer. In totality, there are 14 internal fields and 16 equations. The "brute force" way of obtaining the [A] matrix is to solve the 14 internal fields in terms of e_2 and h_2 using 14 of the 16 equations (leaving the two equations containing e_1 and h_1 out). Then, substitute the solutions in the two remaining equations containing e_1 and h_1. After much unscrambling, the [A] matrix may be obtained. In general, there are $8N+6$ internal fields and $8N+8$ equations, where N represents magnetic and dielectric layer pairs, and the last layer in the multilayered structure is magnetic. In order to put a quantitative perspective to this problem, let's assume $N=100$ pairs. This means that there are 806 internal fields and 808 equations. In order to obtain the [A] matrix, we need to invert an 806×806 matrix! Simplification may be found in the use of the so-called Gaussian elimination procedure discussed in the previous section.

In summary, the electrical properties of a multilayered structure with no exchange coupling between layers may be explained in terms of a "localized" [A] matrix in which the total matrix is the product of localized [A] matrices. However, when exchange coupling between layers is included, the "localized" picture of the [A] matrix is invalid.

One must then treat the whole multilayered structure as a whole entity. Thus, the exchange coupling binds all the layers together as a whole like a "glue" between layers. That is the result of the exchange coupling and that is the essence of magnetism. MATLAB® codes are provided in the appendix for the calculation of the A matrix.

PROBLEMS

12.1 A distributive impedance element may be represented in a transmission line by the following 2×2 matrix:

$$\begin{bmatrix} V_1 \\ I_1 \end{bmatrix} = \begin{bmatrix} 1 & 0 \\ \dfrac{1}{Z} & 1 \end{bmatrix} \begin{bmatrix} V_2 \\ I_2 \end{bmatrix}.$$

Calculate S_{11} and S_{21} assuming that the characteristic impedance of the transmission line equals Z_0, where S_{11} is the reflection coefficient and S_{21} the transmission coefficient.

12.2 An electromagnetic wave is normally incident upon a film characterized by a characteristic impedance Z and effective propagation constant k. Calculate S_{11} and S_{21}.

12.3 Assume symmetrical excitation of the same film as in the above example. Calculate S_{11} and S_{21}.

12.4 For most cases of interest, the resultant 2×2 matrix is of the form:

$$\begin{bmatrix} V_1 \\ I_1 \end{bmatrix} = \begin{bmatrix} 1+\delta & ja\delta \\ j\dfrac{\delta}{a} & 1-\delta \end{bmatrix} \begin{bmatrix} V_2 \\ I_2 \end{bmatrix}.$$

You can represent the above matrix in this form:

$$\begin{bmatrix} \cos kl & jZ\sin kl \\ \dfrac{j\sin kl}{Z} & \cos kl \end{bmatrix}$$

12.5 Re-derive coefficients a_{11}, a_{12}, a_{21}, and a_{22} for H in the film plane.

12.6 If the magnetic motion is coupled to the elastic motion, would you expect the transfer function matrix to be a 2×2 matrix? Explain.

12.7 Write the transfer function matrix for a superconducting film. Define all the variables in the matrix.

12.8 A superconducting wire is placed at the center of a waveguide. The equivalent circuit of the wire-waveguide system is R, L, and C. The capacitance C is introduced to represent the gap between the end of the wire and the waveguide walls. R and L represent the superconducting properties of the wire. Calculate the transmission coefficient for the microwave energy propagating past the wire.

12.9 Repeat Problem 12.8 except that the wire is replaced by a magnetic film covering the cross section in the waveguide.

12.10

  a. Show that

$$\begin{bmatrix} \cos kl & \sin kl \\ -\sin kl & \cos kl \end{bmatrix}^n = \begin{bmatrix} \cos nkl & \sin nkl \\ -\sin nkl & \cos nkl \end{bmatrix}.$$

  b. Also show that

$$\begin{bmatrix} \cos kl & jZ\sin kl \\ \dfrac{j\sin kl}{Z} & \cos kl \end{bmatrix}^n = \begin{bmatrix} \cos nkl & jZ\sin nkl \\ \dfrac{j\sin nkl}{Z} & \cos nkl \end{bmatrix}.$$

12.11 Find eigenvalues of the matrix in Problems 12.10a and b.

12.12 The subject matters of nanowires and magneto-impedances are of great interest. How would you utilize the concept of matrix representation of wave propagation in nanowires and magneto-impedance?

    No solution is provided.

APPENDIX 12.A

CALCULATION OF TRANSMISSION LINE PARAMETERS FROM [A] MATRIX

The A matrix contains matrix elements a_{11}, a_{12}, a_{21}, and a_{22} (see text for derivation of these parameters). In a single layer transmission line, parameters like the characteristic impedance, Z_{eff}, and propagation constant, k_{eff}, may be calculated from

$$[A] = \begin{bmatrix} a_{11} & a_{12} \\ a_{21} & a_{22} \end{bmatrix} = \begin{bmatrix} \cos k_{eff}t & jZ_{eff}\sin k_{eff}t \\ \dfrac{j\sin k_{eff}t}{Z_{eff}} & \cos k_{eff}t \end{bmatrix}.$$

Using the identity $a_{11}a_{22} - a_{12}a_{21} = 1$, it is simple to show that

$$Z_{eff} = \sqrt{a_{12}/a_{21}}$$

and

$$e^{jk_{eff}t} = a_{11} + \sqrt{a_{12}a_{21}}, \text{ where}$$

$$Z_{eff} = \sqrt{\frac{\mu_{eff}}{\varepsilon_{eff}}} \text{ and}$$

$$k_{eff} = \omega\sqrt{\mu_{eff}\varepsilon_{eff}}$$

Alternatively, one may express a's in terms of transmission line parameters μ_{eff} and ε_{eff} as follows:

$$\mu_{eff} = \frac{k_{eff}Z_{eff}}{\omega}$$

and

$$\varepsilon_{eff} = \frac{k_{eff}}{\omega Z_{eff}}.$$

For magnetic metals like permalloy, it is convenient to introduce an effective conductivity, σ_{eff}, in terms of ε_{eff} or

$$\sigma_{eff} = j\omega\varepsilon_{eff}.$$

For symmetrical excitation of a thin film, it can be shown that the surface impedance, Z_S, as calculated in Chapter 11 (see Appendix), may be expressed in terms of a's or Z_{eff} and k_{eff}.

$$Z_S = \frac{a_{11}-1}{a_{21}}$$

or

$$Z_S = jZ_{eff}\tan\frac{k_{eff}t}{2}.$$

We have assumed in the above derivations that

$$a_{11} = \cos\left(k_{eff}t\right),$$

$$a_{12} = jZ_{eff}\sin\left(k_{eff}t\right),$$

$$a_{21} = j\sin\left(k_{eff}t\right)/Z_{eff}, \text{ and}$$

$$a_{22} = a_{11}.$$

We have numerically calculated Z_{eff}, k_{eff}, μ_{eff}, ε_{eff}, σ_{eff}, and Z_S using MATLAB® program (see computer programs below). We have assumed the same parameters as in Chapter 11 for the calculation of Z_S in permalloy. The external applied field is in the film plane (Figures 12A.1–12A.5).

In the above calculations, frequency is fixed at 9.53 GHz. For details of the numerical calculations, the MATLAB® computer program is provided below:

```
%Permalloy in plane FMR
clc;
clear all;
close all;
global w mu0 Qi1s Qi2s Qi3s k1s k2s k3s h1r h1l h2r h2l h3r
h3l Z1s Z2s Z3s
eps
g=2.1;
fourpiMs=10000;
%Hext=1103.3;
dH=30;
sigma=0.7e5;
A=1.14e-6;
mu0=4*pi*10^-9;
f=9.53e9;
e0=1/(36*pi)*10^-11;
gamma=g.*0.8805e7;
omega=2.*pi*f-j.*(dH.*0.924525e7);
w=omega;
topn=j.*fourpiMs.*w/gamma;
H0=10:1:3000;
L=length(H0);
H1=H0;
H2=H0+fourpiMs;
e=-j.*sigma./omega;
M=fourpiMs./(4.*pi);
k0=sqrt(omega.^2.*e.*mu0);
a1=1;
a2=(2.*H0+fourpiMs)./(2.*A./M) -k0.^2;
a3=(H0.*(H0+fourpiMs)-(omega./gamma).^2)./(4.*A.^2./M.^2) -
2.*(H0+fourpiMs)./(2.*A./M).*k0.^2;
a4=((omega./gamma).^2-(H0+fourpiMs).^2)./
(4.*A.^2./M.^2).*k0.^2;
a=a2;
b=a3;
c=a4;
[k1_sq,k2_sq,k3_sq]=solve3order(a,b,c);
k0_sq=-j.*omega.*sigma.*mu0;%k0squared
```

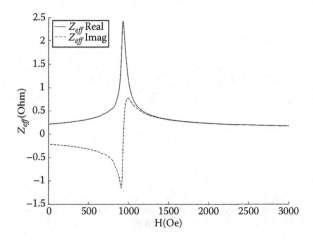

FIGURE 12A.1 Characteristic impedance of 0.3 μm permalloy film plotted as a function of H.

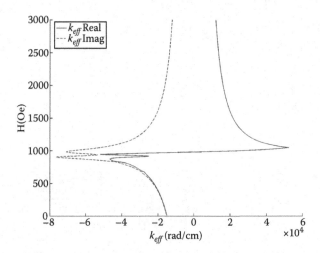

FIGURE 12A.2 Effective propagation constant of 0.3 μm permalloy film.

```
k1 =-sign(imag(k1_sq)).*sqrt(k1_sq);
k2 =-sign(imag(k2_sq)).*sqrt(k2_sq);
k3 =-sign(imag(k3_sq)).*sqrt(k3_sq);
figure(1);%k vs H
hold on;
plot(real(k1),H0,'b','Linewidth',2);
```

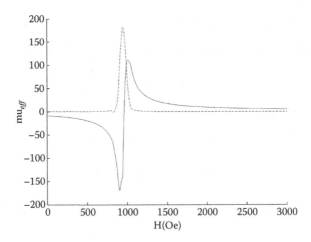

FIGURE 12A.3　Effective permeability of permalloy plotted as a function of H.

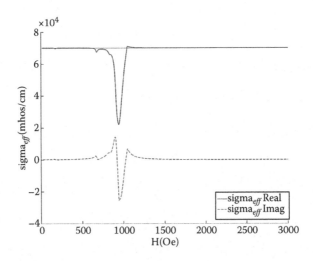

FIGURE 12A.4　Effective conductivity of permalloy plotted as a function of H.

```
plot(imag(k1),H0,'b--','Linewidth',2);
plot(real(k2),H0,'r','Linewidth',2);
plot(imag(k2),H0,'r--','Linewidth',2);
plot(real(k3),H0,'g','Linewidth',2);
plot(imag(k3),H0,'g--','Linewidth',2);
% plot(H0,real(k3),'g','Linewidth',2);
% plot(H0,imag(k3),'g--','Linewidth',2);
% plot(H0,abs(real(k4)),'cyan','Linewidth',2);
```

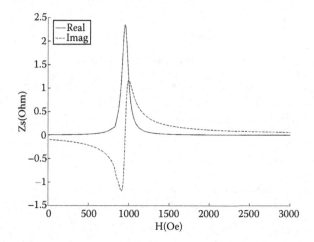

FIGURE 12A.5 Z_S of 0.3 μm permalloy film plotted as a function of H. The above plot was obtained using matrix elements of the [A] matrix. The point here is that it gives exactly the same results as the surface impedance calculated from electromagnetic boundary conditions formulated in Chapter 11.

```
% plot(H0,imag(k4),'cyan--','Linewidth',2);
ylabel('H0(Oe)','Fontsize',24);
%xlabel('Im(k) & Re(k) (cm^-^1)','Fontsize', 24);
xlabel('k(rad/cm)','Fontsize', 24);
%legend('Re(k)>0','Im(k)<0',2);
h=gca;
set(h,'Fontsize',22);
title('Permalloy Dispersion Diagram','FontSize',24);
hold off;
% Calculating the surface impedance of a permalloy film with
magnetization
% Parallel to the plane. Theta_k=pi/2.
% CGS units.
%clc;
%clear all;
%close all;
%global d V
%global w mu0 eps Xi1s Xi2s Xi3s k1s k2s k3s Z1s Z2s Z3s h1r
h1l h2r h2l h3r
h3l
eps0=1/(36*pi)*10^-11; % F/cm
Z0=sqrt(mu0/eps0);
eps=-j*sigma./w; % F/cm
Mu_r=1+fourpiMs.*H2./(H1.*H2-(w./gamma)^2);
% Calculate the surface impedance, only consider resonant
modes
```

```
d=0.3E-4; % cm
Di1=(H1+2.*(A./M).*k1_sq).*(H2+2.*(A./M).*k1_sq) - (w./
gamma).^2;
Di2=(H1+2.*(A./M).*k2_sq).*(H2+2.*(A./M).*k2_sq) - (w./
gamma).^2;
Di3=(H1+2.*(A./M).*k3_sq).*(H2+2.*(A./M).*k3_sq) - (w./
gamma).^2;
Xi1=topn./Di1;
Xi2=topn./Di2;
Xi3=topn./Di3;
Qi1=(1-k1_sq./k0_sq);
Qi2=(1-k2_sq./k0_sq);
Qi3=(1-k3_sq./k0_sq);
Ks_0=0; Ks_d=0;
n_minus_2pi=0;
n_plus_2pi=0;
indicator=1;
for i=1:L
S=[1 1 1 1 1 1;…
exp(-j*k1(i)*d) exp(j*k1(i)*d) …
exp(-j*k2(i)*d) exp(j*k2(i)*d) …
exp(-j*k3(i)*d) exp(j*k3(i)*d); …
Xi1(i)*(Ks_0+j*k1(i)*A) Xi1(i)*(Ks_0-j*k1(i)*A) …
Xi2(i)*(Ks_0+j*k2(i)*A) Xi2(i)*(Ks_0-j*k2(i)*A) …
Xi3(i)*(Ks_0+j*k3(i)*A) Xi3(i)*(Ks_0-j*k3(i)*A);…
Xi1(i)*(Ks_d+j*k1(i)*A)*exp(-j*k1(i)*d)Xi1(i)*(Ks_d-
j*k1(i)*A)*exp(j*k1(i)*d) …
Xi2(i)*(Ks_d+j*k2(i)*A)*exp(-j*k2(i)*d)Xi2(i)*(Ks_d-
j*k2(i)*A)*exp(j*k2(i)*d) …
Xi3(i)*(Ks_d+j*k3(i)*A)*exp(-j*k3(i)*d)Xi3(i)*(Ks_d-
j*k3(i)*A)*exp(j*k3(i)*d); …
j.*k1(i)*Qi1(i) -j.*k1(i)*Qi1(i) …
j.*k2(i)*Qi2(i) -j.*k2(i)*Qi2(i) …
j.*k3(i)*Qi3(i) -j.*k3(i)*Qi3(i); ….
j.*k1(i)*Qi1(i)*(exp(-j*k1(i)*d))
-j.*k1(i)*Qi1(i)*(exp(j*k1(i)
*d))…
j.*k2(i)*Qi2(i)*(exp(-j*k2(i)*d))
-j.*k2(i)*Qi2(i)*(exp(j*k2(i)
*d))…
j.*k3(i)*Qi3(i)*(exp(-j*k3(i)*d))
-j.*k3(i)*Qi3(i)*(exp(j*k3(i)
*d))];
%N=[1e-6 0 0 0 0 0; 0 1e-6 0 0 0 0; 0 0 1e-6 0 0 0; 0 0 0
1e-6 0 0; 0 0 0 0
1e-6 0; 0 0 0 0 0 1e-6];
b=[1 1 0 0 0 0]'; % boundary condition
%b=N*b;
%S=N*S;
```

```
h_vec(:,i)=inv(S)*b;
Z1(i)=k1(i)/(omega*eps);
Z2(i)=k2(i)/(omega*eps);
Z3(i)=k3(i)/(omega*eps);
SP=[Z1(i)*exp(-j*k1(i)*d)  -Z1(i)*exp(j*k1(i)*d)...
Z2(i)*exp(-j*k2(i)*d)  -Z2(i)*exp(j*k2(i)*d)...
Z3(i)*exp(-j*k3(i)*d)  -Z3(i)*exp(j*k3(i)*d); ....
exp(-j*k1(i)*d) exp(j*k1(i)*d) ...
exp(-j*k2(i)*d) exp(j*k2(i)*d) ...
exp(-j*k3(i)*d) exp(j*k3(i)*d); ...
Xi1(i)*(Ks_0+j*k1(i)*A) Xi1(i)*(Ks_0-j*k1(i)*A)...
Xi2(i)*(Ks_0+j*k2(i)*A) Xi2(i)*(Ks_0-j*k2(i)*A)...
Xi3(i)*(Ks_0+j*k3(i)*A) Xi3(i)*(Ks_0-j*k3(i)*A);...
Xi1(i)*(Ks_d+j*k1(i)*A)*exp(-j*k1(i)*d) Xi1(i)*(Ks_d-
j*k1(i)*A)*exp(j*k1(i)*d) ...
Xi2(i)*(Ks_d+j*k2(i)*A)*exp(-j*k2(i)*d) Xi2(i)*(Ks_d-
j*k2(i)*A)*exp(j*k2(i)*d) ...
Xi3(i)*(Ks_d+j*k3(i)*A)*exp(-j*k3(i)*d) Xi3(i)*(Ks_d-
j*k3(i)*A)*exp(j*k3(i)*d); ...
j.*k1(i)*Qi1(i)  -j.*k1(i)*Qi1(i) ...
j.*k2(i)*Qi2(i)  -j.*k2(i)*Qi2(i) ...
j.*k3(i)*Qi3(i)  -j.*k3(i)*Qi3(i); ....
j.*k1(i)*Qi1(i)*(exp(-j*k1(i)*d)) -
j.*k1(i)*Qi1(i)*(exp(j*k1(i)*d)) ...
j.*k2(i)*Qi2(i)*(exp(-j*k2(i)*d)) -
j.*k2(i)*Qi2(i)*(exp(j*k2(i)*d)) ...
j.*k3(i)*Qi3(i)*(exp(-j*k3(i)*d)) -
j.*k3(i)*Qi3(i)*(exp(j*k3(i)*d))];
% SP=N*SP;
B_mat=B_matrix(SP, Z1(i), Z2(i), Z3(i));
a11(i)=B_mat(1, 1);
D=B_mat(1, 1)*B_mat(2, 2)-B_mat(1, 2)*B_mat(2, 1);
Z_square=B_mat(1, 2)/B_mat(2, 1);
Z(i)=sqrt(Z_square);
% calculation of alpha and beta: keff=beta+jalpha
ZS(i)=(B_mat(1,1)-1)/B_mat(2,1);
emcoupl(i)=(ZS(i))/Z(i);
C(i)=B_mat(1,1)+(B_mat(1,2))/Z(i);%1/S21,matched imp.
%C=exp(jwt)=1/S21(S11=0)
thet(i)=angle(C(i));
beta(i)=thet(i)/d;
alpha(i)=-log(abs(C(i)))/d;
% if beta(i)>0
% beta(i)=-2*pi/d+beta(i);
% end
if real(C(i))<0
R(i)=imag(C(i))/real(C(i));
beta(i)=-atan(R(i))/d;
end
```

```
keff(i)=beta(i)+j*alpha(i);
emcoupl(i)=j*tan(keff(i)*(d/2));
% emcoupl(i)=ZS(i)/Z(i);
% ZS(i)=j.*Z(i)*tan(keff(i)*(d/2));
% ZS(i)=B_mat(1,1);
Mueff(i)=(keff(i)*Z(i))/(w.*mu0);
epseff(i)=(keff(i))/(Z(i)*w);
sigma_eff=epseff.*j.*w;
Zeff(i)=sqrt((Mueff(i)*mu0)./epseff(i));
Zs_0(i)=Z1(i)*(h_vec(1,i) -h_vec(2,i))+Z2(i)*(h_vec(3,i)
-h_vec(4,i))+ Z3(i)*(h_vec(5,i) -h_vec(6,i));
Zs_d(i)=Z1(i)*(h_vec(1,i)*exp(-j*k1(i)*d)-h_
vec(2,i)*exp(j*k1(i)
*d))+Z2(i)*(h_vec(3,i)*exp(-j*k2(i)*d)-h_
vec(4,i)*exp(j*k2(i)*d))+...
Z3(i)*(h_vec(5,i)*exp(-j*k3(i)*d)-h_vec(6,i)*exp(j*k3(i)*d));
% Check the result by integrating the poynting vector.
Qi1s=Qi1(i); Qi2s=Qi2(i); Qi3s=Qi3(i); k1s=k1(i);
k2s=k2(i);
k3s=k3(i); Z1s=Z1(i); Z2s=Z2(i);Z3s=Z3(i);
h1r=h_vec(1,i); h1l=h_vec(2,i); h2r=h_vec(3,i); h2l=h_vec
(4,i); h3r=h_vec(5,i);h3l=h_vec(6,i);
Integ(i)=quadgk(@myfun4,0,d); % 2Zs(i)
Integ1(i)=quadgk(@myfun5,0,d)/d; % Effective permeability
by volume
integration
% Integ2(i)=quadgk(@myfun6,0,d)/(Integ(i)/2)^2/d;% Effective
permittivity by volume integration
% Integ3(i)=quadgk(@myfun3,0,d);
Integ2(i)=quadgk(@myfun6,0,d)/d;
end
figure(2);%char. impedance,Z, vs H (circ. pol.)
hold on;
plot(H0,real(Z),'r','Linewidth',2);
plot(H0,imag(Z),'r--','Linewidth',2);
% plot(H0, -real(Integ/2),'b','LineWidth',1);
% plot(H0, -imag(Integ/2),'b--','LineWidth',1);
xlabel('H(Oe)','Fontsize',24);
ylabel('Zeff(Ohm)','Fontsize', 24);
% axis([2500 4500 -100 100]);
h=gca;
set(h,'Fontsize',22);
title('characteristic impedance of 0.3 \mum Permalloy Film',
'FontSize',24);
legend('Zeff Real','Zeff Imag',2);
% legend('y=0 Real','y=0 Imag','y=d Real','y=d
Imag','Integ Real',
'Integ Imag',4);
hold off;
```

```
figure(3);%effective k vs H (calc. from 2×2 matrix)
hold on;
plot(real(keff),H0,'r','Linewidth',2);
plot(imag(keff),H0,'r--','Linewidth',2);
ylabel('H(Oe)','Fontsize',24);
xlabel('keff(rad/cm)','Fontsize', 24);
h=gca;
set(h,'Fontsize',22);
title('Propogation constant of 0.3 \mum Permalloy
Film','FontSize',24);
legend('keff real','keff imag',2);
hold off;
%figure(3);%effective k vs H (calc. from 2×2 matrix)
%hold on;
%plot(H0,beta,'r','Linewidth',2);
%plot(H0,alpha,'r--','Linewidth',2);
%xlabel('H(Oe)','Fontsize',24);
%ylabel('k(rad/cm)','Fontsize', 24);
%h=gca;
%set(h,'Fontsize',22);hold off;
%title('Propogation constant of 0.3 \mum Permalloy
Film','FontSize',24);
%legend('beta','alpha',2);
figure(4);% surface impedance, Zs, vs H (from 4×4 matrix)
hold on;
plot(H0, real(Zs_0),'r','Linewidth',2);
plot(H0, imag(Zs_0),'r--','Linewidth',2);
% plot(Ho,phi,'r','Linewidth',2);
xlabel('H(Oe)','Fontsize',24);
ylabel('Zs(Ohm)','Fontsize', 24);
h=gca;
set(h,'Fontsize',22);
title('Zs_0 of 0.3 \mum Permalloy Film','FontSize',24);
legend('real','Imag',2);
hold off;
figure (5);%surface impedance,ZS,vs H (from 2×2 matrix)
hold on;
plot(H0,real(ZS),'b','Linewidth',2);
plot(H0,imag(ZS),'b--','Linewidth',2);
xlabel('H(Oe)','Fontsize',24);
ylabel('ZS(Ohm)','Fontsize', 24);
h=gca;
set(h,'Fontsize',22);
title('ZS of 0.3 \mum Permalloy Film','FontSize',24);
legend('real','Imag',2);
hold off;
figure(6);%mueff vs H (from poynting vector;mu vs H(from
tensor mu)
hold on;
```

```
plot(H0, real(Mueff),'b','LineWidth',1);
plot(H0, -imag(Mueff),'b--', 'LineWidth',1);
%plot(H0, real(Integ1),'b','LineWidth',1);
%plot(H0, imag(Integ1),'b-', 'LineWidth',1);
xlabel('H(Oe)', 'FontSize', 24);
ylabel('mueff', 'FontSize',24);
%legend('Mueff Real','Mueff Imag','MuS Real','MuS Imag',1);
h=gca;
set(h,'Fontsize',22);
title('Permeability ','FontSize',24);
hold off;
ZSM2=(abs(Zs_0)).^2;
MuS=mu0*Integ1;
epsS=Integ2./ZSM2;
sigmaS=epsS.*(j.*w);
figure(7);%epseff vs H (from poynting vector)
hold on;
plot(H0, real(epseff),'b','LineWidth',2);
plot(H0, imag(epseff),'b--', 'LineWidth',2);
xlabel('H(Oe)', 'FontSize', 24);
ylabel('\epsilon_e_f_f(f/cm)', 'FontSize',32);
legend('\epsilon_e_f_f Real','\epsilon_e_f_f Imag',4);
h=gca;
set(h,'Fontsize',30);
title('Effective Average Permittivity','FontSize',24);
hold off;
figure(8);
hold on;
plot(H0, real(sigma_eff),'LineWidth',2);
plot(H0, imag(sigma_eff),'r--','LineWidth',2);
plot(H0, sigma,'r','LineWidth',2);
h=gca;
set(h,'Fontsize',30);
hold off;
xlabel('H(Oe)', 'FontSize', 30);
ylabel('sigma_e_f_f(mhos/cm)', 'FontSize',32);
title('Effective conductivity','FontSize',30);
legend('sigma_e_f_f Real','sigma_e_f_f Imag',4);
figure(9);%Zeff vs H (calculated from sqrt(mu0*mueff/epseff)
hold on;
plot(H0, real(Zeff),'LineWidth',2);
plot(H0, imag(Zeff),'r','LineWidth',2);
h=gca;
set(h,'Fontsize',30);
hold off;
xlabel('H(Oe)', 'FontSize', 30);
ylabel('Z_e_f_f(Ohm)', 'FontSize',32);
title('Z_e_f_f','FontSize',30);
legend('Z_e_f_f Real','Z_e_f_f Imag',4);
```

```
ZSM2 = (abs(Zs_0)).^2;
MuS = mu0*Integ1;
%epsS =-2.*Zs_0./(j.*w.*ZSM2.*d)+MuS./ZSM2;
epsS = Integ2./ZSM2;
%epsS = conj(epsS);
sigmaS = epsS.*j.*w;
Zsv = 0.5.*(j.*w.*d.*(MuS-(conj(epsS).*ZSM2)));% also, Zs
figure(10);%epsS vs H (from poynting vector)
hold on;
plot(H0, real(epsS),'b','LineWidth',2);
plot(H0, imag(epsS),'b—', 'LineWidth',2);
xlabel('H(Oe)', 'FontSize', 24);
ylabel('epsS(f/cm)', 'FontSize',32);
legend('epsS Real','epsS Imag',4);
h=gca;
set(h,'Fontsize',30);
title('Surface Average Permittivity','FontSize',24);
hold off;
figure(11);%sigmaS vs H (see fig. 7)
hold on;
plot(H0, real(sigmaS),'b','LineWidth',2);
plot(H0, imag(sigmaS),'r—','LineWidth',2);
plot(H0, sigmaS,'r','LineWidth',2);
h=gca;
set(h,'Fontsize',30);
hold off;
xlabel('H(Oe)', 'FontSize', 30);
ylabel('sigmaS(mhos/cm)', 'FontSize',32);
title('Surface conductivity','FontSize',30);
legend('sigmaS Real','sigmaS Imag',4);
figure(12);%Zsv vs H (calculated from poynting integral)
hold on;
plot(H0, real(Zsv),'LineWidth',2);
plot(H0, imag(Zsv),'r','LineWidth',2);
h=gca;
set(h,'Fontsize',30);
hold off;
xlabel('H(Oe)', 'FontSize', 30);
ylabel('Zsv(Ohm)', 'FontSize',32);
title('Zsv','FontSize',30);
legend('Zsv Real','Zsv Imag',4);
figure(13);%emcoupl(i)
hold on;
plot(H0, real(emcoupl),'b','LineWidth',2);
plot(H0, imag(emcoupl),'b—', 'LineWidth',2);
xlabel('H(Oe)', 'FontSize', 24);
ylabel('emcoupl)', 'FontSize',32);
legend('emcoupl real','emcoupl Imag',4);
h=gca;
```

```
set(h,'Fontsize',30);
title('em coupling','FontSize',24);
hold off;
% function A=A_matrix(SP, Z1, Z2)
% alpha_1_plus=det(SP([2:4], [2:4]))/det(SP);
% alpha_1_minus=-det(SP([2:4], [1, 3, 4]))/det(SP);
% alpha_2_plus=det(SP([2:4], [1, 2, 4]))/det(SP);
% alpha_2_minus=-det(SP([2:4], [1, 2, 3]))/det(SP);
%
%
% beta_1_plus=-det(SP([1, 3, 4], [2:4]))/det(SP);
% beta_1_minus=det(SP([1, 3, 4], [1, 3, 4]))/det(SP);
% beta_2_plus=-det(SP([1, 3, 4], [1, 2, 4]))/det(SP);
% beta_2_minus=det(SP([1, 3, 4], [1, 2, 3]))/det(SP);
%
% a11=j*Z1*(alpha_1_plus-alpha_1_minus)
+j*Z2*(alpha_2_plus-alpha
_2_minus);
%
% a12=j*Z1*(beta_1_plus-beta_1_minus)
+j*Z2*(beta_2_plus-beta
_2_minus);
%
%
a21=(alpha_1_plus+alpha_1_minus)+(alpha_2_plus+alpha_2_minus);
%
%
a22=(beta_1_plus+beta_1_minus)+(beta_2_plus+beta_2_minus);
%
% A=[a11, a12; a21, a22];
function B=B__matrix(SP, Z1, Z2, Z3)
alpha_1_plus=det(SP([2:6], [2:6]))/det(SP);
alpha_1_minus=-det(SP([2:6], [1, 3, 4,5,6]))/det(SP);
alpha_2_plus=det(SP([2:6], [1, 2, 4,5,6]))/det(SP);
alpha_2_minus=-det(SP([2:6], [1, 2, 3,5,6]))/det(SP);
alpha_3_plus=det(SP([2:6], [1, 2, 3,4,6]))/det(SP);
alpha_3_minus=-det(SP([2:6], [1:5]))/det(SP);
beta_1_plus=-det(SP([1, 3, 4,5,6], [2:6]))/det(SP);
beta_1_minus=det(SP([1, 3, 4,5,6], [1, 3, 4,5,6]))/det(SP);
beta_2_plus=-det(SP([1, 3, 4,5,6], [1, 2, 4,5,6]))/det(SP);
beta_2_minus=det(SP([1, 3, 4,5,6], [1, 2, 3,5,6]))/det(SP);
beta_3_plus=-det(SP([1, 3, 4,5,6], [1, 2,3, 4,6]))/det(SP);
beta_3_minus=det(SP([1, 3, 4,5,6], [1:5]))/det(SP);
a11=Z1*(alpha_1_plus-alpha_1_minus) +Z2*(alpha_2_plus-
alpha_2_minus) +
Z3*(alpha_3_plus-alpha_3_minus);
a12=Z1*(beta_1_plus-beta_1_minus) +Z2*(beta_2_plus-beta_2_
minus) +
Z3*(beta_3_plus-beta_3_minus);
a21=(alpha_1_plus+alpha_1_minus) +(alpha_2_plus+alpha_2_
minus) +
```

```
(alpha_3_plus+alpha_3_minus);
a22=(beta_1_plus+beta_1_minus) +(beta_2_plus+beta_2_minus)
+
(beta_3_plus+beta_3_minus);
B=[a11, a12; a21, a22];
```

MICROWAVE RESPONSE TO MICROWAVE CAVITY LOADED WITH MAGNETIC THIN FILM

The reader is referred to Chapter 8 for schematic of the experiment. Again, we assume the same experimental conditions as utilized in the above calculations. The equivalent circuit of the microwave cavity and film is shown in Figure 12A.6.

We need to clarify the manner in which matrix $[A]$ is calculated. In this chapter, the surface electric and magnetic fields are related to each other via the $[A]$ matrix. In this calculation, we are interested in determining the voltage reflected from the resonant cavity, which is proportional to the magnitude of the reflection coefficient from the cavity, $|S_{11}|$. Hence, we need to modify the $[A]$ matrix into one which is appropriate for voltage relationships. In particular, if we assume that the cross dimensions of the film are equal to each other (square cross section of the magnetic film), then the $[A]$ matrix as defined in this chapter is indeed suitable for calculating the voltage response. The cumulative matrix containing the coupling to the cavity, spacing before and after the film, and film itself may now be written as

$$
[C] = \begin{bmatrix} a & 0 \\ 0 & \dfrac{1}{a} \end{bmatrix} \begin{bmatrix} \cos kl & jZ_0 \sin kl \\ \dfrac{j \sin kl}{Z_0} & \cos kl \end{bmatrix}
$$

$$
= \begin{bmatrix} a_{11} & a_{12} \\ a_{21} & a_{22} \end{bmatrix} \begin{bmatrix} \cos kl & jZ_0 \sin kl \\ \dfrac{j \sin kl}{Z_0} & \cos kl \end{bmatrix},
$$

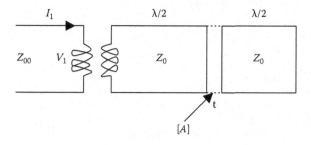

FIGURE 12A.6 Equivalent circuit of microwave cavity system including m.

where

Z_0 is the characteristic impedance of the cavity

$l = (L - t)/2$

L is the length of cavity excited in a TE_{102} resonant mode at 9.53 GHz

k is the propagation constant in the waveguide

From transmission line theory, we write the reflection coefficient as

$$S_{11} = \frac{C_{12} - C_{22}Z_{00}}{C_{12} + C_{22}Z_{00}}.$$

where Z_{00} is the characteristic impedance of the waveguide. First, we have calculated the reflection for an empty cavity and then for various cross sections of the film relative to the cavity's cross section. One may assume an incident voltage amplitude of 1 V in order to obtain order of magnitude from the response. At 9.53 GHz, the FMR field is 955 Oe (see results in "Calculation of Transmission Line Parameters from [A] Matrix" section). So, we fix the field at 955 Oe and calculate $|S_{11}|$ as a function of frequency. The results are shown in Figures 12A.7– 12A.9.

MAT LAB Computer Program:

```
clc
clear all
close all
global w mu0 Qi1s Qi2s Qi3s k1s k2s k3s h1r h1l h2r h2l h3r
h3l Z1s Z2s Z3s
eps d
d=0.3E-4;
f0=9.53e9; % unit: Hz
f1=9.51e9;
f2=9.55e9;
fc=(f1+f2)/2;
P_max=1;
c=3e8;
Q=1500; % Quality factor
length=3./sqrt(5).*c./f0;
lambda_res=length;
beta_res=2.*pi./lambda_res;
% alpha=beta_res./(2.*Q);
alpha=0.03;
e0=1./(36.*pi).*1e-9;
u0=4.*pi.*1e-7;
eta0=120.*pi;
f=9.5e9:1e5:9.6e9;
omiga=2.*pi.*f;
lambda=3./sqrt(5).*c./f;
beta=2*pi./lambda;
gamma_a_av=exp(-2.*alpha.*length);
phi_L=pi;
```

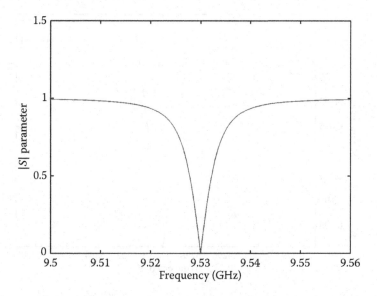

FIGURE 12A.7   Reflection magnitude is plotted as a function of frequency. The resonance frequency of the microwave cavity is at 9.53 GHz, which is typical of an EPR spectrometer.

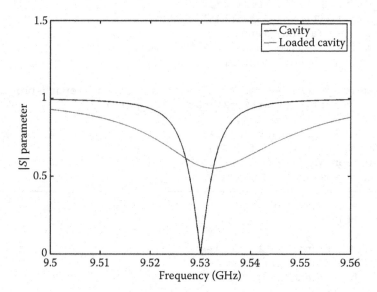

FIGURE 12A.8   Reflection magnitude is plotted as a function of frequency with and without magnetic DC field. The resonance frequency of the cavity is shifted and its Q lowered. The magnetic film is placed at the center of cavity and fills the whole cross section of the waveguide.

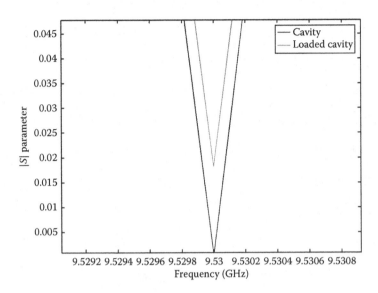

FIGURE 12A.9 Same plot as Figure 10A.8 except that the film fills only 3% waveguide's cross section which is typical of a FMR experimental situation.

```
Z00=3.*eta0./sqrt(5);
Z1_part1=Z00;
Z1_part2=1-j.*f0./f./(2.*Q);
Z1_part3=(1+exp(-2.*alpha.*length))./
(1-exp(-2.*alpha.*length));
Z1_part4=(1-gamma_a_av.^2-2.*j.*gamma_a_
av.*sin(2.*beta.*length-phi_
L))./
(1+gamma_a_av.^2-2.*gamma_a_av.*cos(2.*beta.*length-phi_L));
Z1=Z1_part1.*Z1_part2.*Z1_part3.*Z1_part4;
S11=(Z1-Z00)./(Z1+Z00);
S11_dB=20.*log10(abs(S11));
P=sqrt(1-((f-fc)./(fc-f1)).^2);
P=(P+conj(P))/2;
Pr=P.*(abs(S11)).^2;
%----------------------------------------------
% A matrix method
[aa, bb]=size(f);
H0=955;
Aw=2.5; % unit cm^2
A1 =.08; % unit cm^2
length_B=(length-d)/2;
for ii=1:bb
Z0=Z00.*(1-j.*alpha./beta_res);
k=beta(ii)-j.*alpha;
a=sqrt((1+exp(-2*alpha.*length))./
(1-exp(-2*alpha.*length))./
```

```
(1-j/(2*Q)));
B=[cos(k.*length_B), j.*Z0.*sin(k.*length_B);
j.*sin(k.*length_B)./
Z0, cos(k.*length_B)];
AR=[a, 0; 0, 1./a];
A_H0=cavity_response(H0, f(ii));
C_H0=AR*B*A_H0*B;
S11_C_matrix_H0(ii)=(C_H0(1,2)-C_H0(2, 2).*Z00)./(C_H0
(1, 2)+C_H0(2, 2).*Z00);
% A_0=cavity_response(0, f(ii));
% C_0=AR*B*A_0*B;
% S11_C_matrix_0(ii)=(C_0(1,2)-C_0(2, 2).*Z00)./(C_0(1, 2)
+C_0
(2, 2).*Z00);
S11_magnetic_sample(ii)=(A1*abs(S11_C_matrix_H0(ii))+(Aw-
A1)*abs(S11(ii)))/Aw;
S11_magnetic_sample(ii)=(A1*abs(S11_C_matrix_H0(ii))+(Aw-
A1)*abs(S11(ii)))/Aw;
%S11_magnetic_sample(ii)=S11_C_matrix_H0(ii);
end
figure(1)
f_GHz=f./1e9;
plot(f_GHz, S11_dB, 'k', 'linewidth', 2);
axis([9.5, 9.56, -80, 1])
set(gca, 'FontSize', 25, 'LineWidth', 2.5,
'FontWeight','demi');
xlabel('\it Frequency(GHz)')
ylabel('\it S11(dB)')
set(gca, 'XTick', [9.5:0.01:9.56])
set(gca, 'XTickLabel', {'9.5', '', '9.52', '', '9.54', '',
'9.56'})
%axis([9.5, 9.56, -80, 0])
%set(gca, 'FontSize', 25, 'LineWidth', 2.5, 'FontWeight',
'demi');
%set(gca, 'XTick', [9.5:0.01:9.56])
%set(gca, 'XTickLabel', {'9.5', '', '9.52', '', '9.54',
'','9.56'})
figure(2)
plot(f_GHz, Pr, 'k', 'linewidth', 2);
axis([9.5, 9.56, 0, 1])
set(gca, 'FontSize', 25, 'LineWidth', 2.5, 'FontWeight', 'demi');
xlabel(' Frequency(GHz)')
ylabel(' Reflected Power(W)')
set(gca, 'XTick', [9.5:0.01:9.56])
set(gca, 'XTickLabel', {'9.5', '', '9.52', '', '9.54', '',
'9.56'})
figure(3)
plot(f_GHz, S11, 'k', 'linewidth', 2);
axis([9.5, 9.56, 0, 1])
```

```
set(gca, 'FontSize', 25, 'LineWidth', 2.5, 'FontWeight',
'demi');
xlabel(' Frequency(GHz)')
ylabel(' S11')
set(gca, 'XTick', [9.5:0.01:9.56])
set(gca, 'XTickLabel', {'9.5', '', '9.52', '', '9.54',
'','9.56'})
figure(4)
plot(f_GHz, P, 'k', 'linewidth', 2);
axis([9.5, 9.56, 0, 1])
set(gca, 'FontSize', 25, 'LineWidth', 2.5, 'FontWeight',
'demi');
xlabel(' Frequency(GHz)')
ylabel(' Incident Power (W) ')
set(gca, 'XTick', [9.5:0.01:9.56])
set(gca, 'XTickLabel', {'9.5', '', '9.52', '', '9.54', '',
'9.56'})
figure(5)
plot(f_GHz, (abs(S11)), 'k', f_GHz, (abs(S11_magnetic_
sample)), 'g',…'linewidth', 2);
axis([9.5, 9.56, 0, 1.5])
set(gca, 'FontSize', 25, 'LineWidth', 2.5,
'FontWeight','demi');
xlabel(' Frequency(GHz)')
ylabel(' S parameter ')
legend('cavity', 'loaded cavity');
%set(gca, 'FontSize', 25, 'LineWidth', 2.5, 'FontWeight',
'demi', 'Position',
[0.04, 0.1, 0.9, 0.82]);
```

REFERENCES

S.W. McKnight and C. Vittoria, *Phys. Rev. B*, **36**, 8574, 1987.

F.T. Ruch, D.Z. Barrick, W.D. Stuart, and C.K. Krichbaum, *Radar Cross Section Handbook*, Plenum Press, New York, 1970.

M.E. Van Valkenberg, *Network Analysis*, Prentice-Hall, Inc., Englewood Cliffs, NJ, 1955.

C. Vittoria, *Phys. Rev. B*, **32**, 1679, 1985.

C. Vittoria, *Elements of Microwave Networks*, World Scientific Publishing Co., Singapore, 1998.

SOLUTIONS

12.1 There is a general relation between the *S*-parameters and the distributive impedance [1] (Figure S12.1a).

The *S*-parameters can be written in terms of distributive impedance as follows:

$$(1)\ S_{11} = \frac{A_{12} + A_{11}Z_{0R} - A_{22}Z_{0L} - A_{21}Z_{0R}Z_{0L}}{A_{12} + A_{11}Z_{0R} + A_{22}Z_{0L} + A_{21}Z_{0R}Z_{0L}},$$

$$S_{21} = \frac{2Z_{0R}}{A_{12} + A_{11}Z_{0R} + A_{22}Z_{0L} + A_{21}Z_{0R}Z_{0L}}.$$

Substituting the transfer function matrix given in the problem, we obtain

$$(2)\ S_{11} = \frac{-1}{1 + 2Z/Z_0},\ S_{21} = \frac{1}{1 + Z_0/2Z}.$$

The equivalent circuit is given in Figure S12.1b.

12.2 The transfer function matrix of the film is given by Equation 12.12, which is

$$(a)\ A = \begin{pmatrix} \cos(kd) & jZ\sin(kd) \\ \dfrac{j}{Z}\sin(kd) & \cos(kd) \end{pmatrix},$$

where d is the thickness of the film. Substituting the transfer function matrix into equation (1) of Problem 12.1, we obtain the S-parameters.

$$(b)\ S_{11} = \Gamma_L \frac{e^{-2jkd} - 1}{1 - \Gamma_L^2 e^{-2jkd}},\ S_{21} = e^{-2jkd}(1 + \Gamma_L S_{11}),$$

where

$$(c)\ \Gamma_L = \frac{Z_0 - Z}{Z_0 + Z}.$$

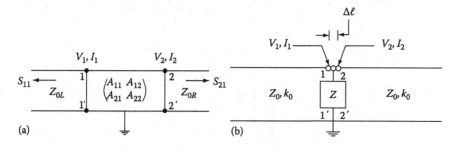

(a)　　　　　　　　　　　　　　　　　(b)

FIGURE S12.1 (a) A two-port microwave network equivalently represented by transfer function matrix and S-parameters. (b) The equivalent circuit of the transfer function matrix given in problem.

12.3 For symmetrical excitation, S_{11} and S_{21} are the same as Problem 12.2 if the network is reciprocal.

12.4 The second form can be approximated as

$$(1) \ A = \begin{pmatrix} \cos(kd) & jZ\sin(kd) \\ \dfrac{j}{Z}\sin(kd) & \cos(kd) \end{pmatrix} \approx \begin{pmatrix} 1 - \dfrac{(kd)^2}{2} & jZkd \\ \dfrac{j}{Z}kd & 1 - \dfrac{(kd)^2}{2} \end{pmatrix}.$$

12.5 As shown in Figure S12.5, the external magnetic field is applied along the z-axis and the electromagnetic waves are plane waves propagating along the y-axis. From Maxwell equation, we obtain

$$(1a) \ Q(k)h_x + 4\pi m_x = 0,$$

$$(1b) \ h_y + 4\pi m_y = 0,$$

$$(1c) \ Q(k)h_z + 4\pi m_z = 0,$$

where $Q(k) = 1 - \dfrac{j\delta^2 k^2}{2}$ and $\delta = \sqrt{\dfrac{c^2}{2\pi\omega\sigma}}$.

From equation of motion of the magnetic moment (without damping term), we obtain the susceptibility matrix:

$$(2) \ \begin{pmatrix} m_x \\ m_y \end{pmatrix} = \frac{M_S}{\Delta} \begin{pmatrix} H_{ext} + \dfrac{2A}{M_S}k^2 & -j\dfrac{\omega}{\gamma} \\ j\dfrac{\omega}{\gamma} & H_{ext} + \dfrac{2A}{M_S}k^2 + 4\pi M_S \end{pmatrix} \begin{pmatrix} h_x \\ h_y \end{pmatrix},$$

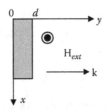

FIGURE S12.5 The geometry of Problem 12.5.

where

$$\Delta = \left(H_{ext} + \frac{2A}{M_S} k^2 \right) \left(H_{ext} + \frac{2A}{M_S} k^2 + 4\pi M_S \right) - \frac{\omega^2}{\gamma^2}.$$

Combining (1a–b) and (3), we obtain two linear equations of h_x and h_y, i.e.,

$$(3) \begin{cases} 0 = \left(\mu_{xx} - \frac{j\delta^2 k^2}{2} \right) h_x - j\frac{\omega}{\gamma} h_y \\[2mm] 0 = j\frac{\omega}{\gamma} h_x + \mu_{yy} h_y \end{cases},$$

where
$$\mu_{xx} = 1 + \frac{4\pi M_S}{\Delta} \left(H_{ext} + \frac{2A}{M_S} k^2 \right) \qquad \text{and}$$

$$\mu_{yy} = 1 + \frac{4\pi M_S}{\Delta} \left(H_{ext} + \frac{2A}{M_S} k^2 + 4\pi M_S \right).$$

Assuming that there exist nonzero solutions of (4), we formally obtain one dispersion relation

$$(4) \ k^2 = -j\frac{2}{\delta^2} \frac{\mu_{xx}\mu_{yy} - \kappa^2}{\mu_{yy}},$$

where $\kappa = \dfrac{4\pi M_S}{\Delta}\dfrac{\omega}{\gamma}$.

(4) is equivalent to (8.21), whereas x and y are interchanged to each other. Since μ_{xx} and μ_{yy} also contain k^2, (4) is actually a cubic equation in k^2. The solutions of (5) are three magnetic modes propagating in the $\pm y$ direction. In linear theory, $m_z = 0$ since the external magnetic field is applied along the z direction. Therefore, from (1c), we obtain the other dispersion relation

$$(5) \ k^2 = -j\frac{2}{\delta^2},$$

which is the skin-depth mode in non-magnetic metals. The h_x of the incident microwave will only excite the magnetic modes and h_z will only excite the skin-depth mode. The transfer function matrix of the skin depth mode can be written out by following Equation 12.12. In this problem, we only concern the magnetic modes. The continuity of h field at $y = 0$ and d can be written as

$$(6a) \ \sum_{i=1}^{3} \left(h_{ix}^+ + h_{ix}^- \right) = h_{0x},$$

$$(6b) \sum_{i=1}^{3} \left(h_{ix}^{+} e^{-jk_i d} + h_{ix}^{-} e^{+jk_i d} \right) = h_{dx},$$

where h_{ix}^{+} and h_{ix}^{-} are the amplitudes of the ith mode propagating in $\pm y$ direction, respectively. The magnetic boundary conditions may be written as

$$(7) \frac{\partial m_x}{\partial n} = 0, \text{ and } A \frac{\partial m_y}{\partial n} - K_s m_y = 0.$$

From (1a–b) and (3), we write

$$(8) \; 4\pi m_x = -Q(k)h_x \text{ and } 4\pi m_y = -P(k)h_x,$$

where $P(k) = -j \dfrac{\kappa}{\mu_{yy}}$.

Therefore, the boundary conditions can be written as

$$(9a) \sum_{i=1}^{3} jk_i Q(k_i)\left(-h_{ix}^{+} + h_{ix}^{-} \right) = 0,$$

$$(9b) \sum_{i=1}^{3} jk_i Q(k_i)\left(-h_{ix}^{+} e^{-jk_i d} + h_{ix}^{-} e^{+jk_i d} \right) = 0,$$

$$(9c) \sum_{i=1}^{3} P(k_i)\left[\left(-jAk_i - K_s \right) h_{ix}^{+} + \left(jAk_i - K_s \right) h_{ix}^{-} \right] = 0,$$

$$(9d) \sum_{i=1}^{3} P(k_i)\left[\left(jAk_i - K_s \right) h_{ix}^{+} e^{-jk_i d} + \left(-jAk_i - K_s \right) h_{ix}^{-} e^{+jk_i d} \right] = 0.$$

From Maxwell equations, we write the e field in the film as

$$(10) \; \vec{e}_i^{\pm} = \pm \frac{c}{4\pi\sigma} k_i h_{ix}^{\pm} e^{j(\omega t - k_i y)} \vec{a}_z.$$

Then, the continuity of e field at the boundary can be written in terms of $h_{ix}^{\pm}$ as

$$(11a) \sum_{i=1}^{3} k_i \left(h_{ix}^{+} - h_{ix}^{-} \right) = e_0,$$

$$(11b) \sum_{i=1}^{3} k_i \left(h_{ix}^{+} e^{-jk_i d} - h_{ix}^{-} e^{jk_i d} \right) = e_d.$$

From (6a–b) and (9a–d), we can uniquely solve $h_{ix}^{\pm}$ in terms of h_0 and h_d. Substituting the solution into (11a–b), we can write e_0 and e_d in terms of h_0 and h_d. Solving (e_d, h_d) from this equation in terms of (e_0, h_0), we finally obtain the transfer function matrix.

12.6 Although the number of modes inside the film increases, the e and h field at both surfaces of the film can still be related by a 2×2 matrix for the normal modes.

12.7 Assuming that the thickness of the superconducting film is d, the transfer matrix can be written as

$$(1) \ A = \begin{pmatrix} \cos kd & jZ \sin kd \\ \dfrac{j}{Z} \sin kd & \cos kd \end{pmatrix},$$

where k and Z are the propagation constant and wave impedance of the electromagnetic wave in the superconductor, which are given by

$$(2) \ jk = \frac{1}{\lambda} \sqrt{1 + 2j \left(\frac{\lambda}{\delta} \right)^2},$$

$$(3) \ Z = \frac{\omega \mu_0}{k}.$$

12.8 The purpose of this problem is to measure the surface resistance of the superconductor [1]. The superconductor wire must be aligned where the E field is maximal in the waveguide for TE_{10} mode (Figure S12.8). The equivalent circuit, in fact, has been shown in Figure S12.1b. The impedance Z can be written in terms of surface resistance (R_s), the capacitance (C), and the inductance (L), i.e.,

$$(1) \ Z = R_s + jX, \text{ where } X = \omega L - \frac{1}{\omega C}.$$

Substituting (1) into (2) of Problem 1, we obtain S_{21}:

$$(2) \ S_{21} = \frac{1}{1 + \dfrac{Z_0}{2 (R_s + jX)}},$$

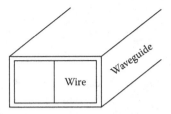

FIGURE S12.8   Superconductivity wire in a waveguide. Note that the wire is not touching the top and bottom of the waveguide.

where Z_0 is the impedance of the TE_{10} mode in the waveguide. At resonance ($\omega^2 = 1/LC$), assuming $Z_0 \gg R_s$, (2) can be simplified as

$$(3)\ S_{21} = \frac{1}{1+\dfrac{Z_0}{R_s}} \approx \frac{2R_s}{Z_0} \ll 1.$$

At off-resonance, R_s is negligible and the norm of S_{21} can be approximated as

$$(3)\ |S_{21}| \approx \frac{1}{\left|1+\dfrac{Z_0}{2X}\right|} \approx \frac{1}{1+\left(\dfrac{Z_0}{2X}\right)^2} \approx 1.$$

12.9 For a magnetic film, if the external field is perpendicular to the film plane, the transfer function matrix is given by Equation 12.10. If the external field is parallel to the film plane, the transfer function matrix is given in Problem 12.5. Once the transfer function matrix is determined, the S-parameters can be calculated as [1]

$$(1)\ S_{11} = \frac{a_{12} + a_{11}Z_0 - a_{22}Z_0 - a_{21}Z_0^2}{a_{12} + a_{11}Z_0 + a_{22}Z_0 + a_{21}Z_0^2},$$

$$(2)\ S_{21} = \frac{2Z_0}{a_{12} + a_{11}Z_0 + a_{22}Z_0 + a_{21}Z_0^2},$$

where Z_0 is usually the characteristic impedance of TE_{10} mode in the waveguide.

12.10 a. Define $x = kl$.
   For $n = 1$,

$$\begin{pmatrix} \cos x & \sin x \\ -\sin x & \cos x \end{pmatrix} = \begin{pmatrix} \cos x & \sin x \\ -\sin x & \cos x \end{pmatrix}.$$

The proposition is true.
Assuming that the proposition is true for $n=m$, i.e.,

$$\begin{pmatrix} \cos x & \sin x \\ -\sin x & \cos x \end{pmatrix}^m = \begin{pmatrix} \cos mx & \sin mx \\ -\sin mx & \cos mx \end{pmatrix}.$$

For $n=m+1$,

$$\begin{pmatrix} \cos x & \sin x \\ -\sin x & \cos x \end{pmatrix}^{m+1} = \begin{pmatrix} \cos mx & \sin mx \\ -\sin mx & \cos mx \end{pmatrix}\begin{pmatrix} \cos x & \sin x \\ -\sin x & \cos x \end{pmatrix}$$

$$= \begin{pmatrix} \cos(m+1)x & \sin(m+1)x \\ -\sin(m+1)x & \cos(m+1)x \end{pmatrix}.$$

Thus, the proposition is true for all values of n.
b. Define $x=kl$.
For $n=1$,

$$\begin{pmatrix} \cos x & jZ\sin x \\ \dfrac{j}{Z}\sin x & \cos x \end{pmatrix} = \begin{pmatrix} \cos x & jZ\sin x \\ \dfrac{j}{Z}\sin x & \cos x \end{pmatrix}.$$

The proposition is true.
Assuming that the proposition is true for $n=m$, i.e.,

$$\begin{pmatrix} \cos x & jZ\sin x \\ \dfrac{j}{Z}\sin x & \cos x \end{pmatrix}^m = \begin{pmatrix} \cos mx & jZ\sin mx \\ \dfrac{j}{Z}\sin mx & \cos mx \end{pmatrix},$$

For $n=m+1$,

$$
\begin{pmatrix} \cos x & jZ\sin x \\ \dfrac{j}{Z}\sin x & \cos x \end{pmatrix}^{m+1} = \begin{pmatrix} \cos mx & jZ\sin mx \\ \dfrac{j}{Z}\sin mx & \cos mx \end{pmatrix} \begin{pmatrix} \cos x & jZ\sin x \\ \dfrac{j}{Z}\sin x & \cos x \end{pmatrix}
$$

$$
= \begin{pmatrix} \cos(m+1)x & jZ\sin(m+1)x \\ \dfrac{j}{Z}\sin(m+1)x & \cos(m+1)x \end{pmatrix}.
$$

Thus, the proposition is true for all values of n.

12.11 The eigenvalues are the roots of the equation

$$(1)\ \det(\lambda E - A) = 0,$$

where E is the identity matrix and A is the matrix in Problem 12.10 part b. Solving (1), we obtain the eigenvalues of A as

$$(2)\ \lambda_{\pm} = e^{\pm jnkl}.$$

Index

Note: **Bold** page numbers refer to tables and *italic* page numbers refer to figures.

Printed in the United States
by Baker & Taylor Publisher Services